Arccos or Cos$^{-1}$
Arcsin or Sin$^{-1}$
Arctan or Tan$^{-1}$ the principal values of inverse circu
Arcsec or Sec$^{-1}$ functions
Arccsc or Csc$^{-1}$
Arccot or Cot$^{-1}$

(3.5)

| | | |
|---|---|---|
| $\angle ABC$ | angle $ABC$ with vertex $B$ | (4.1) |
| $\alpha, \beta, \gamma$, etc. | angles alpha, beta, gamma, etc. | (4.1) |
| $m^\circ(\alpha)$ | degree measure of angle $\alpha$ | (4.1) |
| $m^R(\alpha)$ | radian measure of angle $\alpha$ | (4.1) |
| $x^\circ$ | $x$ degrees | (4.1) |
| $x^R$ | $x$ radians | (4.1) |
| $v$ | linear velocity | (4.2) |
| $\omega$ | angular velocity | (4.2) |
| Trig | any one of the trigonometric functions | (4.4) |
| Circ | any one of the circular functions | (4.4) |
| $\tilde{\alpha}$ | the reference angle of angle $\alpha$ | (4.5) |
| $\Leftrightarrow$ | is equivalent to | (5.1) |
| $\vec{v}$ | geometric vector | (6.4) |
| $\vec{V}$ | the set of geometric vectors | (6.4) |
| $\|\vec{v}\|$ | norm or magnitude of $\vec{v}$ | (6.4) |
| $\alpha$ | alpha, direction angle of $\vec{v}$ (or v) where $-180^\circ < \alpha \le 180^\circ$ | (6.4) |
| $\vec{0}$ | the zero geometric vector | (6.4) |
| $\vec{v}_x$ | geometric vector projection of $\vec{v}$ on the $x$ axis | (6.5) |
| $\vec{v}_y$ | geometric vector projection of $\vec{v}$ on the $y$ axis | (6.5) |
| $\mathbf{v}$ | vector; an ordered pair of real numbers | (7.1) |
| $\mathbf{V}$ | the set of vectors | (7.1) |
| $\mathbf{0}$ | the zero vector; (0, 0) | (7.1) |
| $\|\mathbf{v}\|$ | norm or magnitude of $\mathbf{v}$ | (7.1) |
| $\mathbf{i}$ | the unit vector (1, 0) | (7.2) |
| $\mathbf{j}$ | the unit vector (0, 1) | (7.2) |
| $\mathbf{v}_1 \cdot \mathbf{v}_2$ | inner product of $\mathbf{v}_1$ and $\mathbf{v}_2$ | (7.3) |
| $\rho$ | rho, distance from the origin to a point in the plane | (7.4) |
| $\theta$ | theta, direction angle from the positive $x$ axis to the ray drawn from the origin to a point in a plane | (7.4) |
| $C$ | the set of complex numbers | (8.1) |
| $z$ | complex number; an ordered pair of real numbers | (8.1) |
| $I$ | the set of imaginary numbers | (8.1) |
| $\bar{z}$ | conjugate of $z$ | (8.2) |
| $i$ | the complex number (0, 1) | (8.3) |
| $\rho$ or $|z|$ | absolute value or modulus of $z$ | (8.4) |
| $\theta$ | theta, argument of a complex number | (8.4) |
| $z^\circ$ | the complex number (1, 0) | (8.6) |
| $z^{-n}, n \in N$ | $1/z^n$ | (8.6) |
| $z^{1/n}, n \in N$ | $n$th root of $z$ | (8.6) |
| $\log_b x$ | logarithm to base $b$ of $x$ | (A.1) |
| antilog$_b x$ | antilogarithm to the base $b$ of $x$ | (A.2) |

# TRIGONOMETRY
## An Analytic Approach

# Irving Drooyan & Walter Hadel

*Los Angeles Pierce College, Woodland Hills, California*

# Trigonometry

## An Analytic Approach

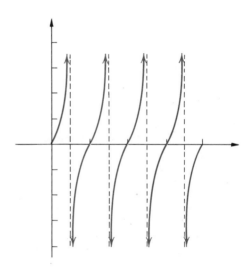

**The Macmillan Company, New York**

**Collier-Macmillan Limited, London**

First Printing

Library of Congress catalog card number: 67-15966

**The Macmillan Company, New York**

**Collier-Macmillan Canada, Ltd., Toronto, Ontario**

Printed in the United States of America

# Preface

T HIS TEXT was written for use in a one-semester course in trigonometry. The point of view and the topics covered reflect recent educational
trends in mathematics. Topics are emphasized that are particularly important
for students who will continue their study of mathematics or enter into fields
which require a strong background in mathematics. These topics include:

1. The periodic nature of the circular functions which relate real numbers
   with real numbers and the trigonometric functions which relate angles
   with real numbers.
2. Relationships among the circular functions and the trigonometric
   functions.
3. Geometric vectors and vectors as ordered pairs of real numbers.
4. Complex numbers.

It is assumed that students undertaking the study of trigonometry have had
prerequisite work in algebra and geometry. However, the modern concepts
and notation which are needed as background are treated in Chapter 1 in
conjunction with a review of the real number system. The emphasis on struc-
ture, evident in Chapter 1, appears throughout the text.

Although the solution of triangles is not stressed, the topic has been included in the text. Furthermore, a treatment of the logarithmic function is
available in Appendix A for use in courses that include this topic.

The problems in the exercise sets are designated as A, B, or C. The B
problems are more challenging than the A problems and provide the instruc-

tor with flexibility in making assignments, depending on the time available and the objectives of the course. The C problems include topics that have not been discussed in the text. In most sections the problems are preceded by a brief explanation. These problems also provide the instructor with flexibility in making assignments. Some exercise sets include B problems but do not include C problems; others include C problems but do not include B problems.

Chapter summaries and review exercise sets follow each chapter.

Answers are provided for the odd-numbered problems; many answers in the form of graphs have also been included. Problems in the text for which graphs are available in the answer section are preceded by the symbol " ■ ".

The authors wish to thank Professors Betty Ribal of Fullerton College and Alexander L. Arning of Manhattanville College of the Sacred Heart and Mr. Tyrus Buquoi of the Los Angeles City School System for their helpful comments and suggestions on the manuscript. Appreciation is also expressed to Mrs. Gertrude Drooyan for typing the manuscript.

IRVING DROOYAN
WALTER HADEL

*Woodland Hills, California*

# Contents

CHAPTER

# 3

## Graphs of the Circular Functions                                      88

CHAPTER

# 4

## Trigonometric Functions                                               122

CHAPTER

# 5

## Identities and Conditional Equations                                  148

CHAPTER

# 6

## Solution of Triangles; Geometric Vectors                  168

CHAPTER

# 7

## Vectors                                                                      204

CHAPTER

# 8

## Complex Numbers                                                     233

# APPENDIXES

# The Real Number System

IN THIS BOOK you will be working with two important sets of numbers; first, the set of real numbers and later, the set of complex numbers. This first chapter is concerned with a review of the basic topics of algebra pertaining to the set of real numbers, using terminology and notation that will be needed in the development of topics in this book.

## 1.1
### Set Notation

Recall that a **set** is simply a well-defined collection of "things." By "well-defined" we mean that it is always possible to determine whether something is or is not in the set. Each "thing" in the set is called a **member** or **element** of the set. The membership can be described by listing the names of the members in braces, { }, or by stating a rule which identifies the members in the set.

**Definition 1.1.** *Two sets are **equal** if and only if they have the same members.*

Customarily, the slant bar, /, drawn through symbols for relations, such as =, indicates negation. Thus,

$$\{3, 4, 5\} = \{5, 3, 4\} \quad \text{but} \quad \{3, 4, 5\} \neq \{4, 5, 6\}.$$

The symbol $\in$ is used to denote membership in a set. For example, you can write

$$3 \in \{3, 4, 5\}$$

to represent the statement "3 is an element of $\{3, 4, 5\}$."

Sets are also designated by means of capital letters, $A$, $B$, $C$, $R$, etc. Thus, if $A = \{2, 3, 4\}$ and $B = \{4, 5, 6\}$, then $3 \in A$ but $3 \notin B$.

An unspecified element of a set is usually denoted by a lower-case italic letter such as $a$, $b$, $c$, $x$, and $y$ or by a lower-case letter from the Greek alphabet such as $\alpha$, $\beta$, $\gamma$, and $\theta$. Such a symbol is called a **variable**; a symbol used to denote a specific element is called a **constant**. Variables are used in conjunction with braces in another symbolism which is often useful in discussing sets. This symbolism, illustrated by

$$\{x \mid x \in A \text{ and } x \in B\},$$

and read "the set of all $x$ such that $x$ is an element of $A$ and $x$ is an element of $B$," is called **set-builder notation**. The vertical line, |, is read "such that." Set-builder notation names a variable and states any conditions on the variable. Thus, if $A = \{2, 3, 4\}$ and $B = \{4, 5, 6\}$, then

$$\{x \mid x \in A \textbf{ or } x \in B\} = \{2, 3, 4, 5, 6\}, \tag{1}$$

because 2, 3, 4, 5, 6 are elements of *one or the other* of the two sets, and

$$\{x \mid x \in A \textbf{ and } x \in B\} = \{4\}, \tag{2}$$

because 4 is an element of *both* sets.

The preceding statements suggest certain operations on sets that will be useful in later sections.

**Definition 1.2.**   *The **union** of two sets $A$ and $B$ is the set of all elements that belong to $A$ or to $B$ or to both.*

The symbol $\cup$ is used to denote this operation; $A \cup B$ is read "the union of $A$ and $B$" or sometimes "$A$ cup $B$." For example, if $A = \{2, 3, 4\}$ and $B = \{4, 5, 6\}$, then

$$A \cup B = \{2, 3, 4, 5, 6\}. \tag{3}$$

Notice that each element in $A \cup B$ is listed only once. Since the symbol 4 denotes only one number, it would be redundant to list the name of the element twice and to write $A \cup B = \{2, 3, 4, 4, 5, 6\}$.

A second useful operation on sets is the following.

**Definition 1.3.**   *The **intersection** of two sets $A$ and $B$ is the set of all elements that belong to both $A$ and $B$.*

The symbol $\cap$ is used to denote this operation; $A \cap B$ is read "the intersection of $A$ and $B$" or sometimes "$A$ cap $B$." For example, if $A = \{2, 3, 4\}$ and $B = \{4, 5, 6\}$, then

$$A \cap B = \{4\}. \tag{4}$$

Notice from examples (1) and (3) above that the operation of *union* is associated with the word "*or*," and from examples (2) and (4) that the operation of *intersection* is associated with the word "*and*."

The set that contains no elements is called the **null set** or the **empty set** and is denoted by the symbol $\emptyset$. The notion of the null set is useful. For example, because the intersection of two sets is a set, if $C = \{2, 3, 4\}$ and $D = \{5, 6, 7\}$ we can write

$$C \cap D = \{2, 3, 4\} \cap \{5, 6, 7\} = \emptyset.$$

Two sets such as $C$ and $D$, which do not have any elements in common, are said to be **disjoint**.

### EXERCISE 1.1

A.

Let $P = \{a, b, c\}$, $S = \{c, a, b\}$, $T = \{b, c\}$, $V = \{a\}$ and $W = \{a, b\}$. Form a true statement by replacing the comma in each of the following pairs with $=$ or $\neq$ where the relationship is between sets, and $\in$ or $\notin$ where the relationship is between an element and a set.

**Examples.** (a) $a, T$.      (b) $P, W$.      (c) $c, P$.

Solutions. (a) $a \notin T$.      (b) $P \neq W$.      (c) $c \in P$.

1. $b, P$.              3. $T, V$.              5. $c, W$.
2. $P, S$.              4. $a, V$.              6. $T, W$.

Let $K = \{j, k, l, m\}$, $L = \{j, k\}$, $M = \{k, l\}$, $N = \{m, n, o\}$, and $P = \{l\}$. List the elements in each set.

**Examples.** (a) $L \cup M$.      (b) $M \cap N$.

Solutions. (a) $\{j, k, l\}$.      (b) $\emptyset$.

7. $K \cup L$.                        12. $P \cup \emptyset$.
8. $K \cap M$.                       13. $(K \cap L) \cap M$.
9. $L \cup N$.                        14. $L \cup (K \cap P)$.
10. $L \cap P$.                      15. $(N \cap K) \cup L$.
11. $L \cap N$.                      16. $(P \cup L) \cap (N \cup M)$.

Let $A = \{1, 3, 5, 7\}$, $B = \{4, 5, 6, 7\}$, $C = \{6, 7, 8, 9\}$, and $D = \{2, 4, 6, 8\}$. List the members of each set.

**Examples.** (a) $\{x \mid x \in A \text{ and } x \in B\}$.      (b) $\{x \mid x \in A\} \cup \{y \mid y \in B\}$.

Solutions. (a) The set that contains all the members that are members in both $A$ and $B$ is $\{5, 7\}$.

         (b) The set that contains all the members that are members of either $A$ or $B$ or both is $\{1, 3, 4, 5, 6, 7\}$.

**17.** $\{x \mid x \in B \text{ and } x \in C\}.$
**18.** $\{x \mid x \in B \text{ or } x \in C\}.$
**19.** $\{y \mid y \in A \text{ or } y \in C\}.$
**20.** $\{y \mid y \in A \text{ and } y \in C\}.$
**21.** $\{a \mid a \in A \text{ or } a \in D\}.$
**22.** $\{a \mid a \in A \text{ and } a \in D\}.$
**23.** $\{b \mid b \in C \text{ and } b \in D\}.$
**24.** $\{b \mid b \in C \text{ or } b \in D\}.$

**25.** $\{x \mid x \in A\} \cap \{y \mid y \in B\}.$
**26.** $\{x \mid x \in B\} \cap \{y \mid y \in C\}.$
**27.** $\{a \mid a \in C\} \cup \{b \mid b \in D\}.$
**28.** $\{a \mid a \in A\} \cup \{a \mid a \in D\}.$
**29.** $\{r \mid r \in A\} \cap \{s \mid s \in D\}.$
**30.** $\{r \mid r \in B\} \cup \{s \mid s \in D\}.$
**31.** $\{a \mid a \in A\} \cup \varnothing.$
**32.** $\{b \mid b \in C\} \cap \varnothing.$

**B.**

Considering the sets $A$, $B$, and $C$ as defined above, state whether each statement is true or false.

**33.** $A \cup B = B \cup A.$
**34.** $A \cap B = B \cap A.$
**35.** $(A \cap B) \cap C = A \cap (B \cap C).$

**36.** $(A \cup B) \cup C = A \cup (B \cup C).$
**37.** $A \cap (B \cup C) = (A \cap B) \cup (A \cap C).$
**38.** $A \cup (B \cap C) = (A \cup B) \cap (A \cup C).$

## 1.2
### The Set of Real Numbers

Recall from your study of algebra that each real number can be associated with one and only one point on a number line and each point on the line can be associated with one and only one real number. Several examples are shown in Figure 1.1. The real number corresponding to a point on the line is called the **coordinate** of the point, and the point is called the **graph** of the number.

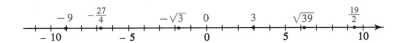

**FIGURE 1.1**

The following sets of numbers that are contained in the set of **real numbers** $R$ will be referred to frequently. The capital letters shown are the ones that are sometimes associated with the respective sets. These are the symbols that we shall use to describe the appropriate set.

1. The set $N$ of **natural numbers**, among whose elements are such numbers as 1, 2, 6, 25, and 624.
2. The set $J$ of **integers**, whose elements consist of the natural numbers, their negatives and zero. Included are such elements as $-24$, $-9$, 0, 7, and 314.
3. The set $Q$ of **rational numbers**, whose elements are all those numbers that can be represented in the form $\dfrac{a}{b}$ or $a/b$, where $a$ and $b$ are integers and $b$ is not zero. Included are such elements as $-\frac{1}{3}$, 0, $\frac{1}{4}$, and 6. Rational numbers can also be described as numbers with *terminating* or *repeating*

*decimal representations.* Thus,

$$\tfrac{1}{4} = 0.25 \qquad \text{and} \qquad -\tfrac{1}{3} = -0.33\ldots,$$

are both rational numbers. Repeating decimals such as $-0.33\ldots$ are sometimes written as $-0.\overline{3}$, $-0.3\overline{3}$, etc., where the bar indicates the repeating digit(s).

4. The set $H$ of **irrational numbers**, whose elements are those numbers whose representations are *nonterminating, nonrepeating decimals.* Included are such elements as $\sqrt{2}$, $\pi$, and $-\sqrt{5}$. An irrational number cannot be represented in the form $a/b$, where $a$ and $b$ are integers.

Radical notation is often useful to represent irrational numbers.

**Definition 1.4.** *For all $a \geq 0$, $\sqrt{a}$ is the nonnegative number such that*

$$\sqrt{a} \cdot \sqrt{a} = a.$$

Frequently, rational numbers are used as approximations for irrational numbers. For example, by using the symbol $\approx$ for "approximately equal to," we write, $\sqrt{2} \approx 1.414$, $\sqrt{3} \approx 1.732$, etc. Table I in Appendix C, page 291, gives rational number approximations for some roots and reciprocals which are irrational numbers. In fact, most of the entries in the other tables in Appendix C are also approximations for irrational numbers.

Figure 1.2 summarizes the relationship between the different sets of numbers noted above. Recall that the set of rational numbers and the set of irrational numbers are disjoint; that is, $Q \cap H = \varnothing$. Furthermore, $Q \cup H = R$.

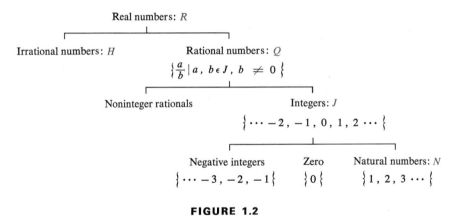

**FIGURE 1.2**

If *every* element of a set $A$ is also an element of a set $B$, then $A$ is said to be a **subset** of $B$.

If the elements of one set are in a *one-to-one correspondence* to the set $\{1, 2\ 3, \ldots, n\}$ for some specified natural number $n$, as shown in Figure 1.3, then the set is said to be a **finite set**. A set that is not the empty set and is not finite is called an **infinite set**. Some infinite sets can be designated by using three dots with some of the members which are listed to establish a pattern as

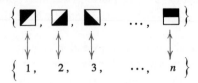

**FIGURE 1.3**

shown for $N$ and $J$ in Figure 1.2. Although the elements in some sets of real numbers cannot be listed to establish a pattern, an order can be established by the following.

**Definition 1.5.**   *For all  a,  b ∈ R,   b is **less than** a if for some positive real number c,*

$$b + c = a.$$

*For such conditions, a is said to be **greater than** b.*

The number line as shown in Figure 1.4 is very helpful in visualizing whether one real number is less than or greater than a second real number.

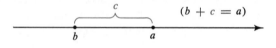

**FIGURE 1.4**

If $b$ is *less than* $a$, then the graph of $b$ lies to the *left* of the graph of $a$. The following symbols are used in connection with the property of order:

$<$   read "is less than";

$\leq$   read "is less than or equal to";

$>$   read "is greater than";

$\geq$   read "is greater than or equal to".

Set-builder notation in conjunction with the above order symbols is especially useful in describing infinite sets of numbers.

**Examples.**   Graph $\{x \mid -2 < x \leq 3, x \in R\}$ on a number line.

*Solution.*

The *closed dot* on the right-hand end of the heavy line on the graph indicates that the *end point is part of the graph,* while the *open dot* on the left indicates that the *end point is not part of the graph.*

The set of real numbers can be partitioned in such a way that if $a \in R$, $a < 0$, $a = 0$, or $a > 0$. Furthermore,

$$\text{if } a < 0, \text{ then } -a > 0,$$

$$\text{if } a = 0, \text{ then } -a = 0,$$

and
$$\text{if } a > 0, \text{ then } -a < 0,$$

where $a + (-a) = 0$. Often we wish to consider the nonnegative member of the pair $a$, $-a$. The special notation $|a|$ is used for this purpose.

**Definition 1.6.**   *For all $a \in R$, and*

$$\text{if } a \geq 0, \text{ then } |a| = a,$$

$$\text{if } a < 0, \text{ then } |a| = -a;$$

$|a|$ *is called the* **absolute value** *of* $a$.

Thus, $|a|$ is always nonnegative.

**Examples.**   (a) $|4| = 4$.        (b) $|-4| = 4$.        (c) $|0| = 0$.

In your study of arithmetic and algebra you paired elements of the set of real numbers over the binary operations of addition and multiplication. You also made assumptions called **axioms,** concerning the properties of these operations and studied the logical consequences of these axioms. For your convenient reference for the study of later sections in this text, axioms and some of their consequences (called **theorems**) which pertain to the set of real numbers are listed in Appendix B, page 288.

Any mathematical system, consisting of a set of elements and two binary operations, in which the equality axioms and the axioms for operations listed in Appendix B are satisfied is said to form a **field**. A system in which the order axioms are also satisfied is said to form an **ordered field**. Thus, *the real number system is an ordered field.* In Chapter 8, you will see that the set of complex numbers which are discussed, also form a field, but not an ordered field.

In your study of arithmetic and algebra, you also paired elements of the set of real numbers over the binary operations of subtraction and division, which were defined in terms of addition and multiplication, respectively.

**Definition 1.7.** *For all $a, b \in R$, the difference*

$$a - b \text{ is the real number } d \text{ such that } b + d = a.$$

It can be shown as a consequence of this definition that if $a, b \in R$, then

$$a - b = a + (-b).$$

The proof of this statement is left as an exercise.

**Examples.**   (a) $7 - 3 = 7 + (-3) = 4$.        (b) $3 - 7 = 3 + (-7) = -4$.

**Definition 1.8.**   *For all a, b ∈ R, (b ≠ 0), the* **quotient**

$$\frac{a}{b} = q \quad \text{such that} \quad b \cdot q = a.$$

It can be shown as a consequence of this definition that if $a, b \in R, (b \neq 0)$, then

$$\frac{a}{b} = a \cdot \frac{1}{b}.$$

The proof of this statement is left as an exercise.

**Examples.**   (a) $\dfrac{2}{3} = 2 \cdot \dfrac{1}{3}.$    (b) $\dfrac{2}{x-5} = 2 \cdot \dfrac{1}{x-5}$, $(x \neq 5).$

### EXERCISE 1.2

**A.**

Let   $A = \{3, -1, 5/6, -\sqrt{7}, 0, -8/11, \sqrt{5}, \sqrt{25}, 4\}$.   Write each of the following sets by using braces and listing the members. The letters $N, J, Q, H,$ and $R$ refer to the sets shown in Figure 1.2.

**Example.**   $\{x \mid x \in A \text{ and } x \in N\}.$

**Solution.**   $\{3, 4, \sqrt{25}\}.$

**1.** $\{x \mid x \in A \text{ and } x \in J\}.$     **3.** $\{x \mid x \in A \text{ and } x \in H\}.$
**2.** $\{x \mid x \in A \text{ and } x \in Q\}.$     **4.** $\{x \mid x \in A \text{ and } x \in R\}.$

Replace the comma in each pair of expressions with the appropriate order symbol, $<, >,$ or $=,$ to form a true statement.

**Examples.**   (a) $3, -6.$        (b) $3, |-6|.$

**Solutions.**   (a) $3 > -6.$        (b) Because $|-6| = 6,$   $3 < |-6|.$

**5.** $5, 7.$                                    **9.** $|-4|, |-7|.$

**6.** $-11, 3.$                                  **10.** $|-5|, 3.$

**7.** $|-4|, 4.$                                 **11.** $\dfrac{-4}{5}, \left|\dfrac{3}{-4}\right|.$

**8.** $|9|, -4.$                                 **12.** $\dfrac{\sqrt{3}}{2}, \dfrac{\sqrt{5}}{3}.$

Graph each set on a number line. Use a separate number line for each problem. Use Table I, page 291 as necessary to find rational number approximations for irrational numbers.

**Example.**  (a) $\{x \mid x \geq 3, x \in J\}$.        (b) $\{x \mid x < 2, x \in R\}$.

*Solution.*   (a)                                (b)

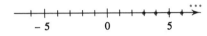

**·13.** $\{-\sqrt{3}, 2, \sqrt{17}\}$.

**14.** $\{-\sqrt{8}, -1, \sqrt{10}\}$.

**15.** $\{x \mid x \geq 5, x \in N\}$.

**16.** $\{x \mid x < 7, x \in J\}$.

**·17.** $\{x \mid x < -2, x \in R\}$.

**18.** $\{x \mid x \geq -1, x \in R\}$.

**19.** $\{x \mid 2 < x \leq 7, x \in N\}$.

**20.** $\{x \mid -1 \leq x < 6, x \in J\}$.

**·21.** $\{x \mid -\sqrt{5} < x < \sqrt{5}, x \in R\}$.

**22.** $\{x \mid -6 \leq x \leq 6, x \in R\}$.

**B.**

**Example.**  $\{x \mid x \leq 2, x \in R\} \cap \{x \mid x > -1, x \in R\}$.

*Solution.*

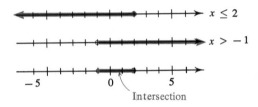

$x \leq 2$

$x > -1$

Intersection

**23.** $\{x \mid x < 4, x \in N\} \cup \{x \mid x \geq 3, x \in N\}$.

**24.** $\{x \mid x > -1, x \in J\} \cup \{x \mid x < -3, x \in J\}$.

**·25.** $\{x \mid x \leq 6, x \in R\} \cap \{x \mid x \geq 1, x \in R\}$.

**26.** $\{x \mid x > -\sqrt{2}, x \in R\} \cap \{x \mid x < \sqrt{8}, x \in R\}$.

**C.**

**27.** Show that $a - b = a + (-b)$.   *Hint:* From Definition 1.7,
$b + (a - b) = a$.

**28.** Show that $a/b = a \cdot (1/b)$   $(b \neq 0)$.   *Hint:* From Definition 1.8,
$b \cdot (a/b) = a$.

# 1.3
## Relations and Functions; Equations in Two Variables

In this book we are going to be interested in associations between elements of one set of numbers and elements of a second set of numbers. These associations are called relations and can be exhibited in several ways. Very often we shall use a representation for a relation that involves set notation.

■ Graphs are included in the answer section.

For example, the set of all pairings that arise when the elements of

$$A = \{1, 2, 3\}$$

are paired with the elements of

$$B = \{5, 10\}$$

is given by

$$A \times B = \{(1, 5), (1, 10), (2, 5), (2, 10), (3, 5), (3, 10)\}.$$

$A \times B$ is called the **Cartesian product** of $A$ and $B$. Each pair in such a set is called an **ordered pair** because the order of listing the elements is important. The pair $(a, b)$ is *not* the same as the pair $(b, a)$. The first number in each pair, called the **first component**, is selected from set $A$ and the second number, called the **second component**, is selected from set $B$. The above example suggests the following.

**Definition 1.9.**    *A **relation** is any set of ordered pairs.*

The set of first components in a set of ordered pairs is called the **domain** of the relation and the set of second components is called the **range.** In the example $A \times B$ above, $A = \{1, 2, 3\}$ is the domain, and $B = \{5, 10\}$ is the range.

**Examples.**    Write the relation $A \times B$ as a set of ordered pairs and specify the domain and range.

(a) $A = \{1, 3\}$, $B = \{2, 4\}$.        (b) $A = \{x \mid x \in R\}$, $B = \{y \mid y \in R\}$.

*Solutions.*
(a) $A \times B = \{(1, 2), (1, 4), (3, 2), (3, 4)\}$; domain is $\{1, 3\}$, range is $\{2, 4\}$.
(b) $A \times B = \{(x, y) \mid x, y \in R\}$; domain is $\{x \mid x \in R\}$, range is $\{y \mid y \in R\}$.

Obviously all the members in Example (b) cannot be listed, but setbuilder notation makes it possible to describe this infinite set of ordered pairs. This Cartesian product is usually designated $R \times R$.

There are certain kinds of relations called functions which have special importance.

**Definition 1.10.** *A **function** is a relation in which each element in the domain is associated with only one element in the range.*

For example, the set of ordered pairs

$$\{(7, 5), (8, 6)\} \tag{1}$$

is a function because each first component is associated with one and only one second component; on the other hand,

$$\{(7, 5), (7, 6)\}$$

is not a function because the same first component is associated with two different second components.

In example (1) a function was exhibited by *listing* the members, in this case, ordered pairs. Functions can also be described by a *rule*. To do this we shall first review some ideas concerning equations and their solutions. Recall from your study of algebra that the ordered pair $(2, 9)$ is called a **solution** of the equation $y = x + 7$ because the replacement of $x$ by the first component $2$ and the replacement of $y$ by the second component $9$ results in a *true statement*, $9 = 2 + 7$. The ordered pair $(2, 9)$ is said to *satisfy* the equation. The set of all such ordered pairs that satisfy an equation is called its **solution set**. The set from which ordered pairs are selected as possible solutions is called the **replacement set**. In algebra, the replacement set is generally $R \times R$.

The equation $y = x + 7$ does not result in a true statement for all ordered pairs from $R \times R$. For example, $(3, 5)$ is not a solution because $5 \neq 3 + 7$. Such equations are called **conditional equations** for the replacement set $R \times R$. Some equations such as

$$x^2 + y^2 = -1$$

*always result in a false statement* when variables are replaced by ordered pairs from the replacement set. For any real number replacements of $x$ and $y$, $x^2$ and $y^2$ are both positive or zero and their sum can never be $-1$. Thus, the solution set in this case is the empty set, $\emptyset$. On the other hand, if an equation such as

$$x + y = y + x$$

*always results in a true statement* for each replacement of $x$ and $y$; the equation is called an **identity**. The solution set for this equation contains all elements in the replacement set $R \times R$.

In order to show that an equation is not an identity, it is only necessary to find one replacement for the variables that will make the equation false. Recall that such a procedure is referred to as showing a **counterexample**.

*Example.*   Show by counterexample that $x^2 - 3y = 3y - x^2$ is not an identity in $R \times R$.

*Solution.*   Replace the variables with any real numbers, say 1 for $x$ and 1 for $y$. Since for these replacements, the left-hand member of the equation equals $1^2 - 3 \cdot 1$, or $-2$, and the right-hand member equals $3 \cdot 1 - 1^2$, or 2,

$$x^2 - 3y \neq 3y - x^2$$

for the ordered pair $(1, 1)$ and therefore the equation is not an identity in $R \times R$.

Now, since equations in two variables express relationships between the variables and serve to pair numbers in the replacement sets of the variables, equations can be used to define functions (and relations). The equations specify which ordered pairs are in the function and which are not; those that are in the function are elements of the solution set. For example,

$$\{(x, y) \mid y = x + 7, x \in \{1, 2, 3\}\} \tag{2}$$

specifies the set

$$\{(1, 8), (2, 9), (3, 10)\}. \quad\begin{array}{l}\text{(Domain)}\\[1ex]\text{(Range)}\end{array}$$

The elements in the range, 8, 9, and 10, are obtained by replacing $x$ in the equation $y = x + 7$ with 1, 2, and 3 in turn and solving for $y$. The Function (2) consists of 3 ordered pairs only.

If the domain of (2) is changed to be the set of real numbers $R$, rather than the finite set $\{1, 2, 3\}$, the function

$$\{(x, y) \mid y = x + 7, x \in R\}$$

consists of an *infinite set* of ordered pairs and obviously all the elements of such a set cannot be listed. *In most of our work the domain of a function shall be the set of real numbers for which both members of the defining equation are themselves defined and in this case the domain shall not be specified.* For example, it is understood that the domain of

$$\{(x, y) \mid y = x + 7\}$$

is the set $R$ because $x + 7$ and therefore $y$ are defined for all real number replacements for $x$. On the other hand, in the function

$$\left\{(x, y) \mid y = \frac{3}{x - 4}\right\},$$

it is understood that $x \in R$ with the restriction that $x \neq 4$, because $y$ is undefined for $x = 4$.

Functions are sometimes designated by names (linear, quadratic, etc.) or by means of a single symbol, generally the symbol $f$. If the discussion includes a consideration of more than one function, other letters such as $g, h, F, \mathscr{F}, P$, etc., are used.

The symbol for the function can be used in conjunction with the variable representing an element in the domain to represent the associated element in the range. For example, $f(x)$ (read "$f$ of $x$" or "the value of $f$ at $x$") is the element in the range of $f$ associated or paired with the element $x$ in the domain. Thus, we can discuss functions defined by an equation such as

$$y = x - 3$$

or

$$f(x) = x - 3,$$

$$g(x) = x - 3, \text{ etc.,}$$

where the symbols $f(x)$, $g(x)$, etc., are playing exactly the same role as $y$. The variable $x$ represents an element in the domain; $y$, $f(x)$, or $g(x)$, etc., represents an element in the range.

The notation, $f(x)$, is especially useful because, by replacing $x$ with a specific real number $a$ in the domain, the notation $f(a)$ will then denote the paired, or corresponding, element in the range of the function.

**Examples.** If $f(x) = 2x - 5$, find the value of:

(a) $f(3)$.      (b) $f(0)$.      (c) $f(-3)$.      (d) $f(a)$.

*Solutions.* Replacing $x$ with 3, 0, $-3$, and $a$ in turn, we have

(a) $f(3) = 2(3) - 5 = 6 - 5 = 1$.
(b) $f(0) = 2(0) - 5 = 0 - 5 = -5$.
(c) $f(-3) = 2(-3) - 5 = -6 - 5 = -11$.
(d) $f(a) = 2(a) - 5 = 2a - 5$.

Function notation makes it convenient for us to consider an operation on two functions, say $f$ and $g$, which leads to new functions which pair $x$ with $f[g(x)]$ or $g[f(x)]$. The new functions are called **composite functions** of $f$ and $g$.

**Example.** If $f$ is defined by $f(x) = x^2 + 1$ and $g$ is defined by $g(x) = 2x + 3$, find $f[g(x)]$ and $g[f(x)]$.

*Solution.* By replacing $x$ in $f(x)$ with $2x + 3$ which is $g(x)$, we obtain

$$f[g(x)] = f(2x + 3) = (2x + 3)^2 + 1$$

$$= 4x^2 + 12x + 9 + 1$$

$$= 4x^2 + 12x + 10.$$

By replacing $x$ in $g(x)$ with $x^2 + 1$, which is $f(x)$, we obtain

$$g[f(x)] = g(x^2 + 1) = 2(x^2 + 1) + 3$$

$$= 2x^2 + 2 + 3$$

$$= 2x^2 + 5.$$

## EXERCISE 1.3

### A.

Write the relation $A \times B$ for the given sets $A$ and $B$.

1. $A = \{1, 2, 3\}$, $B = \{4, 5\}$.
2. $A = \{a, b\}$, $B = \{-2, -1, 0\}$.
3. $A = \{0, 2, 4\}$, $B = \{1\}$.
4. $A = \{3\}$, $B = \{-2, 0, 2\}$.

5. $A = \{x\}$, $B = \{0, 1, 2, 3\}$.
6. $A = \{a, b, c, d\}$, $B = \varnothing$.
7. $A = \{2\}$, $B = \{1, 2, 3, \ldots\}$.
8. $A = \{1, 2, 3, \ldots\}$, $B = \{5\}$.

Specify the domain and the range for each set of ordered pairs and state which relations are also functions.

***Examples.***   (a) $\{(1, 2), (1, 3), (5, 2), (5, 3)\}$.
          (b) $\{(1, 5), (2, 10), (3, 15), (4, 20)\}$.

*Solutions.*   (a) Domain is $\{1, 5\}$ and range is $\{2, 3\}$. Since each element in the domain is paired with both elements of the range, this relation is not a function.

          (b) Domain is $\{1, 2, 3, 4\}$ and range is $\{5, 10, 15, 20\}$. This relation is a function since each element in the domain is paired with one and only one element in the range.

**9.** $\{(0, 2), (0, 4), (3, 2), (3, 4)\}$.
**10.** $\{(3, 1), (9, 1), (27, 1), (8, 2)\}$.
**11.** $\{(2, 0), (4, 0), (8, 0), (16, 1), (32, 1)\}$.
**12.** $\{(3, 5), (3, 6), (3, 7), (3, 8), (3, 9)\}$.
**13.** $\{(0, -1), (1, -2), (2, -3), (3, -4), (4, -5), (5, -6)\}$.
**14.** $\{(1, 2), (2, 3), (3, 4), (4, 5), (5, 6)\}$.

In Problems 15 to 20, rewrite the left-hand member of each equation and show that the equation is an identity in $R \times R$.

**15.** $(x - 2y)(3y + x) = xy - 6y^2 + x^2$.
**16.** $(x + y)^2 - 3y^2 = x^2 + 2xy - 2y^2$.
**17.** $(x + y)^2 - (x - y)^2 = 4xy$.
**18.** $(x + 2y)(x - 2y) - x^2 = -4y^2$.
**19.** $\dfrac{y^2}{4} + \dfrac{x^2 + 2xy}{4} = \dfrac{(x + y)^2}{4}$.
**20.** $(y - 2)^3 = y^3 - 6y^2 + 12y - 8$.

In Problems 21 to 24, show by counterexample that the equations are not identities in $R \times R$.

**21.** $\dfrac{x}{4} + \dfrac{y}{3} = \dfrac{x + y}{12}$.                **23.** $x^2 + y = y - x^2$.

**22.** $\dfrac{x}{y} = \dfrac{x^2}{y^2}$.                **24.** $(x + y)^2 = x^2 + y^2$.

In Problems 25 to 32 specify:

(a) The domain.         (b) The rule for association.
(c) The range.          (d) The relation by listing the members.

***Example.***   $\{(x, y) \mid y = x - 3, x \in \{1, 2, 3\}\}$.

*Solution.*
(a) The domain is $\{1, 2, 3\}$.

(b) The rule for association is $y = x - 3$.

(c) Replacing $x$ with each element of the domain in turn, you can find the associated values of $y$ to be $-2$, $-1$, and 0 respectively. Thus the range is $\{-2, -1, 0\}$.

(d) The relation is $\{(1, -2), (2, -1), (3, 0)\}$.

25. $\{(x, y) \mid y = 2x + 1, x \in \{-2, 0, 2\}\}$.
26. $\{(x, y) \mid xy = 2, x \in \{-2, -1, 1, 2\}\}$.
27. $\{(x, y) \mid 2y + x = 6, x \in \{2, 4, 6, 8\}\}$.
28. $\{(x, y) \mid x = \frac{1}{5}y - 3, x \in \{\frac{1}{5}, \frac{1}{10}, \frac{1}{15}, \frac{1}{20}\}\}$.
29. $\{(x, y) \mid y = 2\sqrt{x^2 + 1}, x \in \{0, \sqrt{3}, \sqrt{8}, \sqrt{15}\}\}$.
30. $\{(x, y) \mid y = 3\sqrt{16 - x^2}, x \in \{-4, -\sqrt{12}, 0, \sqrt{12}, 4\}\}$.
31. $\{(x, y) \mid y = |x|, x \in \{-4, -2, 0, 2, 4\}\}$.
32. $\{(x, y) \mid y = |x - 3|, x \in \{-2, -1, 0, 1, 2\}\}$.

For what replacement values of $x$, $x \in R$, are the following relations undefined?

33. $\left\{(x, y) \mid y = \dfrac{3}{x - 2}\right\}$.

34. $\left\{(x, y) \mid y = \dfrac{2x}{4 - 3x}\right\}$.

State whether or not the given equation defines a function for values $x \in R$ for which $y \in R$.

***Example.*** (a) $y = x^2$.  (b) $y = \pm\sqrt{x^2 + 4}$.

*Solution.* (a) Yes; for each $x$ there is only one value of $y$.
(b) No; for each $x$ there are two values of $y$.

35. $y = -x^3$.

37. $y^2 = x^3$.

36. $y = \pm\sqrt{x^2 + 5}$.

38. $xy = 3$.

In Problems 39 to 42, $f(x) = x^2 - 2$. Find the value of each of the following.

***Example.*** $f(3)$.

*Solution.* Replacing $x$ with 3 in $f(x) = x^2 - 2$ gives

$$f(3) = (3)^2 - 2 = 9 - 2 = 7.$$

39. $f(0)$.  40. $f(-3)$.  41. $f(a)$.  42. $f(a - 1)$.

43. If $f(x) = 2x - 1$ and $g(x) = x + 3$, find $f[g(0)]$.
44. If $f(x) = 2x + 1$ and $g(x) = 3x - 2$, find $g[f(0)]$.
45. If $f(x) = 2x + 1$ and $g(x) = x^2 - 1$, find $f[g(x)]$.
46. If $f(x) = 2x^2 + x$ and $g(x) = x - 1$, find $f[g(x)]$.
47. If $f(x) = x + 3$ and $g(x) = x - 3$, find $f[g(x)]$.
48. If $f(x) = 2x + 1$ and $g(x) = (x - 1)/2$, find $g[f(x)]$.

## 1.4
### Graphs of Relations and Functions

As you probably recall from your study of algebra, an association exists between ordered pairs of real numbers and the points on a geometric plane, sometimes called the **real plane**. If two number lines are drawn perpendicular to each other at their origins as in Figure 1.5, they form a rectangular coordinate system. These perpendicular number lines are called **axes** (singular,

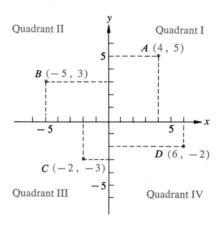

**FIGURE 1.5**

axis). The horizontal line is usually called the **x axis** and the vertical line is usually called the **y axis**. The x axis and the y axis divide the plane into four regions called **quadrants** as shown.

Any point in the plane can be associated with a unique ordered pair of real numbers. For example, the point A, in Figure 1.5, is in Quadrant I and can be associated with the ordered pair (4, 5), where 4 represents the distance of the point to the right of the y axis and 5 represents the distance of the point above the x axis. The points B, C, and D correspond to the ordered pairs (−5, 3), (−2, −3), and (6, −2), respectively.

The *components of an ordered pair* associated with a point in the plane are called the **coordinates** of the point. The first component is called the **abscissa** of the point; the second component is called the **ordinate** of the point. The *point* is called the **graph** of the ordered pair. The entire set of points in the plane is the graph of $R \times R$. An ordered pair is sometimes used as a name for the corresponding point. Thus, we occasionally speak of the point (2, 3), the point (4, −3), or in general, the point (x, y).

Recall that in Section 1.3, a function was defined as a set of ordered pairs. Consequently, a function—or at least a part of a function—can now be displayed as a set of points on the plane. The phrase "part of a function" is used here since the domain or range or both may be the infinite set of real numbers, which cannot be shown on a finite line. For example, consider the function defined by

$$y = x - 4.$$

If arbitrary values are assigned to $x$, say $-3, 0, 3$, and $6$, and the corresponding values for $y$ are computed, the four ordered pairs

$$(-3, -7), (0, -4), (3, -1), \text{ and } (6, 2)$$

are obtained which are solutions to the equation. These points can be located on a coordinate system as shown in Figure 1.6a. These points appear to lie

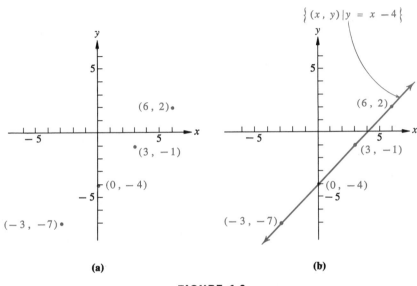

**FIGURE 1.6**

on a straight line, and in fact they do. It can be shown (we shall not do so) that the coordinates of any point on the line (Figure 1.6b) constitute a solution of the first-degree equation in two variables, $y = x - 4$, and conversely, every solution of $y = x - 4$ corresponds to a point on this line. The line is referred to as the *graph of the function* defined by $y = x - 4$, or alternately, as the *graph of the solution set of the equation*, or simply as the *graph of the equation*. Obviously, only part of the line can be displayed, so an arrowhead is placed at both ends to indicate that the graph, or line, continues in both directions indefinitely. Furthermore, since the graphs of first-degree equations of the form

$$y = ax + b$$

are straight lines, it is only necessary to find two ordered pairs in the solution set to sketch the graph. Such equations are also called **linear equations**, and the functions defined by these equations are called **linear functions**.

It can be shown that the ratio of the difference in values of $y$ to the difference in values of $x$ between any two points on a straight line is constant. This ratio is called the **slope** of the line. Thus, the slope designated by $m$ is given by

$$m = \frac{y_2 - y_1}{x_2 - x_1}, \qquad (x_2 \neq x_1).$$

See Figure 1.7a. For example, in Figure 1.7b, the difference in values of $y$ is $6 - 3$, or $3$, and the difference in values of $x$ is $5 - 1$, or $4$, and the slope of the line is $3/4$. In Figure 1.7c, the slope of the *horizontal line* $(y_2 = y_1)$ is equal to *zero* and the slope of the *vertical line* $(x_2 = x_1)$ is *undefined*.

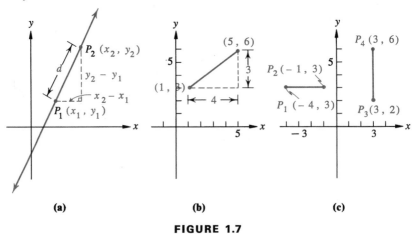

(a)                    (b)                    (c)

**FIGURE 1.7**

If $P_2$ is to the right of $P_1$ as in Figure 1.7a, $x_2 - x_1$ will be greater than zero and the slope will be positive or negative as $y_2 - y_1$ is positive or negative. Since

$$\frac{y_2 - y_1}{x_2 - x_1} = \frac{-(y_2 - y_1)}{-(x_2 - x_1)} = \frac{y_1 - y_2}{x_1 - x_2},$$

the order in which the points are considered is immaterial.

For any two points $P_1$ and $P_2$ on a line, the set of points containing $P_1$ and $P_2$ and all points lying between $P_1$ and $P_2$ is called a **line segment**. We shall designate the line segment by $\overline{P_1 P_2}$.

In addition to the property of slope, a line segment has the important property of length. If we designate the end points of a line segment $\overline{P_1 P_2}$ with the pairs of coordinates $(x_1, y_1)$ and $(x_2, y_2)$, as in Figure 1.7a, the measure or length of the line segment, which we will designate by $l(\overline{P_1 P_2})$ or some variable such as $d$, can be determined by an application of the Pythagorean theorem. From the right triangle shown, we observe that the lengths of the perpendicular sides are $y_2 - y_1$ and $x_2 - x_1$; the square of the distance, or length, of the line segment $P_1 P_2$ is given by

$$d^2 = (x_2 - x_1)^2 + (y_2 - y_1)^2.$$

By considering only the positive square root of the right-hand member, we have that

$$d = \sqrt{(x_2 - x_1)^2 + (y_2 - y_1)^2}. \tag{1}$$

*Example.*   Find the distance $d$ between the pairs of points $(1, 3)$ and $(5, 6)$.

*Solution.*   See Figure 1.7b. Replacing $(x_1, y_1)$ with $(1, 3)$ and $(x_2, y_2)$ with $(5, 6)$ in Equation (1) yields

$$d = \sqrt{(5 - 1)^2 + (6 - 3)^2}$$
$$= \sqrt{16 + 9} = \sqrt{25} = 5.$$

In Equation (1), the positive square root of the right-hand member was taken because the positive real numbers are normally assigned as lengths or measures of line segments. However, sometimes line segments (and as you shall see later, angles) are associated with a direction in which case they are referred to as **directed segments** (and **directed angles**). These are assigned either positive or negative real numbers, depending on the direction of the segment or angle. This notion is particularly useful if the line segment is either parallel to the $x$ axis or to the $y$ axis. For example, in Figure 1.8, where $P_1$ and $P_2$ lie on the same horizontal line, the directed distance (or directed length) between them is considered positive for $\overrightarrow{P_1 P_2}$ and negative for $\overrightarrow{P_2 P_1}$. Similarly, where $P_3$ and $P_4$ lie on the same vertical line, the directed distance between them is considered positive for $\overrightarrow{P_3 P_4}$ and negative for $\overrightarrow{P_4 P_3}$. We use an arrowhead on the bar in the appropriate direction as shown to indicate a directed line segment.

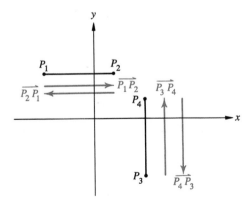

**FIGURE 1.8**

*Examples.*  Find each directed length, given the points

$$P_1(2, 6), \quad P_2(5, 6), \quad P_3(-2, -4), \text{ and } P_4(-2, 3).$$

(a) $l(\overrightarrow{P_1P_2})$.        (b) $l(\overrightarrow{P_2P_1})$.        (c) $l(\overrightarrow{P_3P_4})$.        (d) $l(\overrightarrow{P_4P_3})$.

*Solutions.*

(a) $l(\overrightarrow{P_1P_2}) = 5 - 2 = 3.$

(b) $l(\overrightarrow{P_2P_1}) = 2 - 5 = -3.$

(c) $l(\overrightarrow{P_3P_4}) = 3 - (-4) = 7.$

(d) $l(\overrightarrow{P_4P_3}) = -4 - 3 = -7.$

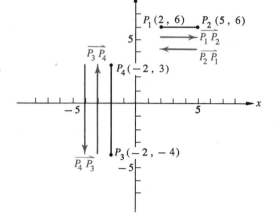

As with first-degree equations in two variables, solutions of higher degree equations in two variables are also ordered pairs, which can be found by arbitrarily assigning values to $x$ and finding the associated values of $y$. For example, in the equation

$$y = x^2 + 2,$$

if $x = -2$, then

$$y = (-2)^2 + 2 = 6,$$

and $(-2, 6)$ is a solution. Similarly, by replacing $x$ with $-1, 0, 1$, and $2$, we obtain $(-1, 3), (0, 2), (1, 3)$, and $(2, 6)$ as additional solutions. Plotting these points on the plane, we have the graph in Figure 1.9a. Apparently, these

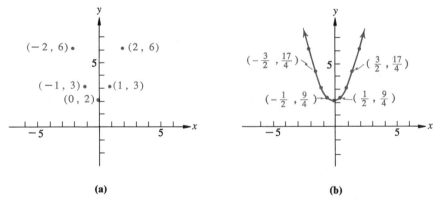

(a)                                                  (b)

**FIGURE 1.9**

points do not lie on a straight line. By plotting additional solutions of $y = x^2 + 2$, such as

$$\left(-\frac{3}{2}, \frac{17}{4}\right), \left(-\frac{1}{2}, \frac{9}{4}\right), \left(\frac{1}{2}, \frac{9}{4}\right), \text{ and } \left(\frac{3}{2}, \frac{17}{4}\right),$$

we have the additional points in Figure 1.9b. These points can now be connected in sequence from left to right by a smooth curve as shown. We assume that this curve is a good approximation to the graph of $y = x^2 + 2$. We have no absolute assurance at this time that a smooth curve results from connecting individual points regardless of the number of individual ordered pairs that we graph. Proving that the graph is indeed a smooth curve requires the use of the concept of continuity from calculus and is not attempted in this book. The statement that a function defined by an equation $y = f(x)$ is continuous in an interval in the domain implies that $y$ exists for each $x \in R$ and that there is a small difference in values of $y$ for a small difference in values of $x$ in the interval.

The curve in Figure 1.9b is an example of a **parabola**. More generally, the graph of any equation of the form

$$y = ax^2 + bx + c, \qquad (a, b, c, x \in R, a \neq 0),$$

is called a **parabola**. Notice that such an equation defines a function since for each $x$, an equation of this form will associate one and only one $y$. Such functions are called **quadratic functions**, the word "quadratic" indicating that the highest power of a variable in the defining equation is 2.

The quadratic equation

$$x^2 + y^2 = r^2,$$

where $r$ is a real number greater than 0, is of special interest. Its graph is a circle with center at the origin and radius of length $r$. However, although this equation defines a relation, it does not define a function.

Whether or not a relation is also a function can be determined by inspecting its graph. For example, consider the typical graphs of $y = ax^2 + bx + c$, $x^2 + y^2 = r^2$, and $y \geq ax^2 + bx + c$ in Figure 1.10. Imagine a vertical line

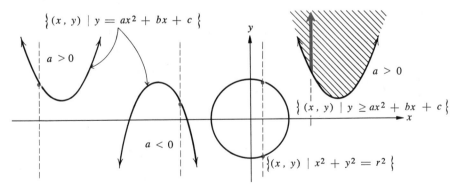

**FIGURE 1.10**

moving across each graph from left to right. Since the line cuts the graph of $y = ax^2 + bx + c$ at *only one point at each position, this equation defines a function*; that is, for each $x$ in this relation, there is only one $y$. Since the line cuts the graphs of $x^2 + y^2 = r^2$ and $y \geq ax^2 + bx + c$ at *more than one point at certain positions, the equation and the inequality do not define functions,* although they do define relations. There are ordered pairs in either relation which have the same first components and different second components.

### EXERCISE 1.4

A.

Find the missing component so that the ordered pair will satisfy the given equation.

**Examples.**   $2y + x = 5$:   (a) $(-1, \ \ )$.          (b) $( \ \ , 0)$.

*Solutions.*   (a) Replacing $x$ with $-1$ in the equation yields

$$2y + (-1) = 5,$$

$$y = 3.$$

Thus, the ordered pair $(-1, 3)$ is a solution of the equation.

(b) Replacing $y$ with 0 in the equation yields

$$2(0) + x = 5,$$

$$x = 5.$$

The ordered pair $(5, 0)$ is a solution of the equation.

**1.** $y = x + 8$:     (a) $(0, \ \ )$.     (b) $( \ \ , 0)$.     (c) $(-5, \ \ )$.
**2.** $y - 2x = 3$:     (a) $(0, \ \ )$.     (b) $( \ \ , 0)$.     (c) $(4, \ \ )$.
**3.** $2y + 4x = 5$:     (a) $(0, \ \ )$.     (b) $( \ \ , 0)$.     (c) $(-\frac{1}{2}, \ \ )$.
**4.** $\frac{1}{2}y = \frac{1}{3}x - 1$:     (a) $(0, \ \ )$.     (b) $( \ \ , 0)$.     (c) $(3, \ \ )$.

Graph each function in $R \times R$.

**•5.** $\{(x, y) \mid y = 2x + 3\}$.                    **8.** $\{(x, y) \mid 3y = x + 6\}$.
**6.** $\{(x, y) \mid y = x - 3\}$.                    **•9.** $\{(x, y) \mid y - x = 0\}$.
**7.** $\{(x, y) \mid y = -\frac{1}{2}x\}$.                    **10.** $\{(x, y) \mid y + x = 0\}$.

**11.** $\{(x, y) \mid y = -2\}$.   *Hint:* Consider $\{(x, y) \mid 0x + y = -2\}$.
**12.** $\{(x, y) \mid y = 5\}$.

Given the points $P_1(4, -3)$, $P_2(10, -3)$, $P_3(5, 1)$, $P_4(5, -5)$ and $P_5(5, -3)$, find each directed length.

**Example.**   $l(\overrightarrow{P_1 P_2}) = 10 - 4 = 6$.

**13.** $l(\overrightarrow{P_2 P_1})$.                    **15.** $l(\overrightarrow{P_4 P_3})$.                    **17.** $l(\overrightarrow{P_5 P_2})$.
**14.** $l(\overrightarrow{P_3 P_4})$.                    **16.** $l(\overrightarrow{P_1 P_5})$.                    **18.** $l(\overrightarrow{P_5 P_4})$.

Find the distance between each of the given pairs of points and find the slope of the line segment joining them.

*Example.* $(3, -2), (-1, -4)$.

*Solution.* Considering $(3, -2)$ and $(-1, -4)$ as the points $P_1$ and $P_2$, respectively, then the distance

$$l(\overrightarrow{P_1P_2}) = \sqrt{(-1-3)^2 + (-4-[-2])^2}$$

$$= \sqrt{(-4)^2 + (-4+2)^2}$$

$$= \sqrt{16+4} = \sqrt{20} = 2\sqrt{5};$$

$$\frac{y_2 - y_1}{x_2 - x_1} = \frac{-4-(-2)}{-1-3} = \frac{-4+2}{-4} = \frac{1}{2}.$$

**19.** $(3, 5), (1, 2)$.                     **21.** $(-2, -6), (2, 6)$.
**20.** $(-3, 4), (1, -1)$.                   **22.** $(-1, 3), (0, 0)$.

Graph each relation in $R \times R$.

*Example.*  $\{(x, y) \mid y = 4 - 3x - x^2\}$.

*Solution.*  Replacing $x$ in the defining equation with some integral values, say $-5, -4, -3, -2, -1, 0, 1,$ and $2$, corresponding values of $y$ are obtained. Thus, some solutions in the relation are $(-5, -6),$ $(-4,0), (-3,4), (-2,6),$ $(-1, 6), (0, 4), (1, 0),$ and $(2, -6)$. The graph of these points and the parabola passing through them are shown in the figure.

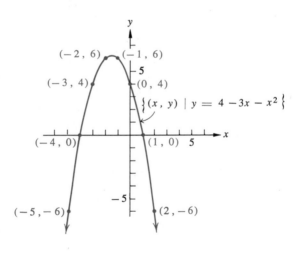

**•23.** $\{(x, y) \mid y = 2x^2 - x\}$.                **26.** $\{(x, y) \mid y = -x^2 + x + 2\}$.
**24.** $\{[x, f(x)] \mid f(x) = -x^2 - 1\}$.        **•27.** $\{(x, y) \mid x^2 + y^2 = 1\}$.
**25.** $\{[x, f(x)] \mid f(x) = x^2 + 3x + 2\}$.    **28.** $\{(x, y) \mid x^2 + y^2 = 9\}$.

By graphical means find approximations to the nearest $\frac{1}{2}$ unit for the elements in the set formed by each intersection.

***Example.***   $\{(x, y)\,|\,y = x^2\} \cap \{(x, y)\,|\,y = 2\}$.

*Solution.*   Graph each set.
The $x$ coordinates of the
points of intersection are approximately $-1.5$ and $1.5$.
Therefore

$\{(x, y)\,|\,y = x^2\}$

$\qquad \cap \{(x, y)\,|\,y = 2\}$

$= \{(\approx -1.5, 2), (\approx 1.5, 2)\}$.

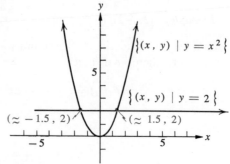

29. $\{(x, y)\,|\,y = x^2 + 5x\} \cap \{(x, y)\,|\,y = 3\}$.
30. $\{(x, y)\,|\,y = 4 - x^2\} \cap \{(x, y)\,|\,y = 2\}$.
■31. $\{(x, y)\,|\,y = x^2 - 3x - 4\} \cap \{(x, y)\,|\,y = 1\}$.
32. $\{(x, y)\,|\,y = x^2 + 5x + 4\} \cap \{(x, y)\,|\,y = -4\}$.
33. $\{(x, y)\,|\,y = x^2 - 6x - 16\} \cap \{(x, y)\,|\,y = 0\}$.
34. $\{(x, y)\,|\,y = x^2 - 7x + 10\} \cap \{(x, y)\,|\,y = 0\}$.
■35. $\{(x, y)\,|\,y = x^2 - 9\} \cap \{(x, y)\,|\,y = x + 1\}$.
36. $\{(x, y)\,|\,y = 4 - x^2\} \cap \{(x, y)\,|\,y = x^2 - 4\}$.

**B.**

Graph each relation in $R \times R$.

***Example.***   $\{(x, y)\,|\,y > x^2 - 5x + 4\}$.

*Solution.*   First graph $y = x^2 - 5x + 4$, shown as a dashed curve in the
figure. Then shade the portion of the plane above the graph of
$y = x^2 - 5x + 4$. *Note:* Had the inequality been $y \geq x^2 - 5x + 4$, the curve
would be a part of the graph of the solution set and would be shown as a solid
curve.

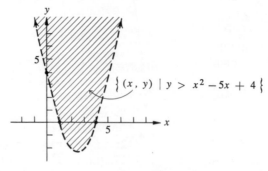

37. $\{(x, y)\,|\,y > x\}$.
38. $\{(x, y)\,|\,y \leq x\}$.
■39. $\{(x, y)\,|\,y \leq 2x - 4\}$.
40. $\{(x, y)\,|\,y > x + 5\}$.

41. $\{(x, y)\,|\,y \geq x^2 - 4\}$.
42. $\{(x, y)\,|\,y \leq 4 - x^2\}$.
■43. $\{(x, y)\,|\,x^2 + y^2 < 9\}$.
44. $\{(x, y)\,|\,x^2 + y^2 \geq 9\}$.

**45.** $\{(x, y) \mid y \le 3 - x\} \cap \{(x, y) \mid y \ge x\}.$
**46.** $\{(x, y) \mid y > 2x - 6\} \cap \{(x, y) \mid y > 2\}.$
**·47.** $\{(x, y) \mid y \ge x^2 - 1\} \cap \{(x, y) \mid y < x + 1\}.$
**48.** $\{(x, y) \mid y \le 4 - x^2\} \cap \{(x, y) \mid y \ge 0\}.$
**49.** $\{(x, y) \mid x^2 + y^2 < 16\} \cap \{(x, y) \mid y > x^2\}.$
**50.** $\{(x, y) \mid x^2 + y^2 \ge 16\} \cap \{(x, y) \mid x^2 + y^2 \le 36\}.$

## 1.5
### Zeros of Functions; Equations in One Variable

In any function, the element(s) in the domain which are paired with the element 0 in the range are called **zeros of the function.** Thus, the zeros of

$$f = \{(-3, 1), (-2, 0), (-1, -2), (0, -4), (1, 0)\}$$

are $-2$ and $1$ because each of these first components is paired with a second component 0. The graph of each such ordered pair lies on the $x$ axis, as

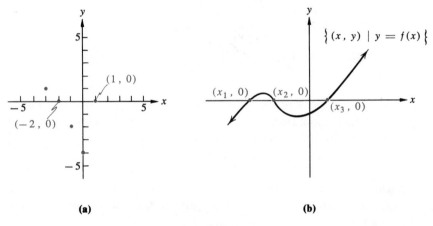

**(a)**                                    **(b)**

**FIGURE 1.11**

shown in Figure 1.11a. For any function defined by an equation

$$y = f(x),$$

the zeros of the function (see Figure 1.11b) are those values of $x$ $(x_1, x_2, x_3)$ for which

$$y = f(x) = 0.$$

These values of $x$ are called **solutions** of the equation in one variable $f(x) = 0$, and the set of such numbers is called the **solution set** of the equation.

Thus, there are three different names for a single idea:

**1.** The zeros of $\{(x, y) \mid y = f(x), x \in R\}$.
**2.** The $x$ intercepts of the graph of $\{(x, y) \mid y = f(x), x \in R\}$.
**3.** Solutions of $f(x) = 0$, $x \in R$.

In many simple cases such as

$$x - 2 = 0,$$

the solution set of an equation in one variable is evident by inspection. However, this is not the case for most first-degree equations. Recall from your work in algebra that you can find solutions of all first-degree equations in one variable by generating **equivalent equations** (a set of equations with the same solution set) until you obtain an equation whose solution set is evident by inspection. For example, consider the equation

$$\frac{2x - 1}{3} = 11.$$

Perhaps the solution set is evident to you simply by inspection. If not, you can generate an equivalent equation (see Appendix B, page 288) by multiplying each member by 3 to obtain

$$3\left(\frac{2x - 1}{3}\right) = 3 \cdot 11,$$

$$2x - 1 = 33,$$

and then adding 1 to each member to obtain the equivalent equation

$$2x = 34.$$

Probably the replacement for $x$ for which this equation is true can now be obtained by inspection. However if necessary, you can multiply each member of the equation by $\frac{1}{2}$ to obtain the equivalent equation

$$x = 17,$$

from which it is now certainly evident that the solution set is {17}.

Since the equations in this example are all equivalent equations, the solution set of

$$\frac{2x - 1}{3} = 11$$

is also {17}.

Recall from your study of algebra that solutions of quadratic equations in one variable are obtained by various methods. For example if the left-hand member of a quadratic equation of the form

$$ax^2 + bx + c = 0, \qquad (a, b, c \in R, a \neq 0)$$

is factorable, the solutions can be obtained from the fact that the left-hand member will equal zero for values of $x$ for which one or both of the factors equals zero (see Appendix B, page 288). Such a procedure of solving a quadratic equation is called *solution by factoring*.

***Example.*** Solve $x^2 + x = 30$.

*Solution.* First write the equation equivalently as

$$x^2 + x - 30 = 0, \tag{1}$$

in which form the right-hand member equals zero. Factoring the left-hand member, yields

$$(x + 6)(x - 5) = 0. \tag{2}$$

The left-hand member equals 0 for values of $x$ for which

$$x + 6 = 0 \quad \text{or} \quad x - 5 = 0 \quad \text{or both.}$$

By inspection you can now observe that the solution set of (2), and therefore also of (1), is $\{5, -6\}$.

Quadratic equations of the form

$$x^2 = a, \quad (a \in R, a \geq 0)$$

can be solved by a method called *extraction of roots*. If the equation has a solution, then $x$ must be a square root of $a$ (see Definition 1.4, page 5). Since each positive real number $a$ has two square roots, the solution set, for $a > 0$, is $\{\sqrt{a}, -\sqrt{a}\}$. If $a = 0$, the solution set is $\{0\}$.

From the method of extraction of roots applied to the equation

$$(x - a)^2 = b, \quad (a, b \in R, b \geq 0),$$

we obtain

$$x - a = \sqrt{b} \quad \text{and} \quad x - a = -\sqrt{b},$$

from which the solution set $\{a + \sqrt{b}, a - \sqrt{b}\}$ is obtained. From this it follows that we can find the solution set of any quadratic equation by first rewriting the equation in the form

$$(x - a)^2 = b.$$

This procedure is called solving a quadratic equation by *completing the square*.

Applying this procedure to the general quadratic equation,

$$ax^2 + bx + c = 0, \quad (a, b, c \in R, a \neq 0),$$

results in the formula

$$x = \frac{-b \pm \sqrt{b^2 - 4ac}}{2a},$$

called the **quadratic formula**, which gives us the solutions of any quadratic equation in the form $ax^2 + bx + c = 0$ expressed in terms of the coefficients

$a, b,$ and $c$ (see Exercise 1.5, Problem 31). The solution set is

$$\left\{ \frac{-b + \sqrt{b^2 - 4ac}}{2a}, \frac{-b - \sqrt{b^2 - 4ac}}{2a} \right\}.$$

The elements in the solution set are real numbers if and only if $b^2 - 4ac \geq 0$.

**Example.**   Solve $x^2 - 5x = 6$ by using the quadratic formula.

*Solution.*   First, rewrite the equation as

$$x^2 - 5x - 6 = 0.$$

Then replacing $a$ with 1, $b$ with $-5$, and $c$ with $-6$ in the quadratic formula yields

$$x = \frac{-(-5) + \sqrt{(-5)^2 - 4(1)(-6)}}{2(1)}$$

or

$$x = \frac{-(-5) - \sqrt{(-5)^2 - 4(1)(-6)}}{2(1)},$$

from which

$$x = \frac{5 + \sqrt{25 + 24}}{2} = 6 \quad \text{or} \quad x = \frac{5 - \sqrt{25 + 24}}{2} = -1,$$

and the solution set is $\{6, -1\}$.

## EXERCISE 1.5

**A.**

Find the set of zeros of each function.

**1.** $\{(-3, 4), (-1, 2), (1, 0)\}.$
**2.** $\{(1, 0), (3, 5), (4, 0)\}.$

**3.** $\{(-1, 0), (0, 1), (1, 2)\}.$
**4.** $\{(-2, 0), (0, 0), (2, 0)\}.$

**Example.**   $\{(x, y) \mid y = x^2 - 3x\}.$

*Solution.*   The values of $x$ for which $y = 0$ are the zeros of the function. Thus, we seek the solution set of the equation

$$x^2 - 3x = 0.$$

Factoring the left member yields

$$x(x - 3) = 0,$$

from which the solution set is $\{0, 3\}$, and the set of zeros of the function above is $\{0, 3\}$.

**5.** $\{(x, y)\,|\,y = 4x - 5\}$.

**6.** $\{(x, y)\,|\,y = 2x + 3\}$.

**7.** $\{(x, y)\,|\,3y + 2 = -x\}$.

**8.** $\{(x, y)\,|\,2y = 3(x - 1)\}$.

**9.** $\{(x, y)\,|\,y = x^2 - 2\}$.

**10.** $\{(x, y)\,|\,y = -x^2 + 4\}$.

**11.** $\{(x, y)\,|\,y = x^2 - 7x + 6\}$.

**12.** $\{(x, y)\,|\,y = -x^2 - 8x + 9\}$.

Find the $x$ intercept(s) of the graph of the function. Graph the function.

**Example.** $\{(x, y)\,|\,y = x^2 - 5x + 4\}$.

*Solution.* The graph of the function will intersect the $x$ axis at the point whose $y$ coordinate is zero. Thus, we seek the solution set of the equation

$$x^2 - 5x + 4 = 0.$$

Factoring the left-hand member yields

$$(x - 4)(x - 1) = 0,$$

whose solution set is $\{1, 4\}$. The $x$ intercepts are 1 and 4. Some additional solutions of the defining equation can be obtained to complete the graph shown in the figure.

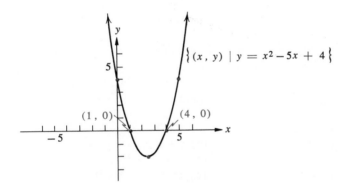

**•13.** $\{(x, y)\,|\,\frac{1}{2}x^2 - 2x = y\}$.

**14.** $\{(x, y)\,|\,y = -2x^2 + 5x\}$.

**15.** $\{(x, y)\,|\,y = x^2 + 3x + 2\}$.

**16.** $\{(x, y)\,|\,2x^2 - x - 3 = y\}$.

In Problems 17 to 20 use the method of extraction of roots to find the solution set.

**17.** $x^2 = 7$.

**18.** $3x^2 = 5$.

**19.** $(x - 3)^2 = 16$.

**20.** $(x + a)^2 = 7$.

Use the quadratic formula to find the solution set of each equation.

**21.** $x^2 - 6 = 5x$.

**22.** $2x^2 - x - 2 = 0$.

**23.** $x^2 + 5x - 2 = 0$.

**24.** $3x^2 + 6x + 1 = 0$.

**25.** $\dfrac{x^2}{3} = \dfrac{x}{2} + \dfrac{3}{2}$.

**26.** $6x = x^2 + 1$.

**C.**

Use graphical methods to solve each of the following inequalities.

*Examples.* (a) $x^2 - 7x + 10 \le 0$.      (b) $x^2 - 7x + 10 > 0$.

*Solution.* (a) Graph the equation $y = x^2 - 7x + 10$. The graphs of all values of $y$ or $x^2 - 7x + 10$ which are less than zero lie *below* the $x$ axis and the graphs of the corresponding values of $x$ are shown in color on the $x$ axis. The solution set of the inequality is $\{x \mid 2 \le x \le 5\}$.

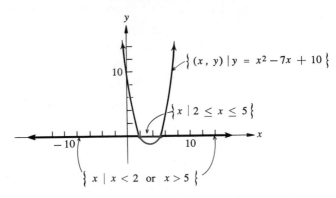

(b) The graphs of all values of $y$ or $x^2 - 7x + 10$ which are greater than zero lie *above* the $x$ axis and the graphs of the corresponding values of $x$ are all points on the $x$ axis not shown in color. The solution set of the inequality is $\{x \mid x < 2\} \cup \{x \mid x > 5\} = \{x \mid x < 2 \text{ or } x > 5\}$.

**·27.** (a) $x + 3 \le 0$.        (b) $x + 3 > 0$.
**28.** (a) $4 - x > 0$.        (b) $4 - x \le 0$.
**29.** (a) $x^2 - 4 \le 0$.        (b) $x^2 - 4 > 0$.
**30.** (a) $x^2 - 3x - 4 \ge 0$.        (b) $x^2 - 3x - 4 < 0$.

**31.** Solve the general quadratic equation $ax^2 + bx + c = 0$, $(a \neq 0)$ for $x$ in terms of $a$, $b$, and $c$ by the method of completing the square.

## 1.6

### Inverse of a Function

If the components of each ordered pair in a given relation are interchanged, the resulting relation and the given relation are called **inverses*** of each other. Thus,

$$\{(1, 3), (5, 7), (11, 13)\} \quad \text{and} \quad \{(3, 1), (7, 5), (13, 11)\}$$

are inverse relations.

---

* Some authors use the word "converses" in this case and reserve the word "inverses" in this sense only for one-to-one functions (see page 33).

The inverse of a relation $\mathscr{R}$ is denoted by $\mathscr{R}^{-1}$ (read "$\mathscr{R}$ inverse" or "the inverse of $\mathscr{R}$"). It is evident from the above example that the domain and range of $\mathscr{R}^{-1}$ are the range and domain, respectively, of $\mathscr{R}$. If $y = \mathscr{R}(x)$ defines a relation, then $x = \mathscr{R}(y)$ defines the inverse of $\mathscr{R}$ or using inverse notation, we have, $y = \mathscr{R}^{-1}(x)$. We have used the script $\mathscr{R}$ for a relation so that it should not be confused with the $R$ which represents the set of real numbers.

***Example.***   Find the inverse of the relation

$$\{(x, y)\,|\,y = 3x - 5\}. \tag{1}$$

*Solution.*   Replacing $x$ with $y$ and $y$ with $x$ in the defining equation in (1), we obtain as the inverse relation

$$\{(x, y)\,|\,x = 3y - 5\}, \tag{2}$$

or, when $y$ is expressed in terms of $x$ in the defining equation,

$$\{(x, y)\,|\,y = \tfrac{1}{3}(x + 5)\}. \tag{2'}$$

Notice that $y = 3x - 5$ from (1) and $x = 3y - 5$ from (2) define inverse relations, because the domains and ranges of the relations are interchanged. However, $x = 3y - 5$ from (2) and $y = \tfrac{1}{3}(x + 5)$ from (2') are equivalent equations and define the same relation.

The graphs of inverse relations are always located symmetrically with respect to the graph of the linear equation $y = x$. To see this, notice the graphs of the ordered pairs $(a, b)$ and $(b, a)$ in Figure 1.12. They are the same

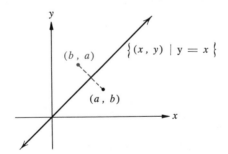

**FIGURE 1.12**

distance from, but on opposite sides of, the line which is the graph of the equation $y = x$. The graph of $y = x$ serves as a reflecting line, or mirror, for these points. Thus, because for every ordered pair $(a, b)$ in the relation $\mathscr{R}$, the ordered pair $(b, a)$ is in $\mathscr{R}^{-1}$, *the graphs of $y = \mathscr{R}(x)$ and $y = \mathscr{R}^{-1}(x)$ are reflections of each other about the graph of $y = x$.*

Using the example above, Figure 1.13 shows the graphs of

$$\{(x, y)\,|\,y = 3x - 5\}$$

and its inverse

$$\{(x, y) \mid x = 3y - 5\} \qquad \text{or} \qquad \{(x, y) \mid y = \tfrac{1}{3}(x + 5)\},$$

together with the graph of $\{(x, y) \mid y = x\}$.

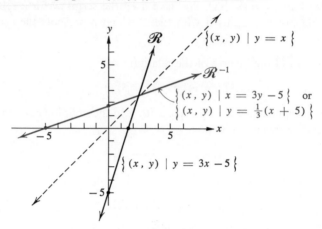

**FIGURE 1.13**

Because every function is a relation, *every function has an inverse, but the inverse is not always a function.* For example, consider the function

$$\{(x, y) \mid y = x^2\}. \tag{3}$$

Its graph, together with the graph of its inverse

$$\{(x, y) \mid x = y^2\} \qquad \text{or} \qquad \{(x, y) \mid y = \sqrt{x}\} \cup \{(x, y) \mid y = -\sqrt{x}\} \tag{4}$$

is shown in Figure 1.14. Since for all but one value of $x$ $(x = 0)$, the inverse

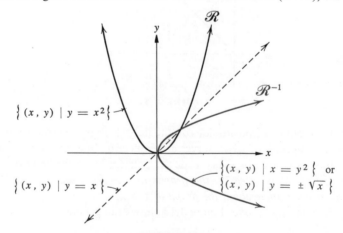

**FIGURE 1.14**

specified by (4) associates two different values of $y$ (one positive and one nega-tive) in the range, the inverse of (3) is not a function. Observe in Figure 1.14 that the graphs of these relations are symmetric with respect to the graph of $y = x$.

In order for a function to have an inverse which is also a function, it must be a *one-to-one function*; that is, each element in the domain of the original function must be associated with one and only one element in its range and each element in its range must be associated with one and only one element in its domain. Many functions have this property. Linear functions are obviously one-to-one functions. So are **exponential functions** defined by equations of the form

$$y = b^x, \qquad (b > 0, b \neq 1, x \in R).$$

Figure 1.15 shows the graphs for the cases where $b = \frac{1}{2}$ and $b = 2$. The inverses of these functions are called **logarithmic functions** and are discussed in detail in Appendix A.

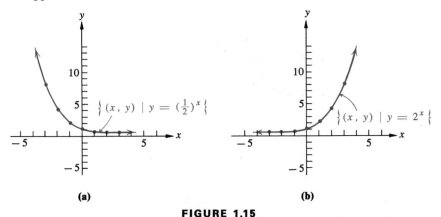

(a)                                      (b)

**FIGURE 1.15**

**EXERCISE 1.6**

A.

Write the inverse of each relation.

1. $\{(0, 1), (1, 3), (2, 5), (3, 7)\}$.

3. $\left\{(0, 1), \left(\frac{\pi}{6}, \frac{1}{2}\right), \left(\frac{\pi}{3}, \frac{\sqrt{3}}{2}\right), \left(\frac{\pi}{2}, 0\right)\right\}$.

2. $\{(-3, 1), (-2, 4), (-1, 7), (0, 2)\}$.

4. $\left\{\left(\frac{\pi}{2}, 1\right), (\pi, 0), \left(\frac{3\pi}{2}, -1\right)\right\}$.

In Problems 5 to 12, write the inverses of the relations showing the de-fining equation with $y$ expressed in terms of $x$. Graph the two relations, including the graph of $y = x$.

**Example.** $\{(x, y) | y = 2x + 6\}$.

(*Solution on the next page.*)

*Solution.* Replacing $x$ with $y$ and $y$ with $x$ in the defining equation $y = 2x + 6$, yields the inverse relation (or function, in this case)

$$\{(x, y) \mid x = 2y + 6\}$$

or, when $y$ is expressed in terms of $x$,

$$\{(x, y) \mid y = \tfrac{1}{2}x - 3\}.$$

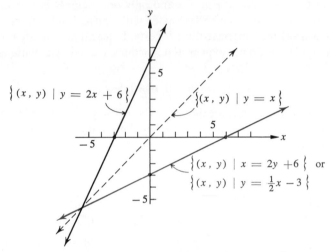

**5.** $\{(x, y) \mid x + 2y = 5\}$.        **9.** $\{(x, y) \mid y = -x^2\}$.

**6.** $\{(x, y) \mid 3x - y = 6\}$.        **10.** $\{(x, y) \mid y = x^3\}$.

**7.** $\{(x, y) \mid y = 6 - 2x\}$.        **11.** $\{(x, y) \mid y = 2x^2 - 2\}$.

**8.** $\{(x, y) \mid y = 3x - 4\}$.        **12.** $\{(x, y) \mid y = 4 - x^2\}$.

**B.**

**\*13.** Graph the function $\{(x, y) \mid y = 1/x\}$ and its inverse. What property does this function and its inverse have?

**14.** Graph the function $\{(x, y) \mid y = 3^x\}$. Graph its inverse. Can you express the inverse function with a defining equation giving $y$ in terms of $x$?

Consider each of the following equations which defines a function $\mathscr{F}$. Find the equation defining $\mathscr{F}^{-1}$ and show that $\mathscr{F}[\mathscr{F}^{-1}(x)] = \mathscr{F}^{-1}[\mathscr{F}(x)] = x$.

**15.** $y = 2x$.        **17.** $y = 2x - 1$.        **19.** $2x - y = 6$.

**16.** $y = x + 3$.        **18.** $y = 4 - 3x$.        **20.** $x - 3y = 3$.

## Chapter Summary

**1.** The following symbols* have been introduced: $\{\ \ \}$, $=$, $\neq$, $\in$, $\notin$, $N$, $J$, $Q$, $H$, $R$, etc., $a$, $b$, $c$, $\alpha$, $\beta$, $\gamma$, $\theta$, etc., $\{x \mid x \ldots\}$, $\cup$, $\cap$, $\varnothing$, $<$, $\leq$, $>$, $\geq$, $|a|$, $A \times B$, $(a, b)$, $R \times R$, $f$, $f(x)$, $\overline{P_1 P_2}$, $l(\overline{P_1 P_2})$, $l(\overrightarrow{P_1 P_2})$, $\mathscr{R}$, $\mathscr{R}^{-1}$.

> \* You should refer to the text for the meanings of these symbols. For convenience, the symbols and their meanings are also listed inside the front cover.

2. A **set** may be formed by *listing* the names of the members or elements, or by a *rule* which determines precisely the elements included in the set, or by *set-builder notation*, $\{x \mid \text{condition on } x\}$.
3. Two sets are **equal** if and only if they have the same members.
4. The following operations on sets lead to new sets:
    (a) the **union** of two sets, $A$ and $B$, is the set of all elements that belong to $A$ or $B$, or to both;
    (b) the **intersection** of two sets, $A$ and $B$, is the set of all elements that belong to both $A$ and $B$; that is, the elements common to both sets;
    (c) the **Cartesian product** of two sets, $A$ and $B$, where $a \in B$ and $b \in B$, is the set $A \times B$ which contains all ordered pairs $(a, b)$ that can be formed.
5. The set that contains no elements is called the **null set**, or **empty set**.
6. Two sets $A$ and $B$ which have no elements in common are said to be **disjoint.** However, if *every* element of $A$ is also an element of $B$, then $A$ is a **subset** of $B$.
7. The **absolute value** of a real number is a nonnegative number defined as follows:
    For all $a \in R$,

    $$\text{if } a \geq 0, \quad \text{then } |a| = a;$$

    $$\text{if } a < 0, \quad \text{then } |a| = -a.$$

8. The real number system forms an ordered field.
9. A **relation** is any set of ordered pairs. A **function** is a relation in which each element in the domain is associated with only one element in the range.
10. The set of all ordered pairs that satisfy an equation in two variables is called the **solution set** of the equation. The set from which ordered pairs are selected as possible solutions is called the **replacement set.**
11. Equations that are not true for all elements in the replacement set of the variables are called **conditional equations**. Equations that result in a true statement for any replacement of the variable are called **identities.**
12. The symbol $f(a)$ represents an element in the range associated with the element $a$ in the domain for the function $f$.
13. The graph of an equation of the form

    $$y = ax + b$$

    is a straight line. The equation defines a **linear function.**
14. The ratio of the difference in the values of $y$ with the difference in the values of $x$ between any two points on a straight line is called the **slope** of the line. If two points are specified by $(x_1, y_1)$ and $(x_2, y_2)$, the slope is given by

    $$m = \frac{y_2 - y_1}{x_2 - x_1}.$$

**15.** For any two points $P_1$ and $P_2$ on a line, the set of points containing $P_1$ and $P_2$ and all points lying between $P_1$ and $P_2$ is called a **line segment** and designated by $\overline{P_1P_2}$ or $\overline{P_2P_1}$. The *length* of a line segment, say $P_1P_2$, is designated by $l(\overline{P_1P_2})$ and its *directed length* by $l(\overrightarrow{P_1P_2})$.

**16.** The length $d$ of a line segment with end points $(x_1, y_1)$ and $(x_2, y_2)$ is given by the formula

$$d = \sqrt{(x_2 - x_1)^2 + (y_2 - y_1)^2}.$$

**17.** The graph of an equation of the form

$$y = a^2x + bx + c, \qquad (a, b, c, x \in R, a \neq 0)$$

is called a **parabola**. The equation defines a **quadratic function.**

**18.** The graph of the quadratic equation

$$x^2 + y^2 = r^2$$

is a circle with its center at the origin of a Cartesian coordinate system and having a radius of length $r$.

**19.** In any function, the elements in the domain which are paired with the element 0 in the range are called **zeros of the function.**

**20.** Sometimes the solution of a first-degree equation in one variable can be determined by inspection. If not, **equivalent equations** can be generated until the solution is evident.

**21.** Solutions of quadratic equations in one variable can be obtained by factoring, by extraction of roots, by completing the square, or by using the quadratic formula

$$x = \frac{-b \pm \sqrt{b^2 - 4ac}}{2a}, \qquad (a \neq 0).$$

**22.** If the components of each ordered pair in a given relation $\mathcal{R}$ are interchanged, the resulting relation $\mathcal{R}^{-1}$ and the given relation are called **inverses** of each other. The domain and range of $\mathcal{R}$ are the range and domain of $\mathcal{R}^{-1}$, respectively.

**23.** The graphs of inverse relations are always located *symmetrically* with respect to the graph of the linear equation $y = x$.

**24.** In order for a given function to have an inverse which is also a function, the given function must be a one-to-one function.

## Chapter Review

A.

Let $A = \{l, m, n\}$, $B = \{m, n, p\}$, $C = \{m, n\}$, and $D = \{n, m, l\}$. Replace the comma in each pair with $=$ or $\neq$ where the relationship is between sets and $\in$ or $\notin$ where the relationship is between an element and a set.

**1.** $p, D.$         **2.** $A, B.$         **3.** $D, A.$         **4.** $n, C.$

Use the sets $A$, $B$, $C$, and $D$ as above, and list the elements in each of the following sets.

**5.** $A \cup B$.
**6.** $B \cap C$.
**7.** $(B \cap C) \cap D$.

**8.** $A \cup \emptyset$.
**9.** $\{x \mid x \in B \text{ or } x \in C\}$.
**10.** $\{x \mid x \in C \text{ and } x \in D\}$.

Let $E = \{5, -3, \frac{7}{8}, -\sqrt{13}, 0, -\frac{1}{5}, \sqrt{6}, 2\}$. Write each of the following sets by listing the members.

**11.** $\{x \mid x \in E \text{ and } x \in J\}$.
**12.** $\{x \mid x \in E \text{ and } x \in H\}$.

**13.** $\{x \mid x \in E \text{ and } x \in Q\}$.
**14.** $\{x \mid x \in E \text{ and } x \in R\}$.

In Problems 15 and 16, graph each set on a number line. Use Table I, page 291 as necessary.

**•15.** $\{-\sqrt{5}, \frac{2}{3}, \sqrt{12}\}$.

**16.** $\{x \mid -\sqrt{8} \le x < 2, x \in R\}$.

Write the relation $A \times B$ for the given sets $A$ and $B$.

**17.** $A = \{a, b, c\}$, $B = \{c, d, e\}$.

**18.** $A = \{1, 3, 5, 7\}$, $B = \{-1, 0\}$.

Specify the domain and range for each set of ordered pairs and state which of the relations is a function.

**19.** $\{(1, -2), (2, -4), (3, -6), (4, -8), (5, -10)\}$.
**20.** $\{(a, -1), (a, 0), (b, -1), (b, 0)\}$.

State whether or not the following equations are identities or conditional equations.

**21.** $(x - 2)(x + 1) = 1$.

**22.** $(x - 1)(x + 1) - x^2 = -1$.

Specify for each set:
(a) the domain,       (b) the rule for association,
(c) the range,        (d) the relation by listing the members.

**23.** $\{(x, y) \mid y = -3x + 4, x \in \{\frac{5}{3}, \frac{7}{3}, 3, \frac{11}{3}\}\}$.
**24.** $\{(x, y) \mid y = \sqrt{4 - x^2}, x \in \{-2, -\sqrt{3}, -1, 0, 1, \sqrt{3}, 2\}\}$.

Specify the replacement values of $x$, $x \in R$, for which $y \in R$.

**25.** $y = \dfrac{x}{(x - 1)(x + 3)}$.

**26.** $y = \dfrac{2}{\sqrt{4 - x^2}}$.

If $f(x) = x^2 - 2x + 4$, find:

**27.** $f(0)$.       **28.** $f(-2)$.       **29.** $f(3)$.       **30.** $f(a + 1)$.

Find the directed length $l(\overrightarrow{P_1 P_2})$ between each pair of points.

**31.** $P_1(4, 9)$; $P_2(4, -2)$.       **32.** $P_1(2, -1)$; $P_2(6, -1)$.

Find the distance between each pair of points and find the slope of the line segment joining them.

**33.** $(-5, -8), (5, 2)$.                    **34.** $(6, 1), (-7, 3)$.

Graph each function in $R \times R$.

**•35.** $\{(x, y) | y = -x + 3\}$.              **36.** $\{(x, y) | y = 6 - 5x - x^2\}$.

By graphical means, find approximations to the nearest $\frac{1}{2}$ unit for the elements in the set formed by each intersection.

**37.** $\{(x, y) | y = 6 - x^2\} \cap \{(x, y) | y = -1\}$.
**38.** $\{(x, y) | y = x^2 - 6x - 16\} \cap \{(x, y) | y = 1\}$.

Find the set of zeros of each function.

**39.** $\{(-5, 1), (-3, 2), (-1, 0), (1, 3), (3, 0), (5, 4)\}$.
**40.** $\{(x, y) | y = x^2 - 8x + 15\}$.

Solve each equation.

**41.** $2x^2 - 5 = 0$.                         **42.** $2x^2 - 9x - 5 = 0$.

Write the inverse of each relation showing the defining equation with $y$ expressed in terms of $x$. Graph each relation, its inverse, and the graph of $y = x$.

**•43.** $\{(x, y) | 5x + 2y = 10\}$.            **44.** $\{(x, y) | y = -2x^2 + 8\}$.

**B.**

In Problems 45 and 46, graph each relation in $R \times R$.

**45.** $\{(x, y) | y > x - 3\}$.               **46.** $\{(x, y) | y \le x^2 - 9\}$.

**47.** Show that if a function $\mathscr{F}$ is defined by $y = x - 5$, then

$$\mathscr{F}[\mathscr{F}^{-1}(x)] = \mathscr{F}^{-1}[\mathscr{F}(x)] = x.$$

**48.** Show that if a function $\mathscr{F}$ is defined by $x + 2y = 8$, then

$$\mathscr{F}[\mathscr{F}^{-1}(x)] = \mathscr{F}^{-1}[\mathscr{F}(x)] = x.$$

**•49.** Graph $\{(x, y) | x^2 + y^2 \le 9\} \cap \{(x, y) | y \ge 1\}$ in $R \times R$.
**50.** Graph $\{(x, y) | y \ge x^2 - 9\} \cap \{(x, y) | y \le 9 - x^2\}$ in $R \times R$.

**C.**

Use graphical methods to solve each inequality.

**•51.** $x^2 - 1 \ge 0$.                        **52.** $x^2 - 4x - 5 < 0$.

# Circular Functions

**I**N CHAPTER 1 we discussed linear, quadratic, and exponential functions. There is another type of function which figures prominently in the work of today's scientists, engineers, and economists. For example, consider the graph of a function in Figure 2.1 which displays the pairings of the

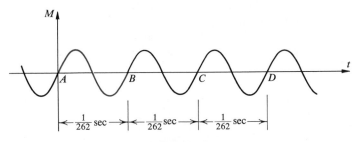

**FIGURE 2.1**

magnitude ($M$) of the displacement of a tuning fork with time ($t$) assuming that there is no damping. Its frequency when tuned to middle $C$ is 262 vibrations per second. Note that the domain of this function can be divided into equal intervals such that the graph in each of the intervals is a duplicate, or repetition, of the graph of the function in any other interval. For example, the graph from $A$ to $B$ is a duplicate of the graph from $B$ to $C$, $C$ to $D$, etc. Other quantities whose measures are related to time in a similar way include the current in a loop of wire, the displacement of a cork bobbing up and down on the surface of a lake, the displacement of the free end of a spring, and the amplitude of carrier waves in radio transmission.

A function which describes a relationship that has a repetitive pattern is said to be periodic, and each of the equal intervals of the domain is called a *period* of the function. We define such a function as follows.

**Definition 2.1.** *If f is a function with domain a subset of R, x ∈ R, such that for some a ∈ R, a ≠ 0, the value x + a is in the domain and such that*

$$f(x) = f(x + a),$$

*then f is called a **periodic function.** Its period is a.*

If a function has a period $a$, then it is also periodic with periods $2a$, $3a$, $-2a$, $-3a$, etc., and generally, $ka$, $k \in J$, $k \neq 0$. The smallest positive number $a$ for which the function is periodic is referred to as the **fundamental period** of the function. Whenever we refer to the period of a function in this text we mean its fundamental period unless otherwise stated.

The most common periodic functions are the **circular functions** and the **trigonometric functions.** We start our study of circular functions in this chapter. Trigonometric functions will be considered in Chapter 4.

## 2.1
### The Coordinate Function

Consider the unit circle with center at the origin in Figure 2.2, the graph of $\{(x, y) \mid x^2 + y^2 = 1\}$. Since the radius is 1 unit, the circumference $2\pi r$ is equal to $2\pi$. Corresponding with the length $s$ of each arc of the circle with initial point $(1, 0)$, there is an ordered pair of real numbers $(x, y)$ which are the coordinates of the terminal point of the arc. We shall assume that there is a one-to-one correspondence between all such arcs on the unit circle and the elements in the set of real numbers. The correspondence can be visualized as shown in Figure 2.2 where some of the elements in $R$ from 0 to

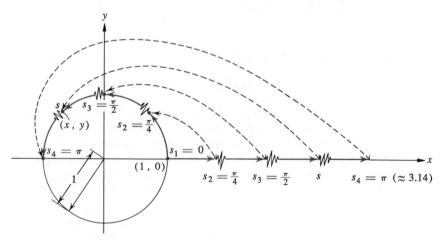

**FIGURE 2.2**

$\pi$ ($\approx 3.14$) associated with line segments on the number line are paired with some arcs on the circumference of the unit circle. The number line can be thought of as being "*wrapped*" about the circumference of the unit circle. Furthermore, if the terminal point of the arc is considered to be in a counter-clockwise direction from the initial point (Figure 2.3), $s$ will be a positive real number, that is, $s > 0$, and if the terminal point is considered to be in a clockwise direction, $s$ will be negative, that is, $s < 0$.

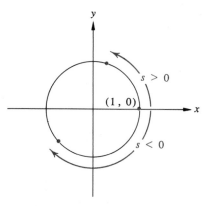

**FIGURE 2.3**

The set of pairings of each arc length $s$ with an ordered pair $(x, y)$ corresponding to the terminal point of the arc with initial point at $(1, 0)$ is called the **coordinate function.** Symbolically we write

$$\textbf{coordinate } (s) = (x, y).$$

The domain of the function is $\{s \mid s \in R\}$ and the range is $\{(x, y) \mid x^2 + y^2 = 1; \ x, y \in R\}$. Since the distance once around the unit circle counterclockwise from $(1, 0)$ back to $(1, 0)$ is $2\pi$, half-way around is $\pi$, one-fourth of the way around is $\pi/2$, etc. Figure 2.4a shows arcs with lengths of $\pi$ and $-\pi$ with the corresponding ordered pair $(-1, 0)$ and Figure 2.4b shows arcs with

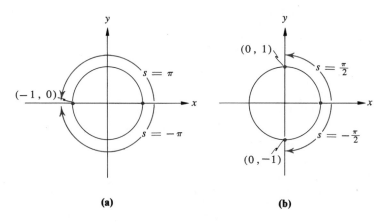

(a)                                    (b)

**FIGURE 2.4**

lengths of $\pi/2$ and $-\pi/2$ with the corresponding ordered pairs $(0, 1)$ and $(0, -1)$, respectively. Elements in the range of the coordinate function for certain other elements in the domain can be obtained by using some ideas from geometry. Since any traversal of an arc of length $2\pi$ or $-2\pi$ results in returning to the same ordered pair $(x, y)$, the coordinate function is periodic.

Thus,

$$\text{coordinate } (0 + 2\pi k, k \in J) = (1, 0),$$

and

$$\text{coordinate } \left(\frac{\pi}{2} + 2\pi k, k \in J\right) = (0, 1).$$

Other special values for coordinate $(s)$ can be found by using the equation of the unit circle and the distance formula, $d^2 = (x_2 - x_1)^2 + (y_2 - y_1)^2$. For $s = \pi/4$, the terminal point of the arc bisects the arc $[\pi/4 = \frac{1}{2} \cdot (\pi/2)]$ from $(1, 0)$ to $(0, 1)$, and at this point $x = y$ (Figure 2.5). Substituting $x$ for $y$ in the equation for the unit circle, $x^2 + y^2 = 1$, we obtain

$$x^2 + x^2 = 1,$$

from which

$$x = \frac{1}{\sqrt{2}} \quad \text{or} \quad x = \frac{-1}{\sqrt{2}}.$$

Since $x = y$,

$$y = \frac{1}{\sqrt{2}} \quad \text{or} \quad y = \frac{-1}{\sqrt{2}}.$$

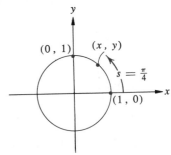

**FIGURE 2.5**

Thus, since both $x$ and $y$ are positive in the first quadrant,

$$\text{coordinate } \left(\frac{\pi}{4} + 2\pi k, k \in J\right) = \left(\frac{1}{\sqrt{2}}, \frac{1}{\sqrt{2}}\right).$$

Now consider coordinate $(\pi/6)$. Figure 2.6 shows the unit circle with $(x, y)$ denoting coordinate $(\pi/6)$. By symmetry of the unit circle, $(x, -y)$ denotes coordinate $(-\pi/6)$. The length of the arc from $(x, -y)$ to $(x, y)$ equals $(\pi/6) + (\pi/6) = \pi/3$ and the length of the arc from $(x, y)$ to $(0, 1)$ equals $(\pi/2) - (\pi/6) = \pi/3$. Because in a circle, equal arcs subtend equal chords, the distance from $(x, -y)$ to $(x, y)$ is the same as the distance from $(x, y)$ to $(0, 1)$. Thus from the distance formula

$$d^2 = (x_2 - x_1)^2 + (y_2 - y_1)^2,$$

we have that

$$(x - x)^2 + (-y - y)^2 = (x - 0)^2 + (y - 1)^2$$

or

$$4y^2 = x^2 + y^2 - 2y + 1.$$

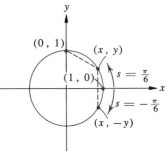

**FIGURE 2.6**

Substituting 1 for $x^2 + y^2$ (from the equation of the unit circle) in the right-hand member yields

$$4y^2 = 1 - 2y + 1,$$
$$4y^2 + 2y - 2 = 0,$$
$$2(2y - 1)(y + 1) = 0,$$

from which

$$y = \tfrac{1}{2} \quad \text{or} \quad y = -1.$$

Because $(x, y)$ is in Quadrant I $(x, y > 0)$, we select $\tfrac{1}{2}$ as the $y$ coordinate. Substituting $\tfrac{1}{2}$ for $y$ in

$$x^2 + y^2 = 1,$$

we have

$$x^2 + (\tfrac{1}{2})^2 = 1,$$

from which

$$x = \pm \frac{\sqrt{3}}{2}.$$

Again selecting the positive value, we have

$$\text{coordinate} \left(\frac{\pi}{6}\right) = \left(\frac{\sqrt{3}}{2}, \frac{1}{2}\right).$$

Figure 2.7 shows

$$\text{coordinate} \left(\frac{\pi}{6}\right) = \left(\frac{\sqrt{3}}{2}, \frac{1}{2}\right)$$

and

$$\text{coordinate} \left(\frac{\pi}{3}\right) = (x, y).$$

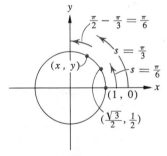

FIGURE 2.7

Observe that, by symmetry, the abscissa and ordinate of $(x, y)$ are the ordinate and abscissa of $\left(\frac{\sqrt{3}}{2}, \frac{1}{2}\right)$, respectively. Therefore,

$$\text{coordinate} \left(\frac{\pi}{3}\right) = \left(\frac{1}{2}, \frac{\sqrt{3}}{2}\right).$$

Figure 2.8 on page 44 summarizes the above results for $s \in \left\{\frac{\pi}{6}, \frac{\pi}{4}, \frac{\pi}{3}\right\}$.

We have now determined the coordinates of the end points of selected arcs on the unit circle, all of whose initial points are $(1, 0)$.

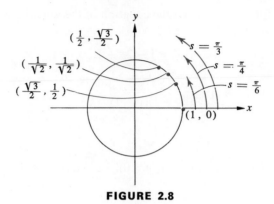

**FIGURE 2.8**

***Examples.*** Find:

(a) coordinate ($9\pi/4$).      (b) coordinate ($25\pi/6$).      (c) coordinate ($19\pi/3$).

*Solutions.*

(a) Since $9\pi/4 = (\pi/4) + 2\pi$,

$$\text{coordinate} \left(\frac{9\pi}{4}\right) = \text{coordinate} \left(\frac{\pi}{4} + 2\pi\right) = \left(\frac{1}{\sqrt{2}}, \frac{1}{\sqrt{2}}\right).$$

(b) Since $25\pi/6 = (\pi/6) + 4\pi$,

$$\text{coordinate} \left(\frac{25\pi}{6}\right) = \text{coordinate} \left(\frac{\pi}{6} + 4\pi\right) = \left(\frac{\sqrt{3}}{2}, \frac{1}{2}\right).$$

(c) Since $19\pi/3 = (\pi/3) + 6\pi$,

$$\text{coordinate} \left(\frac{19\pi}{3}\right) = \text{coordinate} \left(\frac{\pi}{3} + 6\pi\right) = \left(\frac{1}{2}, \frac{\sqrt{3}}{2}\right).$$

Notice that the coordinate function pairs each specified arc length with one and only one coordinate pair. However, with each given ordered pair on the unit circle, there are an infinite number of arcs and real number measures of these arcs. *Therefore, the coordinate function is not a one-to-one function.*

***Examples.*** Find the set of elements in the domain associated with the specified element in the range of the coordinate function.

(a) $\left(\dfrac{1}{\sqrt{2}}, \dfrac{1}{\sqrt{2}}\right)$.      (b) $\left(\dfrac{\sqrt{3}}{2}, \dfrac{1}{2}\right)$.

*Solutions.*   (a) $\{s \mid s = \pi/4 + 2\pi k, k \in J\}$.
              (b) $\{s \mid s = \pi/6 + 2\pi k, k \in J\}$.

### EXERCISE 2.1

A.

**1.** What is the domain of the coordinate function? What is the range?

**2.** Complete the following statement: For all points $(x, y)$ on the unit circle,
$x^2 + y^2 = $ _____?_____ .

Complete the following tables where coordinate $(s) = (x, y)$.

**3.**

| $s$ | $x$ | $y$ |
|---|---|---|
| 0 | 1 | ? |
| $\dfrac{\pi}{2}$ | ? | ? |
| $\pi$ | ? | ? |
| $\dfrac{3\pi}{2}$ | ? | ? |

**4.**

| $s$ | $x$ | $y$ |
|---|---|---|
| $\dfrac{\pi}{6}$ | $\dfrac{\sqrt{3}}{2}$ | ? |
| $\dfrac{\pi}{4}$ | ? | ? |
| $\dfrac{\pi}{3}$ | ? | ? |

Find each ordered pair. Sketch each corresponding arc on a unit circle.

**⋅5.** coordinate $\left(\dfrac{\pi}{6}\right)$.

**6.** coordinate $\left(-\dfrac{\pi}{4}\right)$.

**7.** coordinate $\left(\dfrac{9\pi}{4}\right)$.

**8.** coordinate $\left(\dfrac{13\pi}{6}\right)$.

**⋅9.** coordinate $\left(-\dfrac{13\pi}{6}\right)$.

**10.** coordinate $\left(\dfrac{7\pi}{3}\right)$.

Find the least positive element $s$ in the domain, associated with each specified element, coordinate $(s)$ in the range. Show each corresponding arc on a unit circle.

**11.** $\left(\dfrac{1}{\sqrt{2}}, \dfrac{1}{\sqrt{2}}\right)$.

**12.** $\left(\dfrac{\sqrt{3}}{2}, \dfrac{1}{2}\right)$

**⋅13.** $\left(\dfrac{1}{2}, \dfrac{\sqrt{3}}{2}\right)$.

**14.** $(0, 1)$.

**15.** $(1, 0)$.

**16.** $(-1, 0)$

Specify the quadrant in which the terminal point of the corresponding arc lies.

**Example.** coordinate (5).

**Solution.** The arc length $s$ through one quadrant is $(\pi/2) \approx 1.57$. Thus, since $5/1.57 = 3^+$ $(3\pi/2 < 5 < 2\pi)$, the terminal point of the arc corresponding to coordinate (5) lies in Quadrant IV.

**17.** coordinate (2).

**18.** coordinate $(-3)$.

**19.** coordinate $\left(\dfrac{9}{2}\right)$.

**20.** coordinate $\left(-\dfrac{16}{3}\right)$.

**21.** coordinate $(-2.3)$.

**22.** coordinate $(0.5)$.

**C.**

Two examples of periodic functions are given in Problems 23 and 24, where $[x]$ is the greatest integer not greater than $x$ for each $x \in R$. For example, $[5.91] = [5 + 0.91] = 5$ and $[-1.21] = [-2 + 0.79] = -2$.

**•23.** Graph the function defined by $y = x - [x]$, $-4 \le x < 4$. What is the period of the function?

**24.** Graph the function defined by $y = -2x + [2x]$, $-3 \le x < 3$. What is the period of the function?

## 2.2
### The Cosine and Sine Functions

In the preceding section, the coordinate function was defined as having $\{s \mid s \in R\}$ for its domain and

$$\{(x, y) \mid x^2 + y^2 = 1; x, y \in R\}$$

for its range. The variable $s$ is the measure of an arc on the unit circle whose initial point is $(1, 0)$ and whose terminal point is $(x, y)$.

The components $x$ and $y$ of an element in the range of this function are given special names.

**Definition 2.2.** *For all* $x, y \in R$, $x^2 + y^2 = 1$, *and* coordinate $(s) = (x, y)$, $x$ *is called* **cosine** $s$ *and* $y$ *is called* **sine** $s$. *These are abbreviated as* **cos** $s$ *and* **sin** $s$, *respectively.*

We now pair the real number $s$ with each component of the ordered pair $(x, y)$, or $(\cos s, \sin s)$, and specify two functions

$$\{(s, \cos s) \mid \cos s = x, s \in R\}$$

and

$$\{(s, \sin s) \mid \sin s = y, s \in R\}.$$

These functions are called the **cosine function** and the **sine function**, respectively. Such functions which are defined by using a circle, as above, are called **circular functions.** In this section we shall be concerned primarily with these functions for *special values* of $s$. In later sections we shall consider these functions for all $s \in R$.

Since the point associated with $(x, y)$ or $(\cos s, \sin s)$ is on the unit circle the ranges of these functions are $\{\cos s \mid |\cos s| \le 1\}$ and $\{\sin s \mid |\sin s| \le 1\}$. We can obtain elements in the range for the selected elements in the domain, $s \in \{0, \pi/6, \pi/4, \pi/3, \pi/2, \pi, 3\pi/2\}$ by simply referring to the coordinate function of the preceding section (see Problems 3 and 4, Exercise 2.1). For easy reference, these elements are listed in Table 2.1. You will have need to refer to this table often in this chapter.

Elements in the range of a function are sometimes called **function values**, or **values of the function.** For example, $\sin (\pi/6)$, or $1/2$, is a function value of the sine function for $s = \pi/6$. Similarly, $\cos (\pi/2)$, or $0$, is a function value of the cosine function for $s = \pi/2$.

Since $\cos s$ and $\sin s$ are components of the elements in the range of the

**TABLE 2.1**

| $s$ | $x$ or $\cos s$ | $y$ or $\sin s$ |
|---|---|---|
| $0$ | $1$ | $0$ |
| $\dfrac{\pi}{6}$ | $\dfrac{\sqrt{3}}{2}$ | $\dfrac{1}{2}$ |
| $\dfrac{\pi}{4}$ | $\dfrac{1}{\sqrt{2}}$ | $\dfrac{1}{\sqrt{2}}$ |
| $\dfrac{\pi}{3}$ | $\dfrac{1}{2}$ | $\dfrac{\sqrt{3}}{2}$ |
| $\dfrac{\pi}{2}$ | $0$ | $1$ |
| $\pi$ | $-1$ | $0$ |
| $\dfrac{3\pi}{2}$ | $0$ | $-1$ |

coordinate function, these functions are *periodic with period* $2\pi$. Thus,

$$\cos (s + 2\pi k, \, k \in J) = \cos s$$

and

$$\sin (s + 2\pi k, \, k \in J) = \sin s.$$

These equations can be used to find function values for $\cos s$ and $\sin s$ which differ by $2\pi k$, $k \in J$ from those values of $s$ listed in Table 2.1.

**Examples.**   Find:   (a) $\cos (13\pi/6)$.        (b) $\sin (13\pi/6)$.

*Solutions.*

(a)  Since $13\pi/6 = (\pi/6) + 2\pi$, then

$$\cos \frac{13\pi}{6} = \cos \left( \frac{\pi}{6} + 2\pi \right) = \cos \frac{\pi}{6} = \frac{\sqrt{3}}{2}.$$

(b)  Since $13\pi/6 = (\pi/6) + 2\pi$, then

$$\sin \frac{13\pi}{6} = \sin \left( \frac{\pi}{6} + 2\pi \right) = \sin \frac{\pi}{6} = \frac{1}{2}.$$

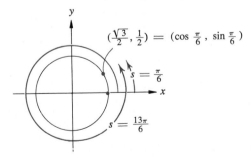

At this time we consider one important relationship between the cosine and sine functions. By substituting $\cos s$ for $x$ and $\sin s$ for $y$ in the equation $x^2 + y^2 = 1$, we have the following result.

**Theorem 2.1.**   *If $s \in R$, then*

$$\cos^2 s + \sin^2 s = 1. \tag{1}$$

Notice that we have used the more convenient symbols $\cos^2 s$ and $\sin^2 s$ for $(\cos s)^2$ and $(\sin s)^2$, respectively. Furthermore, Equation (1) is frequently used in the form $\sin^2 x + \cos^2 x = 1$.

***Example.***   Show that $\cos^2 s + \sin^2 s = 1$ for $s = \pi/3$.

*Solution.*   From Table 2.1, $\cos(\pi/3) = 1/2$ and $\sin(\pi/3) = \sqrt{3}/2$. Therefore,

$$\cos^2 \frac{\pi}{3} + \sin^2 \frac{\pi}{3} = \frac{1}{4} + \frac{3}{4} = 1.$$

***Example.***   If $\sin s = 1/\sqrt{2}$ and $0 < s < \pi/2$, find $\cos s$.

*Solution.*   Substituting $1/\sqrt{2}$ for $\sin s$ in $\sin^2 s + \cos^2 s = 1$, we obtain

$$\left(\frac{1}{\sqrt{2}}\right)^2 + \cos^2 s = 1,$$

$$\tfrac{1}{2} + \cos^2 s = 1,$$

$$\cos^2 s = \tfrac{1}{2},$$

from which

$$\cos s = \frac{1}{\sqrt{2}} \quad \text{or} \quad \frac{-1}{\sqrt{2}}.$$

Since $0 < s < \pi/2$ and $\cos s$ is in the first quadrant, we take the positive value and $\cos s = 1/\sqrt{2}$. Note that this value agrees with the value in Table 2.1.

Other important relationships between the circular functions are discussed in the following sections.

Since the unit circle is symmetric with respect to the horizontal axis, if coordinate $(s) = (x, y)$, then coordinate $(-s) = (x, -y)$. Figure 2.9 illustrates the case for $0 < s < \pi/2$. From Definition 2.2 we have the following.

**Theorem 2.2.**   *If $s \in R$, then*

$$\cos(-s) = \cos s,$$

$$\sin(-s) = -\sin s.$$

***Example.***   Find:   (a) $\sin(-\pi/3)$.       (b) $\cos(-\pi/2)$.

*Solution.*   (a) $\sin(-\pi/3) = -\sin(\pi/3) = -\sqrt{3}/2$.
(b) $\cos(-\pi/2) = \cos(\pi/2) = 0$.

In the field of physics, any oscillatory (periodic) motion satisfying the equation

$$d = A \sin Bt,$$

where $d$ is a measure of displacement, $A$ and $B$ are constants for the particular motion, and $t$ is a measure of time, is called **simple harmonic motion**. Using the function values listed in Table 2.1 and the fact that the sine function is periodic, you can now solve some simple problems pertaining to such motion. There are several such examples in the exercises.

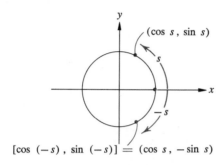

$$[\cos (-s), \sin (-s)] = (\cos s, -\sin s)$$

**FIGURE 2.9**

**EXERCISE 2.2**

A.

Find the given function values.

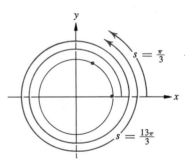

**Example.** $\sin (13\pi/3)$.

**Solution.** $\sin (13\pi/3) = \sin ((\pi/3) + 4\pi)$

$$= \sin \pi/3.$$

From Table 2.1,

$$\sin \frac{\pi}{3} = \frac{\sqrt{3}}{2}.$$

**1.** $\cos \dfrac{25\pi}{6}$.

**2.** $\sin \dfrac{25\pi}{6}$.

**3.** $\sin \dfrac{9\pi}{4}$.

**4.** $\cos \dfrac{9\pi}{4}$.

**5.** $\cos \dfrac{19\pi}{3}$.

**6.** $\sin \dfrac{19\pi}{3}$.

**7.** $\sin \left(-\dfrac{\pi}{4}\right)$.

**8.** $\cos \left(-\dfrac{\pi}{4}\right)$.

**9.** $\cos \left(\dfrac{-7\pi}{3}\right)$.

**10.** $\sin \left(\dfrac{-7\pi}{3}\right)$.

**11.** $\cos (-4\pi)$.

**12.** $\sin (-4\pi)$.

Find the least nonnegative $s$ for which each statement is true.

**Example.** (a) $\cos s = \frac{1}{2}$.      (b) $\cos s = 0$.

**Solution.** The least nonnegative values of $s$ are obtained from Table 2.1.
    (a) $\pi/3$.                          (b) $\pi/2$.

**13.** $\sin s = \dfrac{1}{\sqrt{2}}$.      **15.** $\cos s = 1$.      **17.** $\sin s = \dfrac{\sqrt{3}}{2}$.

**14.** $\cos s = \dfrac{1}{\sqrt{2}}$.      **16.** $\sin s = 0$.      **18.** $\cos s = \dfrac{\sqrt{3}}{2}$.

In Problems 19 to 22, show that $\cos^2 s + \sin^2 s = 1$ for the given arc measures.

**19.** $s = \dfrac{\pi}{4}$.      **20.** $s = \dfrac{\pi}{6}$.      **21.** $s = -\dfrac{\pi}{3}$.      **22.** $s = -\dfrac{\pi}{4}$.

**23.** If $\sin s = 3/5$ and $\cos s > 0$, find $\cos s$.

**24.** If $\cos s = 8/17$ and $\sin s > 0$, find $\sin s$.

**25.** If $\sin s = -5/13$ and $\cos s < 0$, find $\cos s$.

**26.** If $\cos s = 1/\sqrt{2}$ and $\sin s < 0$, find $\sin s$.

**27.** If $\sin s = 1/2$ and $\cos s > 0$, find:

    (a) $\cos s$.            (c) $1/\sin s$.            (e) $\cos s/\sin s$.

    (b) $\sin s/\cos s$.        (d) $1/\cos s$.

**28.** The voltage $E$ in a certain circuit at time $t$ is given as

$$E = RA \sin qt,$$

where $R$, $A$, and $q$ are constants, $E$ is a measure in volts and $t$ is a measure in seconds. Find $E$ if $R = 0.45$, $A = 2.7$, $q = 997\pi$, and $t = 3$ seconds.

**29.** The motion of a cork bobbing on the surface of a pool is described by the equation $d = 2 \sin t$ where $d$ is the measure of a displacement in inches and $t$ is a measure of time in seconds. Find $d$ when $t = 0$, $\pi/2$, $\pi$, $3\pi/2$, and $2\pi$.

**30.** The area $\mathscr{A}$ of a regular polygon of $n$ sides, inscribed in a circle of radius with length $r$, is given by

$$\mathscr{A} = \frac{1}{2} nr^2 \sin \frac{2\pi}{n}.$$

Find the area of a regular polygon of 12 sides inscribed in a circle whose radius has a length of 5 inches.

**31.** The perimeter $p$ of a regular polygon of $n$ sides, inscribed in a circle of radius with length $r$, is given by

$$p = 2nr \sin \frac{\pi}{n}.$$

Find the perimeter if $n = 6$ and $r = 5$ inches.

**B.**

**32.** It is shown in the calculus that an object having simple harmonic motion as described on page 49, has a velocity $v$ given by $v = AB \cos Bt$ and an acceleration $a$ given by $a = -AB^2 \sin Bt$. The motion of a certain spring is expressed by $d = 3 \sin (\pi t/2)$, where $d$ is a measure of displacement in centimeters and $t$ is a measure of time in seconds. (a) Express the velocity and the acceleration of a point on the free end of the spring in terms of $t$. (b) Find the velocity and acceleration when $t = 2, 3,$ and 4 seconds.

# 2.3
## Reduction Formulas for the Cosine Function

In Section 2.2 we considered several special function values $\cos s$ for which the initial point of the associated arc on the unit circle is $(1, 0)$ and the terminal point lies in the first quadrant, on the $x$ axis, or on the $y$ axis. In this section we develop several formulas which will enable you to find special function values $\cos s$ where the terminal point of the associated arc is at other positions on the unit circle. The following theorem is particularly useful to develop such formulas.

**Theorem 2.3.** *If* $s_1, s_2 \in R$, *then*

$$\cos (s_1 + s_2) = \cos s_1 \cos s_2 - \sin s_1 \sin s_2. \qquad (1)$$

*Proof.* The coordinates of the terminal points of the arcs with lengths

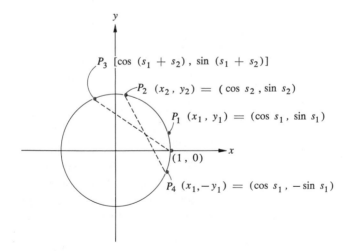

$P_3 \ [\cos (s_1 + s_2), \ \sin (s_1 + s_2)]$

$P_2 \ (x_2, y_2) = (\cos s_2, \sin s_2)$

$P_1 \ (x_1, y_1) = (\cos s_1, \sin s_1)$

$(1, 0)$

$P_4 \ (x_1, -y_1) = (\cos s_1, -\sin s_1)$

**FIGURE 2.10**

$s_1, s_2, -s_1,$ and $s_1 + s_2$, are shown on the unit circle in Figure 2.10. Because the arc measure from $P_1$ to $P_3$ is $s_1 + s_2$ and the arc measure from $P_4$ to $P_2$ is also $s_1 + s_2$, the chords joining these pairs of points are equal in length.

*(Proof continued on the next page.)*

By using the distance formula to express this fact in terms of the coordinates of $P_1, P_2, P_3$, and $P_4$, we have

$$\sqrt{[\cos(s_1 + s_2) - 1]^2 + [\sin(s_1 + s_2) - 0]^2}$$
$$= \sqrt{(\cos s_2 - \cos s_1)^2 + [\sin s_2 - (-\sin s_1)]^2}$$

or, squaring both members,

$$[\cos(s_1 + s_2) - 1]^2 + \sin^2(s_1 + s_2) = (\cos s_2 - \cos s_1)^2 + (\sin s_2 + \sin s_1)^2.$$

Equivalently,

$$\cos^2(s_1 + s_2) - 2\cos(s_1 + s_2) + 1 + \sin^2(s_1 + s_2)$$
$$= \cos^2 s_2 - 2\cos s_1 \cos s_2 + \cos^2 s_1 + \sin^2 s_2 + 2\sin s_1 \sin s_2 + \sin^2 s_1,$$

from which

$$[\cos^2(s_1 + s_2) + \sin^2(s_1 + s_2)] - 2\cos(s_1 + s_2) + 1$$
$$= (\cos^2 s_2 + \sin^2 s_2) + (\cos^2 s_1 + \sin^2 s_1) - 2\cos s_1 \cos s_2 + 2\sin s_1 \sin s_2.$$

Since $\cos^2 s + \sin^2 s = 1$, it follows that

$$1 - 2\cos(s_1 + s_2) + 1 = 1 + 1 - 2\cos s_1 \cos s_2 + 2\sin s_1 \sin s_2,$$

or

$$\cos(s_1 + s_2) = \cos s_1 \cos s_2 - \sin s_1 \sin s_2.$$

A formula for $\cos(s_1 - s_2)$ follows directly from Theorem 2.3 and Theorem 2.2.

**Theorem 2.4.**   *If $s \in R$, then*

$$\cos(s_1 - s_2) = \cos s_1 \cos s_2 + \sin s_1 \sin s_2.$$

*Proof.*   Substituting $-s_2$ for $s_2$ in Equation (1) on page 51 we have

$$\cos[s_1 + (-s_2)] = \cos s_1 \cos(-s_2) - \sin s_1 \sin(-s_2).$$

From Theorem 2.2, $\cos(-s_2) = \cos s_2$ and $\sin(-s_2) = -\sin s_2$. Substituting $\cos s_2$ and $-\sin s_2$ for $\cos(-s_2)$ and $\sin(-s_2)$, respectively, we obtain

$$\cos(s_1 - s_2) = \cos s_1 \cos s_2 + \sin s_1 \sin s_2.$$

These very important relationships for $\cos(s_1 + s_2)$ and $\cos(s_1 - s_2)$ are called the **sum formula** and the **difference formula** for the cosine function, respectively. One use of these expressions is to find additional values for $\cos s$.

***Example.***   Find $\cos(7\pi/12)$.

*Solution.*   Observe that $7\pi/12 = (\pi/3) + (\pi/4)$. Thus, from Theorem 2.3,

$$\cos\frac{7\pi}{12} = \cos\left(\frac{\pi}{3} + \frac{\pi}{4}\right) = \cos\frac{\pi}{3}\cos\frac{\pi}{4} - \sin\frac{\pi}{3}\sin\frac{\pi}{4}.$$

Using Table 2.1 (page 47) or from memory,

$$\cos \frac{7\pi}{12} = \frac{1}{2} \cdot \frac{1}{\sqrt{2}} - \frac{\sqrt{3}}{2} \cdot \frac{1}{\sqrt{2}} = \frac{1 - \sqrt{3}}{2\sqrt{2}}.$$

You can, if you wish, obtain a rational number approximation for $(1 - \sqrt{3})/2\sqrt{2}$ in the above example. Since $\sqrt{3} \approx 1.732$, and $\sqrt{2} \approx 1.414$,

$$\cos \frac{7\pi}{12} \approx \frac{1 - 1.732}{2(1.414)} \approx -0.259.$$

In general we shall leave solutions in radical form.

Although you will learn a more efficient way to find approximations for such function values in Sections 2.7 and 2.8, examples such as these provide you with an opportunity to become familiar with the sum and difference formulas.

Notice that in the above example $\cos (7\pi/12)$ is negative. While $x$ (or $\cos s$) and $y$ (or $\sin s$) are both positive in the first quadrant, this is not true when the terminal point of the arc of length $s$ lies in the other quadrants as shown in Figure 2.11.

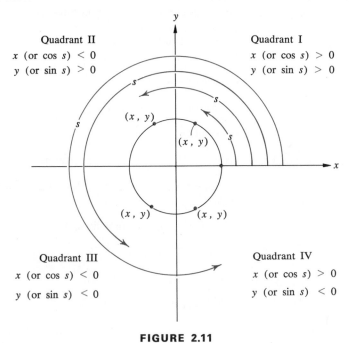

Quadrant II
$x$ (or $\cos s$) $< 0$
$y$ (or $\sin s$) $> 0$

Quadrant I
$x$ (or $\cos s$) $> 0$
$y$ (or $\sin s$) $> 0$

Quadrant III
$x$ (or $\cos s$) $< 0$
$y$ (or $\sin s$) $< 0$

Quadrant IV
$x$ (or $\cos s$) $> 0$
$y$ (or $\sin s$) $< 0$

**FIGURE 2.11**

**Examples.** State whether $\cos s$ is positive or negative and whether $\sin s$ is positive or negative for the following measures $s$.

(a) $3\pi/4$.   (b) $-2\pi/3$.   *(Solutions on the next page.)*

*Solutions.*   (a) If $s = 3\pi/4$, the terminal end of its corresponding arc is in Quadrant II. Thus, $\cos (3\pi/4)$ is negative and $\sin (3\pi/4)$ is positive.

(b) If $s = -2\pi/3$, the terminal end of its corresponding arc is in Quadrant III. Thus, $\cos (-2\pi/3)$ is negative and $\sin (-2\pi/3)$ is negative.

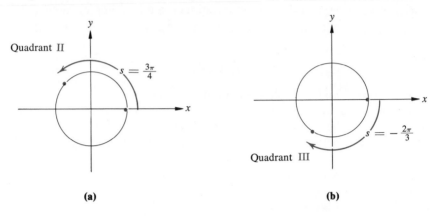

(a)                                                                                (b)

Using the sum and difference formulas, Theorems 2.3 and 2.4, to determine new values for cos $s$ is only one application of these relationships. Of greater importance is the manner in which they are used to show the following.

**Theorem 2.5.**   *If $s \in R$, then*

$$\textbf{I.}\quad \cos (\pi - s) = -\cos s,$$

$$\textbf{II.}\quad \cos (\pi + s) = -\cos s,$$

$$\textbf{III.}\quad \cos (2\pi - s) = \cos s.$$

We shall prove part I. The proofs of II and III are left as exercises.

*Proof of I.* Using the difference formula, Theorem 2.4, and replacing $s_1$ with $\pi$ and $s_2$ with $s$, we obtain

$$\cos (\pi - s) = \cos \pi \cos s + \sin \pi \sin s.$$

Since $\cos \pi = -1$ and $\sin \pi = 0$ (see Table 2.1),

$$\cos (\pi - s) = (-1)(\cos s) + 0(\sin s),$$

$$= -\cos s.$$

The formulas of Theorem 2.5 are called **reduction formulas** for the cosine function because they enable you to write a function value, cos $s$, for a value of $s$ whose associated arc terminates in Quadrant II, III, or IV in terms of cos $\tilde{s}$ where $0 < \tilde{s} < \pi/2$. The arc of length $\tilde{s}$ is called the **reference arc** for arc of length $s$.

The reduction formulas are more easily remembered by an appropriate

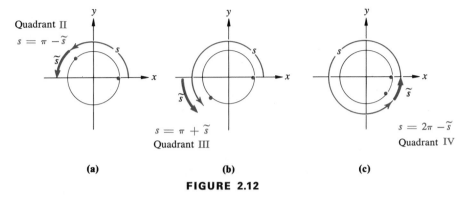

Quadrant II

$s = \pi - \tilde{s}$

$s = \pi + \tilde{s}$

Quadrant III

$s = 2\pi - \tilde{s}$

Quadrant IV

(a)　　　　　　(b)　　　　　　(c)

**FIGURE 2.12**

sketch such as one of those shown in Figure 2.12a, b, and c. The reference arc with measure $\tilde{s}$ is shown in the figure by the heavier arrow. We shall use this designation throughout the text. In each case, it can be seen that, if $s$ is the length of an arc terminating in Quadrants II, III, or IV, then the arc can be visualized either as the sum or as the difference of two arcs, *one of measure $n\pi$, $n \in J$, and the other of measure $\tilde{s}$, where $0 < \tilde{s} < \pi/2$.*

**Examples.** Find each function value. Sketch a figure and show the reference arc in each case.
(a) $\cos (7\pi/6)$.　　　　(b) $\cos (-7\pi/4)$.

*Solutions.* (a) Since $7\pi/6 = \pi + (\pi/6)$, we have from Theorem 2.5 II that

$$\cos \frac{7\pi}{6} = \cos \left( \pi + \frac{\pi}{6} \right) = -\cos \frac{\pi}{6}.$$

From Table 2.1, or from memory, $-\cos (\pi/6) = -\sqrt{3}/2$. A sketch showing the reference arc is helpful.

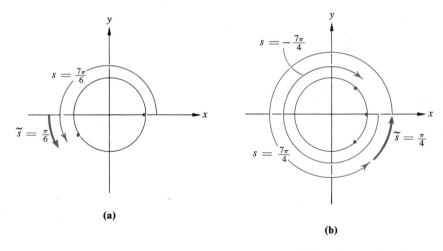

(a)

(b)

*(Solution continued on the next page.)*

(b) From Theorem 2.2, $\cos(-7\pi/4) = \cos(7\pi/4)$. Now since $7\pi/4 = 2\pi - (\pi/4)$, we have from Theorem 2.5 that

$$\cos\left(-\frac{7\pi}{4}\right) = \cos\frac{7\pi}{4} = \cos\left(2\pi - \frac{\pi}{4}\right) = \cos\frac{\pi}{4}.$$

From Table 2.1, or from memory, $\cos(\pi/4) = 1/\sqrt{2}$. Again, the sketch on page 55 is helpful.

The three reduction formulas in Theorem 2.5 are fundamental to much of the work that follows. Other reduction formulas that are sometimes useful are shown in Problems 41 to 44 in the following exercise.

### EXERCISE 2.3

A.

State whether $\cos s$ is positive or negative for each measure, $s$. Use $\pi \approx 3.14$.

1. $\dfrac{7\pi}{6}$.

4. $\dfrac{17\pi}{3}$.

7. $-\dfrac{17\pi}{6}$.

2. $-\dfrac{5\pi}{4}$.

5. $\dfrac{11\pi}{6}$.

8. $-\dfrac{11\pi}{3}$.

3. $-\dfrac{5\pi}{3}$.

6. $\dfrac{23\pi}{4}$.

9. 6.   *Hint*: $2\pi \approx 6.28$. Since $3\pi/2 < 6 < 2\pi$, an arc with measure 6, terminates in Quadrant IV.

10. $-7$.          11. $-4$.          12. 4.

Find each function value. Sketch a figure and show the reference arc in each case.

*13. $\cos\dfrac{3\pi}{4}$.

*17. $\cos\left(-\dfrac{5\pi}{4}\right)$.

*21. $\cos\dfrac{14\pi}{3}$.

14. $\cos\dfrac{7\pi}{6}$.

18. $\cos\left(-\dfrac{5\pi}{6}\right)$.

22. $\cos\dfrac{13\pi}{4}$.

15. $\cos\dfrac{5\pi}{3}$.

19. $\cos\left(-\dfrac{11\pi}{6}\right)$.

23. $\cos\left(-\dfrac{13\pi}{6}\right)$.

16. $\cos\dfrac{4\pi}{3}$.

20. $\cos\left(-\dfrac{2\pi}{3}\right)$.

24. $\cos\left(-\dfrac{11\pi}{3}\right)$.

In Problems 25 to 30 find:
(a) the least positive $s$,
(b) $0 \le s < 2\pi$, for which each statement is true.

**Example.**   $\cos s = -\sqrt{3}/2$.

Solution.   Since $\cos s$ is negative, $\pi/2 < s < \pi$ or $\pi < s < 3\pi/2$.

(a) By Theorem 2.5-I, $\cos s = -\cos(\pi - s)$. Therefore,

$$-\cos(\pi - s) = -\frac{\sqrt{3}}{2},$$

from which

$$\cos(\pi - s) = \frac{\sqrt{3}}{2}.$$

From Table 2.1, or again from memory,

$$\pi - s = \frac{\pi}{6},$$

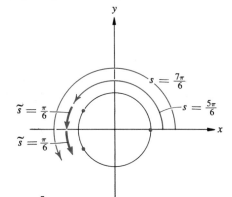

from which the least positive $s$ is given by $\dfrac{5\pi}{6}$.

(b) The reference arc of length $\tilde{s} = \pi/6$ is helpful here. For $\pi < s < 3\pi/2$,

$$s = \tilde{s} + \pi = \frac{\pi}{6} + \pi = \frac{7\pi}{6}.$$

Therefore, for $0 \le s \le 2\pi$, we have $s$ equal to $5\pi/6$ or $7\pi/6$.

**\*25.** $\cos s = \dfrac{1}{2}$.      **27.** $\cos s = -\dfrac{1}{\sqrt{2}}$.      **\*29.** $\cos s = \dfrac{\sqrt{3}}{2}$.

**26.** $\cos s = -\dfrac{1}{2}$.      **28.** $\cos s = \dfrac{1}{\sqrt{2}}$.      **30.** $\cos s = -\dfrac{\sqrt{3}}{2}$.

Complete the tables in Problems 31 and 32.

**31.**

| $s$ | $\dfrac{\pi}{4}$ | $\dfrac{2\pi}{4}$ | $\dfrac{3\pi}{4}$ | $\dfrac{4\pi}{4}$ | $\dfrac{5\pi}{4}$ | $\dfrac{6\pi}{4}$ | $\dfrac{7\pi}{4}$ | $\dfrac{8\pi}{4}$ |
|---|---|---|---|---|---|---|---|---|
| $\cos s$ | ? | ? | ? | ? | ? | ? | ? | ? |

**32.**

| $s$ | $\dfrac{\pi}{6}$ | $\dfrac{2\pi}{6}$ | $\dfrac{3\pi}{6}$ | $\dfrac{4\pi}{6}$ | $\dfrac{5\pi}{6}$ | $\dfrac{6\pi}{6}$ | $\dfrac{7\pi}{6}$ | $\dfrac{8\pi}{6}$ | $\dfrac{9\pi}{6}$ | $\dfrac{10\pi}{6}$ | $\dfrac{11\pi}{6}$ | $\dfrac{12\pi}{6}$ |
|---|---|---|---|---|---|---|---|---|---|---|---|---|
| $\cos s$ | ? | ? | ? | ? | ? | ? | ? | ? | ? | ? | ? | ? |

In Problems 33 and 34, use the sum or the difference formula, Theorem 2.3 or Theorem 2.4, to find each function value.

**33.** $\cos\dfrac{\pi}{12}$.   *Hint:* $\dfrac{\pi}{12} = \dfrac{\pi}{3} - \dfrac{\pi}{4}$.      **34.** $\cos\dfrac{5\pi}{12}$.   *Hint:* $\dfrac{5\pi}{12} = \dfrac{\pi}{6} + \dfrac{\pi}{4}$.

**B.**

**35.** Show that $\cos 2s = \cos^2 s - \sin^2 s$.   *Hint:* $\cos 2s = \cos(s + s)$.
**36.** From the results of Problem 35 show that

$$\cos 2s = 2\cos^2 s - 1 = 1 - 2\sin^2 s.$$

**37.** In Problem 25, find the solution set for $s \in R$.
**38.** In Problem 26, find the solution set for $s \in R$.
**39.** Prove Part II of Theorem 2.5.
**40.** Prove Part III of Theorem 2.5.

In Problems 41 to 44, prove each statement.

**41.** $\cos\left(\dfrac{\pi}{2} - s\right) = \sin s$.

**43.** $\cos\left(\dfrac{3\pi}{2} - s\right) = -\sin s$.

**42.** $\cos\left(\dfrac{\pi}{2} + s\right) = -\sin s$.

**44.** $\cos\left(\dfrac{3\pi}{2} + s\right) = \sin s$.

C.

**45.** Show that $\{(x, y) \mid y = \cos 2x\}$ is periodic with period $\pi$. *Hint*: Use Definition 2.1 and Theorem 2.3.
**46.** Show that $\{(x, y) \mid y = \cos \frac{1}{2}x\}$ is periodic with period $4\pi$.

## 2.4
### Reduction Formulas for the Sine Function

Reduction formulas for the sine function can be obtained in a manner similar to the development of the reduction formulas for the cosine function. First, however, we shall consider two relationships between function values for the cosine and sine.

**Theorem 2.6.** *If $s \in R$, then*

$$\text{I. } \cos\left(\frac{\pi}{2} - s\right) = \sin s,$$

$$\text{II. } \sin\left(\frac{\pi}{2} - s\right) = \cos s.$$

*Proof of I.* From Theorem 2.4,

$$\cos\left(\frac{\pi}{2} - s\right) = \cos\frac{\pi}{2}\cos s + \sin\frac{\pi}{2}\sin s.$$

Replacing $\cos \pi/2$ with $0$ and $\sin \pi/2$ with $1$ (see Table 2.1), we have

$$\cos\left(\frac{\pi}{2} - s\right) = 0\cdot\cos s + 1\cdot\sin s,$$

$$\cos\left(\frac{\pi}{2} - s\right) = \sin s.$$

*Proof of II.* Replacing $s$ with $(\pi/2) - s$ in Theorem 2.6-I, we obtain

$$\cos\left[\frac{\pi}{2} - \left(\frac{\pi}{2} - s\right)\right] = \sin\left(\frac{\pi}{2} - s\right),$$

from which

$$\cos \left[ \left( \frac{\pi}{2} - \frac{\pi}{2} \right) + s \right] = \sin \left( \frac{\pi}{2} - s \right),$$

$$\cos s = \sin \left( \frac{\pi}{2} - s \right).$$

**Examples.**   (a) $\sin \dfrac{\pi}{3} = \cos \left( \dfrac{\pi}{2} - \dfrac{\pi}{3} \right) = \cos \dfrac{\pi}{6}$.

(b) $\cos \dfrac{\pi}{4} = \sin \left( \dfrac{\pi}{2} - \dfrac{\pi}{4} \right) = \sin \dfrac{\pi}{4}$.

Now, consider the sum and difference formulas for the sine function.

**Theorem 2.7.**   *If* $s \in R$, *then*

$$\sin (s_1 + s_2) = \sin s_1 \cos s_2 + \cos s_1 \sin s_2 .$$

*Proof.*   If we replace $s$ with $s_1 + s_2$ in Theorem 2.6-I, we have

$$\sin (s_1 + s_2) = \cos \left[ \frac{\pi}{2} - (s_1 + s_2) \right]$$

$$= \cos \left[ \left( \frac{\pi}{2} - s_1 \right) - s_2 \right],$$

which by Theorem 2.4 can be written equivalently as

$$\sin (s_1 + s_2) = \cos \left( \frac{\pi}{2} - s_1 \right) \cos s_2 + \sin \left( \frac{\pi}{2} - s_1 \right) \sin s_2 .$$

Since by Theorem 2.6, $\cos ((\pi/2) - s_1) = \sin s_1$ and $\sin ((\pi/2) - s_1) = \cos s_1$, we have

$$\sin (s_1 + s_2) = \sin s_1 \cos s_2 + \cos s_1 \sin s_2 .$$

Replacing $s_2$ with $-s_2$ in Theorem 2.7 leads to the following.

**Theorem 2.8.**   *If* $s \in R$, *then*

$$\sin (s_1 - s_2) = \sin s_1 \cos s_2 - \cos s_1 \sin s_2 .$$

The proof is left as an exercise.

Additional values can be found for $\sin s$ by using the sum and difference formulas.

**Example.**   Find $\sin \pi/12$.

*Solution.*   Observe that $\pi/12 = (\pi/3) - (\pi/4)$. Thus, from Theorem 2.8,

$$\sin \frac{\pi}{12} = \sin \left( \frac{\pi}{3} - \frac{\pi}{4} \right) = \sin \frac{\pi}{3} \cos \frac{\pi}{4} - \cos \frac{\pi}{3} \sin \frac{\pi}{4},$$

(*Solution continued on the next page.*)

and using values from Table 2.1 we have

$$\sin \frac{\pi}{12} = \left(\frac{\sqrt{3}}{2}\right) \cdot \left(\frac{1}{\sqrt{2}}\right) - \left(\frac{1}{2}\right) \cdot \left(\frac{1}{\sqrt{2}}\right) = \frac{\sqrt{3}-1}{2\sqrt{2}}.$$

As with the cosine function, of greater importance is the manner in which the sum and difference formulas are used to show the following reduction formulas.

**Theorem 2.9.**   *If* $s \in R$, *then*

$$\text{I.} \quad \sin (\pi - s) = \sin s,$$

$$\text{II.} \quad \sin (\pi + s) = -\sin s,$$

$$\text{III.} \quad \sin (2\pi - s) = -\sin s.$$

We shall prove part I and leave the other two parts as exercises.

*Proof of I.*   Using the difference formula for the sine function, Theorem 2.8, and replacing $s_1$ with $\pi$ and $s_2$ with $s$, we obtain

$$\sin (\pi - s) = \sin \pi \cos s - \cos \pi \sin s.$$

Since $\sin \pi = 0$ and $\cos \pi = -1$, we have

$$\sin (\pi - s) = (0) \cos s - (-1) \sin s = \sin s.$$

The reduction formulas for the sine function, as for the cosine function, are more easily remembered by an appropriate sketch as one of those shown in Figure 2.12a, b, and c (page 55). Recall that in each case, the arc can be visualized either as the sum or as the difference of two arcs, *one of measure* $n\pi$, $n \in J$, *and a reference arc, of length* $\tilde{s}$, *where* $0 < \tilde{s} < \pi/2$.

**Example.**   Find $\sin (3\pi/4)$.

*Solution.*   Since $3\pi/4 = \pi - (\pi/4)$, by Theorem 2.9,

$$\sin \frac{3\pi}{4} = \sin \left(\pi - \frac{\pi}{4}\right) = \sin \frac{\pi}{4} = \frac{1}{\sqrt{2}}.$$

A sketch showing the reference arc is helpful.

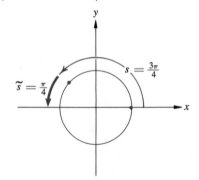

The reduction formulas in Theorem 2.9 like similar reduction formulas for the cosine function in Theorem 2.5 are of fundamental importance. Other reduction formulas for the sine function are shown in Problems 44 to 47 in the following exercise.

Under certain conditions, a relationship exists between function values for the cosine and sine. If two functions $f$ and $g$ exist such that $f(x) = g(y)$ for all

$x, y \in R$, where $x + y = \pi/2$, then $f$ and $g$ are called **cofunctions**. In Exercise 2.3, Problem 41, it was shown that

$$\cos y = \sin x,$$

where $y = (\pi/2) - x$, or $x + y = \pi/2$. Thus, the cosine and the sine are cofunctions. You will see several other examples of cofunctions in Section 2.6.

### EXERCISE 2.4

**A.**

State whether $\sin s$ is positive or negative for each measure, $s$.

**1.** $\dfrac{5\pi}{6}$.  **4.** $\dfrac{15\pi}{6}$.  **7.** $-\dfrac{20\pi}{6}$.

**2.** $-\dfrac{3\pi}{4}$.  **5.** $\dfrac{9\pi}{4}$.  **8.** $-\dfrac{14\pi}{3}$.

**3.** $-\dfrac{7\pi}{3}$.  **6.** $\dfrac{23\pi}{4}$.

**9.** 4.   *Hint*: $\pi \approx 3.14$. Since $\pi < 4 < 3\pi/2$, an arc with measure 4, terminates in Quadrant III.

**10.** $-5$.          **11.** $-7$.          **12.** 6.

Find each function value. Sketch a figure and show the reference arc in each case.

**˙13.** $\sin \dfrac{3\pi}{4}$.  **˙17.** $\sin \left(-\dfrac{7\pi}{4}\right)$.  **˙21.** $\sin \dfrac{5\pi}{3}$.

**14.** $\sin \left(-\dfrac{3\pi}{4}\right)$.  **18.** $\sin \left(-\dfrac{9\pi}{4}\right)$.  **22.** $\sin \left(-\dfrac{8\pi}{3}\right)$.

**15.** $\sin \dfrac{5\pi}{4}$ :  **19.** $\sin \dfrac{7\pi}{3}$.  **23.** $\sin \left(-\dfrac{19\pi}{6}\right)$.

**16.** $\sin \left(-\dfrac{5\pi}{4}\right)$.  **20.** $\sin \dfrac{5\pi}{6}$.  **24.** $\sin \dfrac{10\pi}{3}$.

In Problems 25 to 32, find:
(a) the least positive $s$,
(b) $0 \le s \le 2\pi$, for which each statement is true.

**Example.**  $\sin s = -\frac{1}{2}$.

*Solution.*  Since $\sin s$ is negative, $\pi < s < 3\pi/2$ or $3\pi/2 < s < 2\pi$. The notion of a reference arc is helpful here. Since

$$\sin s = -\sin \tilde{s} = -\tfrac{1}{2},$$

$$\sin \tilde{s} = \tfrac{1}{2}.$$

(*Solution continued on the next page.*)

From Table 2.1, $\tilde{s} = \pi/6$. Since

$$s = \pi + \tilde{s} = \pi + \frac{\pi}{6} = \frac{7\pi}{6},$$

or

$$s = 2\pi - \tilde{s} = 2\pi - \frac{\pi}{6} = \frac{11\pi}{6},$$

(a) the least positive $s$ is $\dfrac{7\pi}{6}$;

(b) for $0 \le s \le 2\pi$, $s$ equals $\dfrac{7\pi}{6}$ or $\dfrac{11\pi}{6}$.

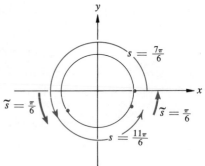

**•25.** $\sin s = \dfrac{1}{2}$.

**•29.** $\sin s = -\dfrac{\sqrt{3}}{2}$.

**26.** $\sin s = -\dfrac{1}{\sqrt{2}}$.

**30.** $\sin s = 0$.

**27.** $\sin s = \dfrac{1}{\sqrt{2}}$.

**31.** $\sin s = 1$.

**28.** $\sin s = \dfrac{\sqrt{3}}{2}$.

**32.** $\sin s = -1$.

Complete the tables in Problems 33 and 34.

**33.**

| $s$ | $\dfrac{\pi}{4}$ | $\dfrac{2\pi}{4}$ | $\dfrac{3\pi}{4}$ | $\dfrac{4\pi}{4}$ | $\dfrac{5\pi}{4}$ | $\dfrac{6\pi}{4}$ | $\dfrac{7\pi}{4}$ | $\dfrac{8\pi}{4}$ |
|---|---|---|---|---|---|---|---|---|
| $\sin s$ | ? | ? | ? | ? | ? | ? | ? | ? |

**34.**

| $s$ | $\dfrac{\pi}{6}$ | $\dfrac{2\pi}{6}$ | $\dfrac{3\pi}{6}$ | $\dfrac{4\pi}{6}$ | $\dfrac{5\pi}{6}$ | $\dfrac{6\pi}{6}$ | $\dfrac{7\pi}{6}$ | $\dfrac{8\pi}{6}$ | $\dfrac{9\pi}{6}$ | $\dfrac{10\pi}{6}$ | $\dfrac{11\pi}{6}$ | $\dfrac{12\pi}{6}$ |
|---|---|---|---|---|---|---|---|---|---|---|---|---|
| $\sin s$ | ? | ? | ? | ? | ? | ? | ? | ? | ? | ? | ? | ? |

In Problems 35 and 36, use the sum or the difference formula, Theorem 2.7 or Theorem 2.8, to find each function value.

**35.** $\sin \dfrac{\pi}{12}$.   *Hint:* $\dfrac{\pi}{12} = \dfrac{\pi}{3} - \dfrac{\pi}{4}$.        **36.** $\sin \dfrac{5\pi}{12}$.   *Hint:* $\dfrac{5\pi}{12} = \dfrac{\pi}{6} + \dfrac{\pi}{4}$.

**B.**

**37.** Show that $\sin 2s = 2 \sin s \cos s$.   *Hint:* $\sin 2s = \sin(s + s)$.
**38.** From the results of Problem 37, show that $\sin s = 2 \sin(s/2) \cos(s/2)$.
**39.** In Problem 25, find the solution set for $s \in R$.
**40.** In Problem 26, find the solution set for $s \in R$.
**41.** Prove Theorem 2.8.
**42.** Prove Part II of Theorem 2.9.
**43.** Prove Part III of Theorem 2.9.

In Problems 44 to 47, prove each statement.

**44.** $\sin\left(\dfrac{\pi}{2} - s\right) = \cos s.$    **46.** $\sin\left(\dfrac{3\pi}{2} - s\right) = -\cos s.$

**45.** $\sin\left(\dfrac{\pi}{2} + s\right) = \cos s.$    **47.** $\sin\left(\dfrac{3\pi}{2} + s\right) = -\cos s.$

**C.**

**48.** Show that $\{(x, y) \mid y = \sin 4x\}$ is periodic with period $\pi/2$.
    *Hint*: Use Definition 2.1 and Theorem 2.7.
**49.** Show that $\{(x, y) \mid y = \sin \frac{1}{4}x\}$ is periodic with period $8\pi$.
**50.** Show that $\{(x, y) \mid y = \sin 2x + \cos \frac{1}{2}x\}$ is periodic with period $4\pi$.

## 2.5
## The Tangent Function

The ratio of the sine and cosine function values is used so often that it is given a special name.

**Definition 2.3.**  *For all $s \in R$   ($s \neq (\pi/2) + k\pi, k \in J$),*

$$\text{tangent } s = \frac{\sin s}{\cos s}.$$

Tangent $s$ *is abbreviated as* **tan** *s.*

The set of all pairings $(s, \tan s)$ make up the tangent function. The restriction on $s$ above assures that $\cos s \neq 0$. In Section 2.8 you shall see that the range of the tangent function is the set of real numbers $R$.

You can find elements in the range of the tangent function for the selected elements $s \in \{0, \pi/2, \pi, 3\pi/2\}$ by referring to the sine and cosine function values in Table 2.1. Notice that

$$\tan 0 = \frac{\sin 0}{\cos 0} = \frac{0}{1} = 0,$$

$$\tan \pi = \frac{\sin \pi}{\cos \pi} = \frac{0}{-1} = 0,$$

and $\tan \pi/2$ and $\tan 3\pi/2$ are undefined since, from Definition 2.3, this is $\tan s$, where $s = (\pi/2) + k\pi, k \in J$. Function values for $s \in \{\pi/6, \pi/4, \pi/3\}$ also can be obtained from the sine and cosine function values in Table 2.1. Thus

$$\tan \frac{\pi}{6} = \frac{\sin (\pi/6)}{\cos (\pi/6)} = \frac{1/2}{\sqrt{3}/2} = \frac{1}{\sqrt{3}},$$

$$\tan \frac{\pi}{4} = \frac{\sin (\pi/4)}{\cos (\pi/4)} = \frac{1/\sqrt{2}}{1/\sqrt{2}} = 1,$$

and

$$\tan \frac{\pi}{3} = \frac{\sin (\pi/3)}{\cos (\pi/3)} = \frac{\sqrt{3}/2}{1/2} = \sqrt{3}.$$

These special function values for the tangent function, along with those for the cosine and sine, are shown in Table 2.2.

**TABLE 2.2**

| $s$ | $\cos s$ | $\sin s$ | $\tan s$ |
|---|---|---|---|
| $0$ | $1$ | $0$ | $0$ |
| $\dfrac{\pi}{6}$ | $\dfrac{\sqrt{3}}{2}$ | $\dfrac{1}{2}$ | $\dfrac{1}{\sqrt{3}}$ |
| $\dfrac{\pi}{4}$ | $\dfrac{1}{\sqrt{2}}$ | $\dfrac{1}{\sqrt{2}}$ | $1$ |
| $\dfrac{\pi}{3}$ | $\dfrac{1}{2}$ | $\dfrac{\sqrt{3}}{2}$ | $\sqrt{3}$ |
| $\dfrac{\pi}{2}$ | $0$ | $1$ | undefined |
| $\pi$ | $-1$ | $0$ | $0$ |
| $\dfrac{3\pi}{2}$ | $0$ | $-1$ | undefined |

The following relationship will enable you to find other function values for the tangent function.

**Theorem 2.10.** *If $s \in R$, then*

$$\tan(-s) = -\tan s.$$

*Proof.* By Definition 2.3 and Theorem 2.2,

$$\tan(-s) = \frac{\sin(-s)}{\cos(-s)} = \frac{-\sin s}{\cos s} = -\tan s.$$

**Example.** Find $\tan(-\pi)$.

*Solution.* From Theorem 2.10,

$$\tan(-\pi) = -\tan \pi.$$

From Table 2.2

$$-\tan \pi = 0.$$

Sum and difference formulas for the tangent function follow from the sum and difference formulas for the sine and cosine functions and Theorem 2.10.

**Theorem 2.11.** *If $s_1, s_2 \in R$, and $s_1, s_2, s_1 + s_2 \neq (\pi/2) + k\pi, k \in J$, then*

$$\tan(s_1 + s_2) = \frac{\tan s_1 + \tan s_2}{1 - \tan s_1 \tan s_2}.$$

*Proof.*   By Definition 2.3,

$$\tan (s_1 + s_2) = \frac{\sin (s_1 + s_2)}{\cos (s_1 + s_2)}.$$

By Theorems 2.3 and 2.7,

$$\tan (s_1 + s_2) = \frac{\sin s_1 \cos s_2 + \cos s_1 \sin s_2}{\cos s_1 \cos s_2 - \sin s_1 \sin s_2}.$$

Dividing the numerator and denominator of the right-hand member by $\cos s_1 \cos s_2$ yields

$$\tan (s_1 + s_2) = \frac{\dfrac{\sin s_1 \cos s_2}{\cos s_1 \cos s_2} + \dfrac{\cos s_1 \sin s_2}{\cos s_1 \cos s_2}}{\dfrac{\cos s_1 \cos s_2}{\cos s_1 \cos s_2} - \dfrac{\sin s_1 \sin s_2}{\cos s_1 \cos s_2}},$$

from which by Definition 2.3 and simplifying the right-hand member, we obtain

$$\tan (s_1 + s_2) = \frac{\tan s_1 + \tan s_2}{1 - \tan s_1 \tan s_2}.$$

By substituting $-s_2$ for $s_2$ in the sum formula, the difference formula for the tangent function is established.

**Theorem 2.12.**   *If $s_1, s_2 \in R$, and $s_1, s_2, s_1 - s_2 \neq (\pi/2) + k\pi, k \in J$, then*

$$\tan (s_1 - s_2) = \frac{\tan s_1 - \tan s_2}{1 + \tan s_1 \tan s_2}.$$

The proof is left as an exercise.

The important reduction formulas for the tangent function follow directly from the sum and difference formulas.

**Theorem 2.13.**   *If $s \in R$, then*

> I.   $\tan (\pi - s) = -\tan s,$
>
> II.   $\tan (\pi + s) = \tan s,$
>
> III.   $\tan (2\pi - s) = -\tan s.$

We shall prove only part I above and leave parts II and III as exercises.

*Proof.*   Using the difference formula for the tangent function, Theorem 2.10, and replacing $s_1$ with $\pi$ and $s_2$ with $s$, we have

$$\tan (\pi - s) = \frac{\tan \pi - \tan s}{1 + (\tan \pi)(\tan s)}.$$

*(Proof continued on the next page.)*

Since, from Table 2.2, $\tan \pi = 0$,

$$\tan (\pi - s) = \frac{0 - \tan s}{1 + (0)(\tan s)} = -\tan s.$$

***Example.*** Find $\tan (3\pi/4)$.

*Solution.* Since $3\pi/4 = \pi - (\pi/4)$, from Theorem 2.13-I,

$$\tan \frac{3\pi}{4} = \tan \left( \pi - \frac{\pi}{4} \right) = -\tan \frac{\pi}{4}.$$

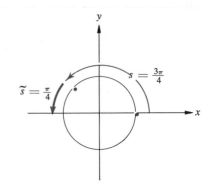

From Table 2.2,

$$-\tan \frac{\pi}{4} = -1.$$

Again, a sketch using the notion of a reference arc helps to interpret the reduction formulas.

***Example.*** Find $\tan (4\pi/3)$.

*Solution.* Since $4\pi/3 = \pi + (\pi/3)$, from Theorem 2.13-II,

$$\tan \frac{4\pi}{3} = \tan \left( \pi + \frac{\pi}{3} \right) = \tan \frac{\pi}{3}.$$

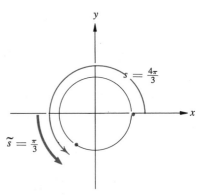

From Table 2.2,

$$\tan \frac{\pi}{3} = \sqrt{3}.$$

Because $\tan s = \sin s / \cos s$, $\tan s > 0$ if the arc corresponding to $s$ terminates in Quadrant I or III since $\sin s$ and $\cos s$ are either both positive or both negative (see Figure 2.11, page 53). $\tan s < 0$ if the arc corresponding to $s$ terminates in Quadrant II or IV since $\sin s$ is positive and $\cos s$ is negative or $\sin s$ is negative and $\cos s$ is positive, respectively.

Observe from Theorem 2.13-II, that

$$\tan s = \tan (\pi + s) = \tan (s + \pi),$$

which means that the *tangent function has period* $\pi$. It can be shown, that the tangent function has no smaller positive period and $\pi$ is its fundamental period. This is left as an exercise.

**EXERCISE 2.5**

**A.**

State whether $\tan s$ is positive or negative for each measure, $s$.

**1.** $\dfrac{5\pi}{6}$.

**2.** $-\dfrac{7\pi}{4}$.

**3.** $-\dfrac{8\pi}{3}$.

**4.** $\dfrac{14\pi}{3}$.

**5.** $\dfrac{11\pi}{6}$.

**6.** $-\dfrac{19\pi}{4}$.

**7.** $-\dfrac{10\pi}{3}$.

**8.** $-\dfrac{16\pi}{3}$.

**9.** 7.

**10.** $-5$.

**11.** $-7$.

**12.** 9.

Find each function value.

**13.** $\tan\left(-\dfrac{\pi}{6}\right)$.

**14.** $\tan\left(-\dfrac{\pi}{4}\right)$.

**15.** $\tan\left(-\dfrac{\pi}{3}\right)$.

**16.** $\tan\left(-\dfrac{\pi}{2}\right)$.

Find each function value. Sketch a figure and show the reference arc in each case.

**▪17.** $\tan\dfrac{5\pi}{4}$.

**18.** $\tan\dfrac{11\pi}{6}$.

**19.** $\tan\dfrac{2\pi}{3}$.

**20.** $\tan\dfrac{4\pi}{3}$.

**▪21.** $\tan\left(-\dfrac{3\pi}{4}\right)$.

**22.** $\tan\left(-\dfrac{7\pi}{6}\right)$.

**23.** $\tan\left(-\dfrac{11\pi}{6}\right)$.

**24.** $\tan\left(-\dfrac{7\pi}{4}\right)$.

**▪25.** $\tan\dfrac{11\pi}{3}$.

**26.** $\tan\dfrac{13\pi}{4}$.

**27.** $\tan\left(-\dfrac{13\pi}{6}\right)$.

**28.** $\tan\left(-\dfrac{7\pi}{3}\right)$.

In Problems 29 to 34, find:
(a) The least positive $s$,
(b) $0 \le s < 2\pi$, for which each statement is true.

***Example.*** $\tan s = -\sqrt{3}$.

*Solution.* Since $\tan s$ is negative, $\pi/2 < s < \pi$ or $3\pi/2 < s < 2\pi$. The notion of a reference arc is helpful here. Since

$$\tan s = -\tan \tilde{s} = -\sqrt{3},$$

$$\tan \tilde{s} = \sqrt{3}.$$

*(Solution continued on the next page.)*

From Table 2.2 or memory, $\tilde{s} = \pi/3$. Since

$$s = \pi - \tilde{s} = \pi - \frac{\pi}{3} = \frac{2\pi}{3},$$

or

$$s = 2\pi - \tilde{s} = 2\pi - \frac{\pi}{3} = \frac{5\pi}{3},$$

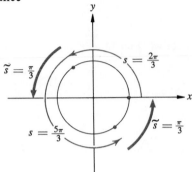

(a) the least positive $s$ is $\dfrac{2\pi}{3}$;

(b) for $0 \le s < 2\pi$, $s$ equals $\dfrac{2\pi}{3}$ or $\dfrac{5\pi}{3}$.

**■29.** $\tan s = 1$.                            **32.** $\tan s = \dfrac{1}{\sqrt{3}}$.

**30.** $\tan s = -1$.                           **■33.** $\tan s = \sqrt{3}$.

**31.** $\tan s = -\dfrac{1}{\sqrt{3}}$.            **34.** $\tan s = 0$.

Complete the tables in Problems 35 and 36.

**35.**

| $s$ | $\dfrac{\pi}{4}$ | $\dfrac{2\pi}{4}$ | $\dfrac{3\pi}{4}$ | $\dfrac{4\pi}{4}$ | $\dfrac{5\pi}{4}$ | $\dfrac{6\pi}{4}$ | $\dfrac{7\pi}{4}$ | $\dfrac{8\pi}{4}$ |
|---|---|---|---|---|---|---|---|---|
| $\tan s$ | ? | ? | ? | ? | ? | ? | ? | ? |

**36.**

| $s$ | $\dfrac{\pi}{6}$ | $\dfrac{2\pi}{6}$ | $\dfrac{3\pi}{6}$ | $\dfrac{4\pi}{6}$ | $\dfrac{5\pi}{6}$ | $\dfrac{6\pi}{6}$ | $\dfrac{7\pi}{6}$ | $\dfrac{8\pi}{6}$ | $\dfrac{9\pi}{6}$ | $\dfrac{10\pi}{6}$ | $\dfrac{11\pi}{6}$ | $\dfrac{12\pi}{6}$ |
|---|---|---|---|---|---|---|---|---|---|---|---|---|
| $\tan s$ | ? | ? | ? | ? | ? | ? | ? | ? | ? | ? | ? | ? |

In Problems 37 and 38, use the sum or the difference formula, Theorem 2.11 or Theorem 2.12, to find each function value.

**37.** $\tan \dfrac{\pi}{12}$.                   **38.** $\tan \dfrac{5\pi}{12}$.

**B.**

**39.** In Problem 29, find the solution set for $s \in R$.
**40.** In Problem 30, find the solution set for $s \in R$.
**41.** Prove Theorem 2.12.
**42.** Prove Part II of Theorem 2.13.
**43.** Prove Part III of Theorem 2.13.

**44.** Show that $\tan 2s = \dfrac{2 \tan s}{1 - \tan^2 s}$.

C.

45. Show that $\pi$ is the fundamental period of the tangent function. *Hint*: Assume that there is a value $a$, $0 < a < \pi$, such that $\tan(s + a) = \tan s$, and show that this implies that $\tan a = 0$, an impossibility.

## 2.6
## The Secant, Cosecant, and Cotangent Functions

The reciprocals of $\sin s$ and $\cos s$, and the ratio $\cos s/\sin s$, are given special names.

**Definition 2.4.** *For all $s \in R$,*

$$\text{I. Secant } s = \frac{1}{\cos s}, \qquad \left(s \neq \frac{\pi}{2} + k\pi, k \in J\right),$$

$$\text{II. Cosecant } s = \frac{1}{\sin s}, \qquad (s \neq k\pi, k \in J),$$

$$\text{III. Cotangent } s = \frac{\cos s}{\sin s}, \qquad (s \neq k\pi, k \in J).$$

Secant $s$, cosecant $s$, *and* cotangent $s$ *are abbreviated as* **sec** $s$, **csc** $s$, *and* **cot** $s$, *respectively*.

**Example.** Find $\sec s$, $\csc s$, and $\cot s$, if $\sin s = \frac{1}{2}$, $0 < s < \pi/2$.

**Solution.** Because $0 < s < \pi/2$, $\cos s$ is positive. Thus

$$\cos s = \sqrt{1 - \sin^2 s} = \sqrt{1 - \tfrac{1}{4}} = \frac{\sqrt{3}}{2},$$

$$\sec s = \frac{1}{\cos s} = \frac{1}{\sqrt{3/2}} = \frac{2}{\sqrt{3}},$$

$$\csc s = \frac{1}{\sin s} = \frac{1}{1/2} = 2,$$

and

$$\cot s = \frac{\cos s}{\sin s} = \frac{\sqrt{3/2}}{1/2} = \sqrt{3}.$$

The set of all pairings $(s, \sec s)$ is called the **secant function**. Because $\sec s$ is the reciprocal of $\cos s$, the domain of the secant function is the set of real numbers, except those values of $s$ for which $\cos s = 0$; namely, $\{s \mid s \in R, s \neq (\pi/2) + k\pi, k \in J\}$. Because $|\cos s| \leq 1$ for all $s \in R$, the range of the secant function is $\{\sec s \mid |\sec s| \geq 1\}$. Because the cosine function has a period of $2\pi$, the secant function also has a period of $2\pi$; that is,

$$\sec(s + 2\pi) = \sec s.$$

***Example.***   Find sec $(13\pi/6)$.

*Solution.*   Because $13\pi/6 = 2\pi + (\pi/6)$,

$$\sec \frac{13\pi}{6} = \sec \left(\frac{\pi}{6} + 2\pi\right) = \sec \frac{\pi}{6}.$$

From Definition 2.4 and Table 2.2,

$$\sec \frac{\pi}{6} = \frac{1}{\cos (\pi/6)} = \frac{1}{\sqrt{3}/2} = \frac{2}{\sqrt{3}}.$$

The set of all pairings $(s, \csc s)$ is called the **cosecant function**. Since $\csc s$ is the reciprocal of $\sin s$, the domain of the cosecant function is the set of real numbers, except those values of $s$ for which $\sin s = 0$; namely, $\{s \,|\, s \in R, s \neq k\pi, k \in J\}$. Since $|\sin s| \leq 1$ for all $s \in R$, the range of the cosecant function is $\{\csc s \,|\, |\csc s| \geq 1\}$. Since the sine function has a period of $2\pi$, the cosecant function also has a period of $2\pi$.

The set of all pairings $(s, \cot s)$ is called the **cotangent function**. Since $\cot s = \cos s/\sin s$, the domain of the cotangent function is the set of real numbers, except those values of $s$ for which $\sin s = 0$; namely, $\{s \,|\, s \in R, s \neq k\pi, k \in J\}$. In Section 2.8 you shall see that the range of the cotangent function, like that of the tangent function, is $R$. Also, like the tangent function, the fundamental period of the cotangent function is $\pi$. Since

$$\frac{\cos s}{\sin s} = \frac{1}{\dfrac{\sin s}{\cos s}}, \qquad \text{for } \sin s, \cos s \neq 0,$$

then

$$\cot s = \frac{1}{\tan s}, \qquad \text{for } s \in R, s \neq \frac{k\pi}{2}, k \in J.$$

We can find selected values for $\sec s$, $\csc s$, and $\cot s$ by using the function values for the sine, cosine, and tangent functions in Table 2.2. For example,

$$\sec 0 = \frac{1}{\cos 0} = \frac{1}{1} = 1,$$

and

$$\cot \frac{\pi}{6} = \frac{1}{\tan (\pi/6)} = \frac{1}{1/\sqrt{3}} = \sqrt{3}.$$

Demonstrations for other function values listed in Table 2.3 are left for you to verify. For convenient reference, function values $\cos s$, $\sin s$, and $\tan s$ for the same values $s$ are also included in the table.

**TABLE 2.3**

| $s$ | $\cos s$ | $\sin s$ | $\tan s$ | $\sec s$ | $\csc s$ | $\cot s$ |
|---|---|---|---|---|---|---|
| $0$ | $1$ | $0$ | $0$ | $1$ | undefined | undefined |
| $\dfrac{\pi}{6}$ | $\dfrac{\sqrt{3}}{2}$ | $\dfrac{1}{2}$ | $\dfrac{1}{\sqrt{3}}$ | $\dfrac{2}{\sqrt{3}}$ | $2$ | $\sqrt{3}$ |
| $\dfrac{\pi}{4}$ | $\dfrac{1}{\sqrt{2}}$ | $\dfrac{1}{\sqrt{2}}$ | $1$ | $\sqrt{2}$ | $\sqrt{2}$ | $1$ |
| $\dfrac{\pi}{3}$ | $\dfrac{1}{2}$ | $\dfrac{\sqrt{3}}{2}$ | $\sqrt{3}$ | $2$ | $\dfrac{2}{\sqrt{3}}$ | $\dfrac{1}{\sqrt{3}}$ |
| $\dfrac{\pi}{2}$ | $0$ | $1$ | undefined | undefined | $1$ | $0$ |
| $\pi$ | $-1$ | $0$ | $0$ | $-1$ | undefined | undefined |
| $\dfrac{3\pi}{2}$ | $0$ | $-1$ | undefined | undefined | $-1$ | $0$ |

The following theorem, which is similar to the theorems in Sections 2.3, 2.4, and 2.5 for $\cos s$, $\sin s$, and $\tan s$ are derived directly from their reciprocals.

**Theorem 2.14.**   *If $s \in R$, then*

$$\textbf{I.}\quad \sec(-s) = \sec s,$$

$$\textbf{II.}\quad \csc(-s) = -\csc s,$$

$$\textbf{III.}\quad \cot(-s) = -\cot s.$$

We shall prove part I and leave parts II and III as exercises.

*Proof of I.*    $\sec(-s) = \dfrac{1}{\cos(-s)} = \dfrac{1}{\cos s} = \sec s.$

***Example.***   Find $\csc(-\pi/3)$.

*Solution.*   From Theorem 2.14-II and Table 2.3,

$$\csc\left(-\frac{\pi}{3}\right) = -\csc\frac{\pi}{3} = -\frac{2}{\sqrt{3}}.$$

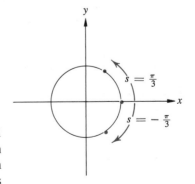

Reduction formulas for the secant and cotangent functions also are obtained from the reciprocal of their respective function values. Proofs of the following theorems are left as exercises.

**Theorem 2.15.**   *If $s \in R$, then*

$$\text{I. sec } (\pi - s) = -\sec s,$$

$$\text{II. sec } (\pi + s) = -\sec s,$$

$$\text{III. sec } (2\pi - s) = \sec s.$$

**Theorem 2.16.**   *If $s \in R$, then*

$$\text{I. csc } (\pi - s) = \csc s,$$

$$\text{II. csc } (\pi + s) = -\csc s,$$

$$\text{III. csc } (2\pi - s) = -\csc s.$$

**Theorem 2.17.**   *If $s \in R$, then*

$$\text{I. cot } (\pi - s) = -\cot s,$$

$$\text{II. cot } (\pi + s) = \cot s,$$

$$\text{III. cot } (2\pi - s) = -\cot s.$$

***Examples.***   Find:
(a) $\sec (3\pi/4)$.          (b) $\csc (7\pi/6)$.          (c) $\cot (7\pi/4)$.

*Solutions.*   The notion of a reference arc is helpful.
(a) $\sec (3\pi/4) = \sec (\pi - (\pi/4))$. From Theorem 2.15-I and Table 2.3,

$$\sec \left( \pi - \frac{\pi}{4} \right) = -\sec \frac{\pi}{4} = -\sqrt{2}.$$

(b) $\csc (7\pi/6) = \csc (\pi + (\pi/6))$. From Theorem 2.16-II and Table 2.3,

$$\csc \left( \pi + \frac{\pi}{6} \right) = -\csc \frac{\pi}{6} = -2.$$

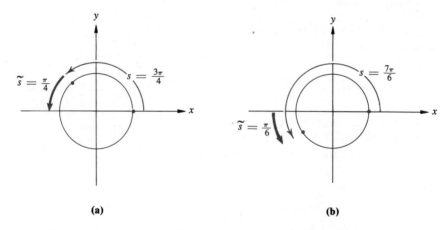

(a)                                              (b)

(c) $\cot(7\pi/4) = \cot(2\pi - (\pi/4))$. From Theorem 2.17-III and Table 2.3,

$$\cot\left(2\pi - \frac{\pi}{4}\right) = -\cot\frac{\pi}{4} = -1.$$

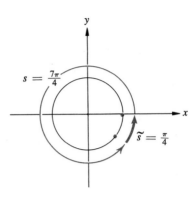

Whether particular function values $\sec s$, $\csc s$, and $\cot s$, are negative or positive real numbers can be readily determined from their definitions and the particular value of $s$. You can verify that these results together with similar information for $\sin s$, $\cos s$, and $\tan s$ summarized in Table 2.4 are correct.

**TABLE 2.4**

| Quadrant | cos $s$ or sec $s$ | sin $s$ or csc $s$ | tan $s$ or cot $s$ |
|----------|--------------------|--------------------|--------------------|
| I        | +                  | +                  | +                  |
| II       | −                  | +                  | −                  |
| III      | −                  | −                  | +                  |
| IV       | +                  | −                  | −                  |

**EXERCISE 2.6**

**A.**

State whether each function value is positive or negative.

**1.** $\cot\dfrac{5\pi}{3}$.

**2.** $\sec\dfrac{11\pi}{6}$.

**3.** $\csc\left(-\dfrac{5\pi}{4}\right)$.

**4.** $\cot\left(-\dfrac{11\pi}{6}\right)$.

**5.** $\sec\left(-\dfrac{9\pi}{4}\right)$.

**6.** $\csc\dfrac{14\pi}{3}$.

**7.** $\sec\dfrac{13\pi}{4}$.

**8.** $\cot\left(-\dfrac{19\pi}{6}\right)$.

**9.** $\cot 5$.

**10.** $\sec(-4)$.

**11.** $\csc 3.5$.

**12.** $\cot(-7.3)$.

Find each function value. Sketch a figure and show each reference arc.

**•13.** $\cot \dfrac{\pi}{4}$.

**•17.** $\cot \dfrac{5\pi}{6}$.

**•21.** $\cot \dfrac{19\pi}{6}$.

**14.** $\sec \dfrac{\pi}{6}$.

**18.** $\csc \dfrac{3\pi}{4}$.

**22.** $\sec \dfrac{10\pi}{3}$.

**15.** $\csc \left(-\dfrac{\pi}{3}\right)$.

**19.** $\csc \left(-\dfrac{2\pi}{3}\right)$.

**23.** $\csc \left(-\dfrac{11\pi}{4}\right)$.

**16.** $\sec \left(-\dfrac{\pi}{2}\right)$.

**20.** $\sec \left(-\dfrac{7\pi}{4}\right)$.

**24.** $\cot \left(-\dfrac{23\pi}{6}\right)$.

In Problems 25 to 30, find:
(a) The least positive $s$,
(b) $0 \le s < 2\pi$, for which each statement is true.

**•25.** $\sec s = \sqrt{2}$.

**28.** $\sec s = 1$.

**26.** $\cot s = -\sqrt{3}$.

**•29.** $\cot s = -1$.

**27.** $\csc s = -2$.

**30.** $\csc s = \dfrac{2}{\sqrt{3}}$.

**B.**

**31.** Show that $\tan ((\pi/2) - s) = \cot s$ and $\cot ((\pi/2) - s) = \tan s$; hence the tangent and cotangent are cofunctions. (See page 61.)
**32.** Show that $\tan ((\pi/2) + s) = -\cot s$ and $\cot ((\pi/2) + s) = -\tan s$.
**33.** In Problem 25, find the solution set for $s \in R$.
**34.** In Problem 26, find the solution set for $s \in R$.
**35.** Show that the secant and the cosecant are cofunctions.
**36.** Prove Theorem 2.14, part II.
**37.** Prove Theorem 2.14, part III.     **39.** Prove Theorem 2.16.
**38.** Prove Theorem 2.15.                **40.** Prove Theorem 2.17.

**C.**

Circular function values can be directly associated with the lengths of certain line segments. For example, in the figure at the right, $\overline{OA}$ and $\overline{OB}$ are radii of a unit circle, $\overrightarrow{BN}$ is perpendicular to $\overrightarrow{AT}$ and $\overrightarrow{LM}$, and $\overrightarrow{BN}$ and $\overrightarrow{AT}$ are tangents to the circle. Since the coordinates of the point $P$ are $(\cos s, \sin s)$, then

$$l(\overrightarrow{OQ}) = \cos s$$

and

$$l(\overrightarrow{QP}) = \sin s.$$

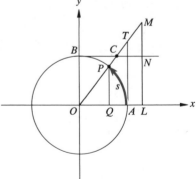

**41.** Show that $l(\overrightarrow{AT}) = \tan s$.
**42.** Show that $l(\overrightarrow{OT}) = \sec s$.
**43.** Show that $l(\overrightarrow{OC}) = \csc s$.
**44.** Show that $l(\overrightarrow{BC}) = \cot s$.

Any function satisfying the condition that $f(-x) = f(x)$ for all $x$ in the domain is called an even function. Any function satisfying the condition that $f(-x) = -f(x)$ for all $x$ in the domain is called an odd function.

***Examples.***   (a) $f(x) = x^2$ defines an even function because

$$f(-x) = (-x)^2 = x^2 = f(x).$$

(b) $f(s) = \sin s$ defines an odd function because

$$f(-s) = \sin (-s),$$

and by Theorem 2.2,

$$\sin (-s) = -\sin (s) = -f(s).$$

Which of the functions defined by the following equations are even and which are odd?

**45.** $f(x) = x^3$.                        **49.** $f(s) = \sec s$.
**46.** $f(x) = x^4 - x^2$.                  **50.** $f(s) = \csc s$.
**47.** $f(s) = \cos s$.                     **51.** $f(s) = \cot s$.
**48.** $f(s) = \tan s$.

## 2.7
## Tables of Function Values

In previous sections you obtained function values for the six circular functions for certain selected elements in their respective domains. In this section you will find ways to approximate function values for other real numbers $s$ where $0 < s < \pi/2$. Because the circular functions are periodic and reduction formulas are available, these values in turn will enable you to obtain additional function values.

Until this section we have been using the variable $s \in R$ exclusively as an element in the domain of each of the six circular functions and associated this variable with the length of an arc on the unit circle. Other variables can also be used; in particular, $x$ is generally used for this purpose. Because this is general practice, we will also use $x$ in this way from now on even though the variable $x$ will now have been used in two different ways. Recall that we previously used $x$ (and will continue to do so) as the first component of an ordered pair $(x, y)$ associated with a point in the geometric plane.

Table III*, page 294, lists function values, to four significant figures, for the circular functions ($0 \leq x \leq 1.57$) at intervals of 0.01. Note that $1.57 \approx \pi/2$. *Function values are, in general, irrational numbers and the four-place decimals shown in the table are mostly rational number approximations.* For example, the value for tan 0.91, located in Table III in the column labeled

---

* The entries in Table III which refer to measures of angles will be used in later sections.

"tan" horizontally opposite 0.91 in the column labeled $x$, is 1.286 and we write tan 0.91 $\approx$ 1.286. Notice further in the table that the function values sin $x$, tan $x$, and sec $x$, increase while cos $x$, cot $x$, and csc $x$ decrease for increasing values of $x$ where $0 \leq x \leq 1.57$.

***Examples.***   Find:

(a) sin 0.21.                    (b) sec 1.42.                    (c) cos 0.43.

*Solutions.*   From Table III,

(a) sin 0.21 $\approx$ 0.2085.      (b) sec 1.42 $\approx$ 6.657.      (c) cos 0.43 $\approx$ 0.9090.

Table III can also be used to find some elements in the domain of the circular functions for selected elements in the range. For example, given that cos $x$ = 0.9820, an approximation for the least positive value of $x$ such that cos $x$ = 0.9820 is 0.19.

Table III consists of function values only for real number values of $x$ at intervals of 0.01. However, in calculus it is shown that the circular functions are continuous over the intervals for which they are defined. As discussed in Section 1.4, page 21, this implies that, in these intervals, function values exist for every $x \in R$, and also that for small differences between two values of $x$ there are small differences between the corresponding function values.

To find approximations for function values for real number values of $x$ at smaller intervals than 0.01, such as sin 0.714, you can also read the entry in the table nearest the function value you seek. Thus from Table III,

$$\text{sin } 0.714 \approx \text{sin } 0.71 \approx 0.6518.$$

Generally, such approximations will be adequate for your purpose in this text. However, there may be times when you wish a closer approximation. Then you can either refer to tables with greater detail or you can use a method called **linear interpolation** illustrated in the following examples. The method of linear interpolation assumes that differences in function values are directly proportional to the differences of the elements in the domain over small intervals. In fact, the differences are not directly proportional and you are simply obtaining closer rational number approximations for the function values than could be obtained by reading directly from Table III. A geometric interpolation of the proportion used in linear interpolation is discussed in Chapter 3 in conjunction with the graphs of the cosine and sine functions.

***Example.***   Find sin 0.714.

*Solution.*   From Table III we find that sin 0.71 $\approx$ 0.6518 and sin 0.72 $\approx$ 0.6594. The following arrangement of the data helps us set up the direct proportion that we assume:

$$0.010 \left\{ 0.004 \left\{ \begin{array}{l} \text{sin } 0.710 \approx 0.6518 \\ \text{sin } 0.714 \approx \quad ? \\ \text{sin } 0.720 \approx 0.6594 \end{array} \right\} d \right\} 0.0076 \quad (d = \text{difference}).$$

We have that

$$\frac{0.004}{0.010} \approx \frac{d}{0.0076},$$

from which

$$d \approx \frac{4}{10}(0.0076) \approx 0.0030.$$

Thus, $\sin 0.714 \approx 0.6518 + 0.0030 = 0.6548.*$

You can find the function value for a number expressed in terms of $\pi$ by first substituting a rational approximation for $\pi$ (3.14 or 3.142) and then using Table III.

**Example.**  Find $\cos(\pi/8)$.

*Solution.*  Because $\pi \approx 3.142$, $\pi/8 \approx 0.393$. From Table III, $\cos 0.390 \approx 0.9249$ and $\cos 0.400 \approx 0.9211$. By arranging the data for interpolation, we obtain

$$0.010\left\{0.003\left\{\begin{array}{l}\cos 0.390 \approx 0.9249 \\ \cos 0.393 \approx \quad ? \\ \cos 0.400 \approx 0.9211\end{array}\right\}d\right\}0.0038.$$

We have that

$$\frac{0.003}{0.010} \approx \frac{d}{0.0038},$$

from which

$$d \approx \frac{3}{10}(0.0038) \approx 0.0011.$$

Thus

$$\cos\frac{\pi}{8} \approx \cos 0.393 \approx 0.9249 - 0.0011 = 0.9238.$$

Notice that we *subtracted* the difference, 0.0011, because in this interval *values for* $\cos x$ *decrease as* $x$ *increases.*

Interpolation can also be used to find approximations to elements in the domain for function values which lie between the values given in the table.

**Example.**  Find an approximation for the least positive real number $x$ such that $\tan x = 0.3369.$                                  *(Solution on the next page.)*

---

* In any such chain of statements involving both the symbols " = " and " ≈," it should be understood that either relationship is only valid for the expressions on either side of the particular symbol.

*Solution.* From Table III, $\tan 0.320 \approx 0.3314$ and $\tan 0.330 \approx 0.3425$. By arranging the data for interpolation, we obtain

$$0.010 \left\{ d \left\{ \begin{array}{l} \tan 0.320 \approx 0.3314 \\ \tan x \quad\;\; \approx 0.3369 \\ \tan 0.330 \approx 0.3425 \end{array} \right\} 0.0055 \right\} 0.0111.$$

We have that

$$\frac{d}{0.010} \approx \frac{0.0055}{0.0111},$$

from which

$$d \approx \frac{55}{111} (0.010) \approx 0.005.$$

Thus,

$$x \approx 0.320 + 0.005 = 0.325.$$

Although at this time you may not have the mathematical background to understand completely how the function values for Table III are obtained, it may be of interest to you to become acquainted with the kinds of formulas, developed in calculus, which are used to produce this table. Two of these formulas are discussed in the following exercise.

### EXERCISE 2.7

**A.**

In Problems 1 to 50, use Table III, page 294. Find an approximation for each function value.

**1.** $\sin 0.10$.
**2.** $\cos 0.26$.
**3.** $\tan 0.38$.
**4.** $\csc 0.54$.
**5.** $\sec 0.79$.
**6.** $\cot 0.91$.
**7.** $\cos 1.12$.
**8.** $\sin 1.29$.

Find an approximation for the least positive element in the domain of the circular function for the given element in the range.

**9.** $\sec x = 1.007$.
**10.** $\cot x = 3.478$.
**11.** $\tan x = 0.4228$.
**12.** $\csc x = 1.883$.
**13.** $\sin x = 0.7243$.
**14.** $\cos x = 0.5978$.
**15.** $\cot x = 0.5090$.
**16.** $\sec x = 3.375$.

Use linear interpolation to find an approximation for each function value. Use $\pi \approx 3.142$.

**17.** $\sin 0.255$.
**18.** $\tan 0.536$.
**19.** $\sec 0.904$.
**20.** $\cos 1.283$.
**21.** $\csc 1.097$.
**22.** $\cot 1.562$.

**23.** $\sin \dfrac{\pi}{8}$.  *Hint:* $\dfrac{\pi}{8} \approx \dfrac{3.142}{8} \approx 0.393$.

**24.** $\cos \dfrac{3\pi}{8}$.

**26.** $\csc \dfrac{4\pi}{9}$.

**28.** $\cot \dfrac{5\pi}{12}$.

**25.** $\tan \dfrac{\pi}{9}$.

**27.** $\sec \dfrac{\pi}{12}$.

Use linear interpolation to find an approximation for the least positive element in the domain for each given function value.

**29.** $\cot x = 3.333$.

**31.** $\cos x = 0.5849$.

**33.** $\tan x = 7.325$.

**30.** $\csc x = 1.850$.

**32.** $\sec x = 4.200$.

**34.** $\sin x = 0.1329$.

**C.**

In calculus it is shown that for all $x \in R$ and $n \in N$,

$$\cos x = 1 - \frac{x^2}{2!} + \frac{x^4}{4!} + \cdots + (-1)^{n-1} \frac{x^{2n-2}}{(2n-2)!} + \cdots \tag{1}$$

and

$$\sin x = x - \frac{x^3}{3!} + \frac{x^5}{5!} + \cdots + (-1)^{n-1} \frac{x^{2n-1}}{(2n-1)!} + \cdots, \tag{2}$$

where

$$2! = 1 \cdot 2, \quad 3! = 1 \cdot 2 \cdot 3, \quad 4! = 1 \cdot 2 \cdot 3 \cdot 4, \quad \text{etc.}$$

The expression in the right member of each equation is called an **infinite power series in** $x$. By taking any finite number of terms in the power series to obtain an approximation for the function value, *the error that exists is less than the value of the next term in the series.*

*Example.* Find an approximation for $\cos 0.3$ correct to four decimal places.

*Solution.* Replacing $x$ in Equation (1) by 0.3 yields

$$\cos 0.3 = 1 - \frac{(0.3)^2}{2!} + \frac{(0.3)^4}{4!} - \frac{(0.3)^6}{6!} + \cdots$$

$$= 1 - \frac{0.09}{2} + \frac{0.0081}{24} - \frac{0.000729}{720} + \cdots$$

$$= 1 - 0.045 + 0.00034 - 0.000001 + \cdots.$$

Since we wish accuracy to four decimal places, and since the error, by taking the first three terms is less than the value of the fourth term (0.000001), then correct to at least four decimal places,

$$\cos 0.3 \approx 1 - 0.045 + 0.00034 = 0.9553.$$

Note that this is the value for $\cos 0.3$ shown in Table III.

Use Equation (1) or (2) above to find each function value correct to four decimal places.

**35.** sin 0.01.             **37.** cos 0.50.             **39.** sin 1.00.
**36.** cos 0.02.             **38.** sin 0.60.             **40.** cos 1.10.

## 2.8

### Function Values for $x \in R$

You can find function values for the six circular functions for values of $x$ (or $s$) which are not in the interval $0 \le x \le \pi/2$ by using an appropriate reduction formula from Sections 2.3 to 2.6 and Table III. The notion of a reference arc which you have been using can also be helpful. The method of finding any circular function value of a real number $x$ (or $s$) associated with an arc on the unit circle whose initial point is $(1, 0)$, is summarized as follows:

1. Find the analogous function value of the real number $\tilde{x}$ (or $\tilde{s}$) associated with the reference arc.
2. Prefix the algebraic sign, $+$ or $-$, depending upon whether the original function value is positive or negative. Refer to Table 2.4 (page 73) as necessary.

   The real number $\tilde{x}$ is a number between 0 and $\pi/2$ such that $x = k\pi + \tilde{x}$ or $x = k\pi - \tilde{x}$, $k \in J$. See Figure 2.12, page 55. *Notice that either the initial point or the terminal point of the reference arc lies on the x axis.*

*Examples.*   Find an approximation for:   (a) sin 2.45.   (b) csc 5.84.

*Solutions.*   (a) Sketch a figure showing the reference arc. Since $\pi \approx 3.14$,

$$\tilde{x} \approx 3.14 - 2.45 = 0.69.$$

Since the arc of length 2.45 terminates in Quadrant II, sin 2.45 is positive. Therefore,

$$\sin 2.45 = \sin \tilde{x} \approx \sin 0.69 \approx 0.6365.$$

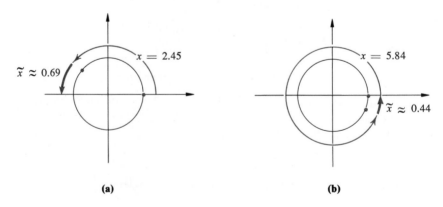

(a)                                                    (b)

(b) Sketch a figure showing the reference arc. Since $2\pi \approx 6.28$,

$$\tilde{x} \approx 6.28 - 5.84 = 0.44.$$

Since the arc of length 5.84 terminates in Quadrant IV, csc 5.84 is negative. Therefore,

$$\csc 5.84 = -\csc \tilde{x} \approx -\csc 0.44 \approx -2.348.$$

**Example.** Find an approximation for cos 1.956 by using linear interpolation.

*Solution.* Sketch a figure showing the reference arc. Let $\pi \approx 3.142$. Thus, $\tilde{x} \approx 3.142 - 1.956 = 1.186$. Since the arc of length 1.956 terminates in Quadrant II, cos 1.956 is negative. Therefore,

$$\cos 1.956 = -\cos \tilde{x} \approx -\cos 1.186.$$

The data arranged for interpolation appears as

$$0.010 \left\{ 0.006 \left\{ \begin{array}{l} \cos 1.180 \approx 0.3809 \\ \cos 1.186 \approx \quad ? \\ \cos 1.190 \approx 0.3717 \end{array} \right\} d \right\} 0.0092.$$

Then

$$\frac{0.006}{0.010} \approx \frac{d}{0.0092},$$

from which

$$d \approx \frac{6}{10}(0.0092) \approx 0.0055.$$

Thus,

$$\cos 1.956 \approx -\cos 1.186 \approx -(0.3809 - 0.0055) = -0.3754.$$

*Notice that because the cosine function value is decreasing as x increases, the difference 0.0055 was subtracted to obtain the answer.*

As you have seen, for most elements in the range of a circular function, you can find two corresponding elements in its domain over the interval 0 to $2\pi$. Furthermore, since the functions are periodic, you can find an infinite number of elements in the domain for $\{x \mid x \in R\}$.

**Examples.** Given that sin $x = 0.4706$. Find an approximation for:
(a) The least positive $x$.       (b) $0 \le x < 2\pi$.

*Solutions.* (a) An approximation for the least positive $x$ can be found directly from Table III. If sin $x = 0.4706$, $x \approx 0.49$.

(b) Since sin $x$ is positive, and 0.49 is the length $\tilde{x}$ of the reference arc, we have that $x \approx \pi - \tilde{x}$, from which $x \approx 3.14 - 0.49 = 2.65$ and for $0 \le x < 2\pi$, we have that $x$ approximately equals 0.49 or 2.65.

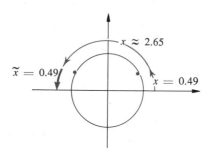

The entries in Table III, the reduction formulas, and linear interpolation enable you to find approximate function values for the six circular functions for values of $x \in R$.

Since the range of the cosine function is $\{\cos x \mid |\cos x| \le 1\}$ and the range of the sine function is $\{\sin x \mid |\sin x| \le 1\}$, the absolute value of the quotients

$$\frac{\sin x}{\cos x} = \tan x \qquad \left( x \ne \frac{\pi}{2} + k\pi, k \in J \right)$$

and

$$\frac{\cos x}{\sin x} = \cot x \qquad (x \ne k\pi, k \in J)$$

can be made as large as we like by assigning values to their respective de-nominators sufficiently close to zero. Furthermore, since these quotients can be positive or negative, the range of the tangent function and the range of the cotangent function are both the set of real numbers $R$. Observe from Table III that tan $x$ increases (becomes greater) and cot $x$ decreases (becomes smaller) as $x$ takes on values from 0 to 1.57.

## EXERCISE 2.8

A.

Use reduction formulas to express each of the following in terms of the same function value of $x$.

*Examples.*    (a) sec $(\pi + x)$.          (b) tan $(2\pi - x)$.

*Solutions.*    From the appropriate reduction formulas:

       (a) sec $(\pi + x) = -\sec x$.      (b) tan $(2\pi - x) = -\tan x$.

| | | |
|---|---|---|
| **1.** sin $(\pi - x)$. | **5.** csc $(2\pi - x)$. | **9.** sec $(2\pi - x)$. |
| **2.** cos $(\pi + x)$. | **6.** cot $(\pi + x)$. | **10.** tan $(\pi - x)$. |
| **3.** tan $(\pi + x)$. | **7.** cos $(2\pi - x)$. | **11.** cot $(2\pi - x)$. |
| **4.** sec $(\pi - x)$. | **8.** sin $(\pi + x)$. | **12.** csc $(\pi + x)$. |

Find approximate function values for each of the following. Use $\pi \approx 3.14$. Make a sketch showing the reference arc in each case.

| | | |
|---|---|---|
| **▪13.** sin 1.76. | **19.** cos $(-0.26)$. | **▪25.** tan 2.71. |
| **14.** cos 3.81. | **20.** tan $(-3.57)$. | **26.** sec 3.97. |
| **15.** tan 4.91. | **▪21.** sec $(-2.55)$. | **27.** csc $(-5.02)$. |
| **16.** sec 4.23. | **22.** csc $(-1.27)$. | **28.** cot $(-6.14)$. |
| **▪17.** csc 6.06. | **23.** cot 10.13. | **▪29.** sin 7.48. |
| **18.** cot 2.91. | **24.** sin 9.30. | **30.** cos 10.48. |

Find function values to four decimal places for each of the following. Use $\pi \approx 3.142$ and employ linear interpolation as necessary. Make a sketch showing the reference arc in each case.

| | | |
|---|---|---|
| **31.** sin 1.724. | **▪33.** tan 5.685. | **35.** csc $(-2.889)$. |
| **32.** cos 4.333. | **34.** sec 7.497. | **36.** cot $(-3.118)$. |

For each function value in Problems 37 to 46, find an approximation for:
(a) The least positive $x$.      (b) $0 \leq x < 2\pi$.

| | | |
|---|---|---|
| **▪37.** sin $x = 0.2280$. | **▪41.** csc $x = 1.116$. | **44.** csc $x = -3.179$. |
| **38.** cos $x = 0.0807$. | **42.** cot $x = 0.9331$. | **▪45.** cot $x = -0.8785$. |
| **39.** tan $x = 1.398$. | **43.** sec $x = -4.372$. | **46.** sin $x = -0.9168$. |
| **40.** sec $x = 1.266$. | | |

**B.**

**47.** In Problem 37, find the solution set for $x \in R$.
**48.** In Problem 38, find the solution set for $x \in R$.
**49.** In Problem 39, find the solution set for $x \in R$.
**50.** In Problem 40, find the solution set for $x \in R$.

In each of the following, state whether the function value increases or decreases as $x$ increases in the specified interval.

| | | |
|---|---|---|
| **51.** sin $x$,   $1.57 \leq x \leq 3.14$. | | **55.** tan $x$,   $1.57 < x \leq 3.14$. |
| **52.** sin $x$,   $3.14 \leq x \leq 4.71$. | | **56.** tan $x$,   $3.14 \leq x < 4.71$. |
| **53.** cos $x$,   $3.14 \leq x \leq 4.71$. | | **57.** cot $x$,   $4.71 \leq x < 6.28$. |
| **54.** cos $x$,   $4.71 \leq x \leq 6.28$. | | **58.** sec $x$,   $1.57 < x \leq 3.14$. |

C.

**59.** The displacement, $d$, of a vibrating spring under certain conditions is given by

$$d = \frac{1}{50}\sin 50t - t\cos 50t,$$

where $d$ is in feet and $t$ is in seconds. Find an approximation for the displacement when $t = 0.3$ seconds. Use $\pi \approx 3.14$.

**60.** The current $i$ at the end of $t$ seconds in a certain experiment is given by the formula,

$$i = 42\sin 96\pi t + 10\sin 445\pi t,$$

where $i$ is in amperes and $t$ is in seconds. Find an approximation for $i$ when $t = 0.52$ seconds. Use $\pi \approx 3.14$.

## Chapter Summary

**1.** Functions with the property that $f(x) = f(x + a)$, with domain a subset of $R$ and for some $a \in R$, $(a \neq 0)$, $x + a$ is an element in the domain, are said to be **periodic.** The smallest positive number $a$ for which a function is periodic is called its **fundamental period.**

**2.** The coordinate function,

$$\textbf{coordinate } (s) = (x, y),$$

pairs the length $s$ of each arc with initial point $(1, 0)$ on the circumference of a unit circle with an ordered pair $(x, y)$ in a Cartesian coordinate system. The domain of the coordinate function is $\{s \mid s \in R\}$ and its range is $\{(x, y) \mid x^2 + y^2 = 1; x, y \in R\}$.

**3.** Special names are given to the components in the range of the coordinate function:

$$x = \textbf{cosine } s \qquad \text{(abbreviated } \textbf{cos } s\text{)},$$

and

$$y = \textbf{sine } s \qquad \text{(abbreviated } \textbf{sin } s\text{)}.$$

**4.** The cosine function

$$\{(s, \textbf{cos } s) \mid \textbf{cos } s = x, s \in R\}$$

and the sine function

$$\{(s, \textbf{sin } s) \mid \textbf{sin } s = y, s \in R\}$$

pair the real number $s$ with each component of $(x, y)$, or $(\cos s, \sin s)$, respectively. *The ranges are* $\{\cos s \mid |\cos s| \leq 1\}$ *and* $\{\sin s \mid |\sin s| \leq 1\}$, *respectively. The functions are periodic with fundamental period* $2\pi$.

5. Elements in the range of the cosine and sine functions, commonly called **function values**, can be obtained by geometric methods for special elements in the domain, $0$, $\pi/6$, $\pi/4$, $\pi/3$ and $\pi/2$. Since these functions are periodic with period $2\pi$, $f(x + 2\pi)$ can be found when $f(x)$ is known.

6. Some important relationships pertaining to function values follow from the definitions for the sine and cosine functions. Unless otherwise specified, $s$, $s_1$, and $s_2$ are elements of $R$ in the formulas in this and the following paragraphs.

   (a) $\sin^2 s + \cos^2 s = 1$,
   (b) $\cos(-s) = \cos s$,
   (c) $\sin(-s) = -\sin s$,
   (d) $\cos(s_1 + s_2) = \cos s_1 \cos s_2 - \sin s_1 \sin s_2$,   (sum formula),
   (e) $\cos(s_1 - s_2) = \cos s_1 \cos s_2 + \sin s_1 \sin s_2$,   (difference formula).

7. Special cases of the two formulas, 6d and 6e above, lead to the following reduction formulas:

   (a) $\cos(\pi - s) = -\cos s$,
   (b) $\cos(\pi + s) = -\cos s$,
   (c) $\cos(2\pi - s) = \cos s$.

8. Two additional relationships follow from formulas 6d and 6e:

   (a) $\cos\left(\dfrac{\pi}{2} - s\right) = \sin s$,   (b) $\sin\left(\dfrac{\pi}{2} - s\right) = \cos s$.

9. The following formulas which apply to the sine function follow from the formulas in 8 above:

   (a) $\sin(s_1 + s_2) = \sin s_1 \cos s_2 + \cos s_1 \sin s_2$,   (sum formula),
   (b) $\sin(s_1 - s_2) = \sin s_1 \cos s_2 - \cos s_1 \sin s_2$,   (difference formula),
   (c) $\sin(\pi - s) = \sin s$,
   (d) $\sin(\pi + s) = -\sin s$,  } (reduction formulas).
   (e) $\sin(2\pi - s) = -\sin s$,

10. The name **tangent** $s$ is given to the ratio of $\sin s$ and $\cos s$. Furthermore

$$\left\{(s, \tan s) \,\middle|\, \tan s = \frac{\sin s}{\cos s}, \ \left(s \neq \frac{\pi}{2} + k\pi, k \in J\right)\right\}$$

is called the **tangent function**. *Its range is $R$; its fundamental period is $\pi$.*

11. The following formulas apply to the tangent function.

   (a) $\tan(-s) = -\tan s$,
   (b) $\tan(s_1 + s_2) = \dfrac{\tan s_1 + \tan s_2}{1 - \tan s_1 \tan s_2}$,   (sum formula),
   (c) $\tan(s_1 - s_2) = \dfrac{\tan s_1 - \tan s_2}{1 + \tan s_1 \tan s_2}$,   (difference formula),
   (d) $\tan(\pi - s) = -\tan s$,
   (e) $\tan(\pi + s) = \tan s$,   } (reduction formulas).
   (f) $\tan(2\pi - s) = -\tan s$,

12. Three additional circular functions, **secant**, **cosecant**, and **cotangent** are defined as follows:

(a) $\left\{(s, \sec s) \mid \sec s = \dfrac{1}{\cos s}, \left(s \neq \dfrac{\pi}{2} + k\pi, k \in J\right)\right\},$

(b) $\left\{(s, \csc s) \mid \csc s = \dfrac{1}{\sin s}, (s \neq k\pi, k \in J)\right\},$

(c) $\left\{(s, \cot s) \mid \cot s = \dfrac{\cos s}{\sin s}, (s \neq k\pi, k \in J)\right\}.$

Ranges for these functions and their periods are determined from their defining equations.

13. Reduction formulas for sec $s$, cosec $s$, and cot $s$, can be derived directly from the defining equations of their respective functions.

14. Table III, page 294, can be used to find approximations for function values for the six circular functions for some $x$ (or $s$) $\in R$, $0 < x < 1.57$.

15. Function values for real number values of $x$ (or $s$) at smaller intervals than 0.01 listed in Table III, can be obtained by referring to tables with greater detail or by using *linear interpolation*.

16. Any circular function value of any real number $x$ (or $s$) for which the function value is defined can be approximated by using the appropriate reduction formula in conjunction with Table III. A sketch of a unit circle showing the reference arc whose length has been designated as $\tilde{x}$ (or $\tilde{s}$) is usually helpful.

### Chapter Review

A.

1. Specify the domain and range of the coordinate function.
2. Complete the following:

(a)   coordinate $(0 + 2\pi k, k \in J) = $ _____?_____ .

(b)   coordinate $\left(\dfrac{3\pi}{2} + 2\pi k, k \in J\right) = $ _____?_____ .

3. Find each ordered pair.
(a) coordinate $(\pi/3)$.          (b) coordinate $(-\pi/6)$.

4. Find the element $s$, $0 < s < 2\pi$, in the domain associated with each specified element, coordinate $(s)$, in the range.

(a) $\left(\dfrac{\sqrt{3}}{2}, -\dfrac{1}{2}\right).$          (b) $\left(-\dfrac{1}{\sqrt{2}}, \dfrac{1}{\sqrt{2}}\right).$

5. Using $\pi \approx 3.14$, determine the quadrant in which the terminal point of the associated arc lies.
(a) coordinate (2.5).          (b) coordinate $(-15/2)$.

6. Find the function values.

(a) $\cos \dfrac{5\pi}{3}$.    (b) $\sin \left(\dfrac{-5\pi}{4}\right)$.    (c) $\tan \left(\dfrac{-14\pi}{3}\right)$.

7. Find the least positive $s$ for which the following statements are true.

(a) $\cos s = 0$.    (b) $\sin s = \dfrac{\sqrt{3}}{2}$.

8. Show that $\cos^2 s + \sin^2 s = 1$ for the following arc measures.

(a) $s = \dfrac{\pi}{3}$.    (b) $s = -\dfrac{\pi}{6}$.

9. If $\sin s = -4/5$ and $\cos s < 0$, find:
(a) $\cos s$.    (b) $\tan s$.    (c) $\sec s$.

10. State which of the six circular function values are positive and which are negative for the following measures:

(a) $\dfrac{5\pi}{4}$.    (b) $-\dfrac{16\pi}{3}$.    (c) 6.

11. Find the least positive $s$ for which each statement is true.
(a) $\cos s = -1$.    (b) $\sin s = -\frac{1}{2}$.    (c) $\tan s = -\sqrt{3}$.

12. Use the appropriate sum or difference formulas to show the following.
(a) $\cos (\pi - s) = -\cos s$.    (b) $\tan (\pi + s) = \tan s$.

13. Use Table III to find an approximation for:
(a) $\cos 0.74$.    (b) $\sin 1.15$.

(c) $\tan \dfrac{3\pi}{8}$. Use linear interpolation ($\pi \approx 3.142$).

14. Use Table III to find an approximation for the least positive element $x$ in the domain of the circular function for the given element in the range.
(a) $\cos x = 0.9856$.    (b) $\tan x = -1.162$.
(c) $\csc x = 2.607$. Use linear interpolation ($\pi \approx 3.142$).

15. Use Table III to find an approximation for each of the following:
(a) $\cos 5.05$.    (b) $\sin 3.86$.
(c) $\tan 6.489$. Use linear interpolation ($\pi \approx 3.142$).

B.

16. Find approximations for the solution set of $\cos x = 0.4085$, $x \in R$.
17. Find approximations for the solution set of $\tan x = -1.459$, $x \in R$.
18. Find approximations for the solution set of $\sec x = 1.805$, $x \in R$. Use linear interpolation ($\pi \approx 3.142$).

C.

19. Use Equation (1), page 79, to find a value for $\cos 0.04$ correct to four decimal places.
20. Use Equation (2), page 79, to find a value for $\sin 0.07$ correct to four decimal places.

# 3

# Graphs of the Circular Functions

T HE PERIODIC CHARACTERISTICS of the circular functions can be investigated further by studying their graphs.

In this chapter you will graph the circular functions (and combinations of these with nonperiodic functions) in the Cartesian coordinate system. In Section 7.4 you will graph similar periodic functions in a different coordinate system.

## 3.1
### Sine and Cosine Functions

In Chapter 2 we used the variables $s$ and $x$ as real numbers associated with arc lengths on a unit circle. In this section we use the variable $x$ for a real number associated with the length of a line segment on the $x$ axis of a rectangular coordinate system. In Section 2.1 we thought of the number line as *winding* on the circumference of a unit circle. Now line segments on the $x$ axis of a Cartesian coordinate system can be thought of as obtained by *unwinding* the arc that has the same length (see Figure 2.2).

The graphs of the sine and cosine functions, defined by the equations $y = \sin x$ and $y = \cos x$, consist of the set of points corresponding to $(x, \sin x)$ and $(x, \cos x)$ which satisfy the respective equations. Although Table III is available to find values for $\sin x$ and $\cos x$ for many values of $x$, function values which can be obtained from Table 2.3, page 71, and the appropriate reduction formulas are sufficient to obtain a good approximation to the graphs of these functions for $0 \leq x \leq 2\pi$. Because the graphs of functions defined by $y = \sin x$ or $y = \cos x$ are called **sine waves** or **sinusoidal waves**, we first consider the graph of

$$\{(x, y) \mid y = \sin x\}. \tag{1}$$

In order not to distort the form of the graph of (1) and the following graphs, we shall use similar units of measurement on the $x$ axis and $y$ axis.

Using the special values for $x$ in the interval 0 to $2\pi$ from Table 2.3 we obtain some ordered pairs in the function which are plotted on the plane in Figure 3.1a. The fact that the sine function is continuous implies that its

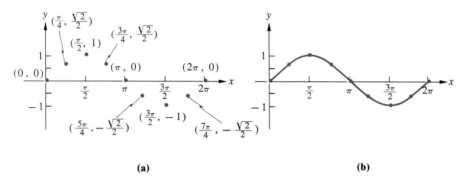

(a)                                                             (b)

**FIGURE 3.1**

graph contains no "breaks." Furthermore, it is shown in calculus that function values increase as $x$ increases from 0 to $\pi/2$ and then decrease as $x$ increases from $\pi/2$ to $3\pi/2$, etc. Therefore we connect the points in Figure 3.1a with a smooth curve to produce Figure 3.1b. Now, because the sine function is periodic with period $2\pi$; that is,

$$\sin (x + 2\pi) = \sin x,$$

we repeat the pattern in Figure 3.1b in both directions and obtain Figure 3.2 which is part of the graph of $y = \sin x$ for all $x \in R$. The graph over one period is called a **cycle** of the wave.

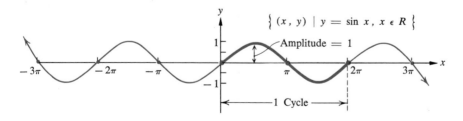

**FIGURE 3.2**

**Definition 3.1.** *The absolute value of half the difference of the maximum and minimum ordinates of a wave which is the graph of a periodic function is called the* **amplitude** *of the wave. This number is also called the amplitude of the function.*

Thus the amplitude of the wave in Figure 3.2, or of function (1) on page 89, is

$$|\tfrac{1}{2}[1 - (-1)]| = 1.$$

Several important notions of the sine function which we have previously discussed are also evident from the graph. Observe that the range of the function is

$$\{y \mid -1 \le y \le 1\}.$$

Also observe that the zeros of the function, values of $x$ for which $\sin x = 0$ and which are associated with the points where the curve intersects the $x$ axis, are $k\pi$, $k \in J$.

We obtain the graph of the cosine function,

$$\{(x, y) \mid y = \cos x\}, \tag{2}$$

in the same way that we obtained the graph of the sine function. Some ordered pairs in the function for $0 \le x \le 2\pi$ are obtained using Table 2.3, page 71, or Table III (or from memory), and the points associated with these ordered pairs are plotted on the plane in Figure 3.3a. In this case function values decrease as $x$ increases from 0 to $\pi$ and they increase as $x$ increases from $\pi$ to $2\pi$. We connect the points in Figure 3.3a with a smooth curve to produce Figure 3.3b.

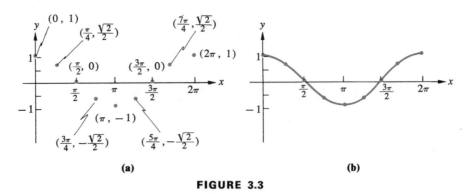

(a)                                                                (b)

**FIGURE 3.3**

Since the cosine function is periodic with period $2\pi$,

$$\cos (x + 2\pi) = \cos x,$$

we repeat this pattern in Figure 3.3b and obtain Figure 3.4 which is part of the graph of $y = \cos x$ for all $x \in R$. From the graph, observe that the range

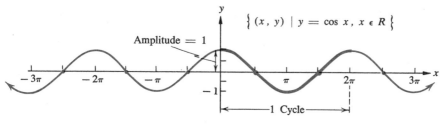

**FIGURE 3.4**

of the function is

$$\{y \mid -1 \le y \le 1\},$$

and therefore the amplitude of the wave is 1. Furthermore, observe that the zeros of the function which are associated with the points where the curve crosses the x axis are $(\pi/2) + k\pi, k \in J$.

In Section 2.7 you sometimes used linear interpolation to find closer approximations for elements in the range or the domain of circular functions than are available in Table III. We can now interpret geometrically the proportions we obtained. For example, consider a part of the graph of

$$y = \sin x,$$

which is shown in Figure 3.5 using an exaggerated curvature to illustrate the principle involved. Consider the value of $\sin 0.217$ which equals $l(\overrightarrow{MN})$, the

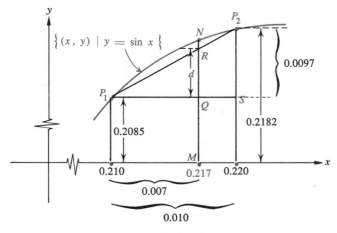

**FIGURE 3.5**

ordinate to the curve at $x = 0.217$. Since we cannot obtain this value because 0.217 is not an entry in Table III, we shall use $l(\overrightarrow{MR})$, which we can find from values available in the table, as an approximation for $l(\overrightarrow{MN})$. Line segments $\overline{P_2S}$ and $\overline{NQ}$ are perpendicular to $\overline{P_1S}$. For the purpose of interpolation we use the numerals 0.210 for 0.21 and 0.220 for 0.22. The values 0.2085 and 0.2182 are read directly from the table for sin 0.210 and sin 0.220, respectively.

Recall from Section 1.4 that the slope of $\overrightarrow{P_1P_2}$ is given by $l(\overrightarrow{SP_2})/l(\overrightarrow{P_1S})$ and that the slope of $\overrightarrow{P_1R}$ is given by $d/l(\overrightarrow{P_1Q})$. Since

$$l(\overrightarrow{SP_2}) \approx 0.2182 - 0.2085 = 0.0097, \quad l(\overrightarrow{P_1S}) = 0.010, \quad l(\overrightarrow{P_1Q}) = 0.007,$$

and

$$\text{slope of } \overrightarrow{P_1P_2} = \text{slope of } \overrightarrow{P_1R},$$

we have

$$\frac{0.0097}{0.010} \approx \frac{d}{0.007},$$

and $d = 7/10 \ (0.0097) \approx 0.0068$. Thus,

$$l(\overrightarrow{MN}) \approx l(\overrightarrow{MR}) \approx 0.2085 + 0.0068$$
$$= 0.2153.$$

## EXERCISE 3.1

A.

(a) Graph the function defined by each equation by plotting selected points over one period and then repeating the pattern.
(b) Specify the zeros of the function.

**Example.**  $y = 2 \sin x$.

*Solution.*  (a) Obtain some ordered pairs in the function defined by the equation and plot the points as shown. Connect the points with a smooth curve.

(b) Zeros: $k\pi, k \in J$.

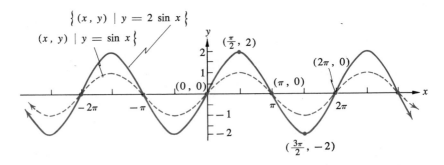

**1.** $y = \sin(-x)$.

**2.** $y = \cos(-x)$.

**3.** $y = -\sin x$.

**4.** $y = -\cos x$.

**5.** $y = 3 \sin x$.

**6.** $y = 2 \cos x$.

**7.** $y = \frac{1}{2} \cos x$.

**8.** $y = -\frac{1}{3} \sin x$.

**Example.** $y = \cos 2x$.

*Solution.* (a) Obtain some ordered pairs in the function defined by the equation and plot the points as shown. Connect the points with a smooth curve.
(b) Zeros: $(\pi/4) + k\pi$ and $(3\pi/4) + k\pi, \, k \in J$.

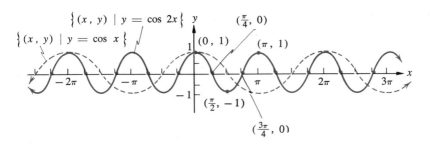

**▪9.** $y = \sin 2x$.

**11.** $y = \cos \frac{1}{2}x$.

**▪13.** $y = \sin \left( x - \dfrac{\pi}{2} \right)$.

**10.** $y = \sin \frac{1}{2}x$.

**12.** $y = \cos \frac{1}{3}x$.

**14.** $y = \cos \left( x + \dfrac{\pi}{4} \right)$.

**B.**

Graph two cycles of each of the following.

**Example.** $y = \cos (\pi x/3)$.

*Solution.* Obtain some ordered pairs in the function defined by the equation and plot the points as shown. Connect the points with a smooth curve in the form of a sine wave. In this case, the selection of rational values for elements, $x$, in the domain, facilitates obtaining elements, $y$, in the range. Label the $x$ axis in rational units. Some ordered pairs in the function are shown in the figure.

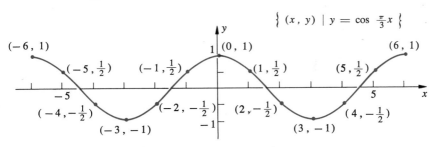

**15.** $y = \sin \pi x$.

**▪17.** $y = \cos \dfrac{\pi}{2} x$.

**16.** $y = \cos \pi x$.

**18.** $y = \sin \dfrac{\pi}{2} x$.

In each of the following, state whether values for sin $x$ and cos $x$, which are approximated by linear interpolation, are greater than or less than their actual values for the given interval.

**19.** sin $x$,   $\dfrac{\pi}{2} < x < \pi$.

**20.** sin $x$,   $\pi < x < \dfrac{3\pi}{2}$.

**21.** cos $x$,   $\dfrac{\pi}{2} < x < \pi$.

**22.** cos $x$,   $\dfrac{3\pi}{2} < x < 2\pi$.

**23.** sin $x$,   $-\dfrac{\pi}{2} < x < 0$.

**24.** cos $x$,   $-\dfrac{3\pi}{2} < x < -\pi$.

## 3.2
### General Sine Waves

Functions defined by equations of the form

$$y = A \sin B(x + C) \qquad \text{or} \qquad y = A \cos B(x + C),$$

where $A$, $B$, $C$, $x$, $y \in R$ and $A$, $B \neq 0$ always have sine waves for their graphs. You probably observed in Exercise 3.1 that depending upon the values for $A$, $B$, and $C$, these sine waves have *different amplitudes, different periods,* and *different horizontal positions* with respect to the origin. To facilitate graphing such functions, we shall now consider these factors in more detail.

First consider how the value of $A$ in the defining equation affects the graph of the function. For each value of $x$, each ordinate of the graph of $y = A \sin x$ is just $A$ times the ordinate of the graph of $y = \sin x$. The effect of $A$ then is to multiply the amplitude of the graph of $y = \sin x$ by $|A|$. Also, the graph of $y = A \cos x$ is related to the graph of $y = \cos x$ in a similar way.

*Example.*   Graph $y = 2 \cos x$, $-2\pi \le x \le 4\pi$.

*Solution.*   (1) Sketch the graph of $y = \cos x$, $0 \le x \le 2\pi$, as a reference.

(2) Since the amplitude of the graph of $y = 2 \cos x$ is 2, sketch the desired graph on the same coordinate system over the same interval by making each ordinate 2 times the ordinate of the graph of $y = \cos x$.

(3) Repeat the cycle obtained in (2) over the interval $-2\pi \le x \le 4\pi$.

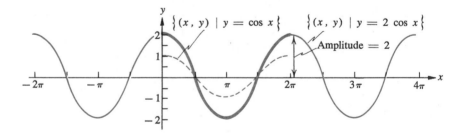

The first cycle is shown sketched with a heavier line for emphasis.

Now consider how different values for $B$ affect the graph of either $y = \sin Bx$ or $y = \cos Bx$. Since a complete cycle of the graph of $y = \sin x$ (or $y = \cos x$) occurs over the period $2\pi$, a complete cycle of the graph of $y = \sin Bx$ (or $y = \cos Bx$) is obtained as $Bx$ increases from 0 to $2\pi$. Therefore, the period of either the sine function or the cosine function is the solution of the equation

$$Bx = 2\pi,$$

or

$$x = \frac{2\pi}{B}.$$

In order to specify the period by a positive number, we use the absolute value of $B$. This argument establishes the following.

**Theorem 3.1.** *The **period** $P$ of a periodic function defined by an equation of the form $y = \sin Bx$ is given by*

$$P = \frac{2\pi}{|B|}.$$

***Example.*** Graph $y = \sin 3x$, $0 \leq x \leq 2\pi$.

*Solution.* (1) First sketch the graph of $y = \sin x$, $0 \leq x \leq 2\pi$, as a reference.

(2) Since $P = 2\pi/|B| = 2\pi/3$, sketch a complete cycle of the desired graph on the same coordinate system over the interval $0 \leq x \leq 2\pi/3$.

(3) Repeat the cycle obtained in (2) over the interval $0 \leq x \leq 2\pi$.

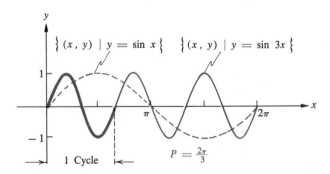

Finally, consider how different values for $C$ affect the graph of either $y = \sin (x + C)$ or $y = \cos (x + C)$. For $x = -C$,

$$\sin (x + C) = \sin 0 = 0,$$

and, for any $x_1$, the value of $\sin (x + C)$ at $x_1 - C$ will be the same as the value of $\sin x_1$ at $x_1$. That is,

$$\sin [(x_1 - C) + C] = \sin x_1.$$

For this reason, the graph of $y = \sin(x + C)$ is shifted $C$ units to the left of the graph of $y = \sin x$ if $C > 0$ and shifted $|C|$ units to the right of the graph of $y = \sin x$ if $C < 0$ as shown in Figure 3.6.

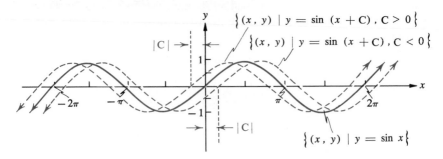

**FIGURE 3.6**

**Definition 3.2.**   *The number $|C|$ is called the **phase shift** of the graph of periodic functions defined by equations of the form $y = \sin B(x + C)$. This number is also called the phase shift of the corresponding function.*

**Example.**   Graph $y = \sin\left(x - \dfrac{\pi}{3}\right)$,      $-2\pi \le x \le 4\pi$.

*Solution.*   (1) First sketch the graph of $y = \sin x$, $0 \le x \le 2\pi$, as a reference.

(2) Since $C = -\pi/3$, the graph of $y = \sin x$ is shifted $|-\pi/3|$ units to the right to obtain the desired graph.

(3) Sketch the graph over the interval $-2\pi \le x \le 4\pi$.

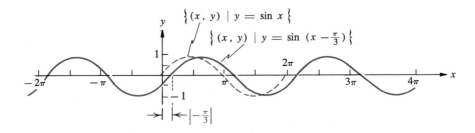

You can now use all the information about the effect of $A$, $B$, and $C$ on the graphs of functions defined by the equation $y = A \sin B(x + C)$ or by the equation $y = A \cos B(x + C)$ to facilitate making such graphs. Such an example is shown in the exercises.

If the period of a graph, $P$, is a positive rational number instead of a multiple of $\pi$ which we encountered in the examples above, it is helpful to use selected rational values as elements of the domain and to label the x axis in rational units to facilitate sketching the graph.

**Example.**  Graph $y = \frac{1}{2} \cos \pi x$, $-5 \le x < 5$.

*Solution.*  (1) Since the phase shift, $|C|$, is zero, compare this equation with

$$y = A \cos Bx.$$

It is evident that the graph of $y = \frac{1}{2} \cos \pi x$ has

(a) Amplitude,     $|A| = \frac{1}{2}$.

(b) Period,     $P = \dfrac{2\pi}{|B|} = \dfrac{2\pi}{\pi} = 2$.

(2) Sketch the graph, marking the x axis in rational units.

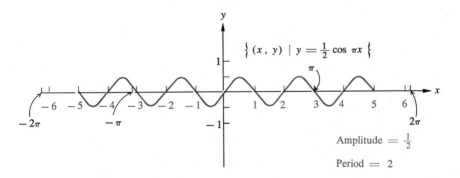

EXERCISE 3.2

**A.**

Sketch the graph of each equation over the interval $-2\pi \le x \le 2\pi$.

**Example.**   $y = 3 \cos 2x$.

*Solution.*  (1) Compare the equation with $y = A \cos Bx$. It is evident that the graph of $y = 3 \cos 2x$ has

(a) Amplitude,     $|A| = 3$.

(b) Period,     $P = \dfrac{2\pi}{|B|} = \dfrac{2\pi}{2} = \pi$.

(2) Sketch the graph as shown on page 98.

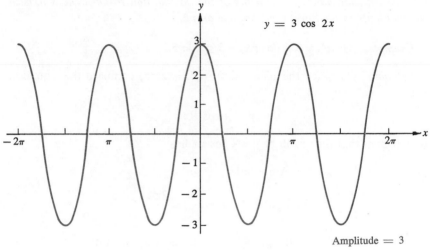

$$y = 3 \cos 2x$$

Amplitude $= 3$

Period $= \pi$

**▪1.** $y = 3 \sin x.$     **▪5.** $y = \frac{1}{2} \sin x.$     **▪9.** $y = \sin \frac{1}{2}x.$

**2.** $y = 2 \cos x.$     **6.** $y = -\frac{1}{3} \cos x.$     **10.** $y = -\cos \frac{1}{3}x.$

**3.** $y = -2 \cos x.$     **7.** $y = \cos 2x.$     **11.** $y = 2 \sin 2x.$

**4.** $y = -5 \sin x.$     **8.** $y = \sin 3x.$     **12.** $y = \frac{1}{3} \cos \frac{1}{2}x.$

*Hint*:   In Problems 13 to 16 use selected rational values for elements, $x$, in the domain. Mark the $x$ axis in rational units.

**▪13.** $y = \cos \pi x.$                          **▪17.** $y = \cos (x + \pi).$

**14.** $y = \sin \dfrac{\pi}{2} x.$                 **18.** $y = \sin \left( x - \dfrac{\pi}{2} \right).$

**15.** $y = \dfrac{1}{2} \sin \dfrac{\pi}{3} x.$    **19.** $y = 2 \sin \left( x - \dfrac{\pi}{4} \right).$

**16.** $y = 2 \cos \dfrac{\pi}{4} x.$               **20.** $y = \dfrac{1}{2} \cos \left( x + \dfrac{\pi}{6} \right).$

**B.**

*Example.*   Graph $y = 2 \cos \left( 3x + \dfrac{\pi}{2} \right).$

*Solution.*   (1) First rewrite the equation by factoring 3 from the expression in parentheses:

$$y = 2 \cos 3 \left( x + \dfrac{\pi}{6} \right).$$

(2) Compare this equation with $y = A \cos B(x + C).$

It is evident that the graph of $y = 2 \cos 3\left(x + \dfrac{\pi}{6}\right)$ has

(a) Amplitude, $|A| = 2$.

(b) Period, $P = \dfrac{2\pi}{|B|} = \dfrac{2\pi}{3}$.

(c) Phase shift is $\pi/6$ units to the left of the graph of $y = 2 \cos 3x$.

(3) Sketch the graph.

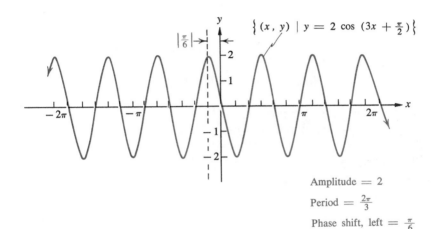

$$\{(x, y) \mid y = 2 \cos (3x + \tfrac{\pi}{2})\}$$

Amplitude $= 2$

Period $= \dfrac{2\pi}{3}$

Phase shift, left $= \dfrac{\pi}{6}$

**21.** $y = 2 \cos \left(2x + \dfrac{\pi}{2}\right)$.

**22.** $y = \dfrac{1}{2} \sin \left(\dfrac{1}{2} x - \dfrac{\pi}{4}\right)$.

**23.** $y = \dfrac{1}{2} \sin \left(\dfrac{1}{2} x + \dfrac{\pi}{6}\right)$.

**24.** $y = 2 \cos (2x - \pi)$.

**25.** Show graphically that $\sin x = \cos \left(x - \dfrac{\pi}{2}\right) = -\cos \left(x + \dfrac{\pi}{2}\right)$.

*Hint:* Graph

$$y = \sin x, \quad y = \cos \left(x - \dfrac{\pi}{2}\right), \quad \text{and} \quad y = -\cos \left(x + \dfrac{\pi}{2}\right).$$

**26.** Show graphically that $\cos x = -\sin \left(x - \dfrac{\pi}{2}\right) = \sin \left(x + \dfrac{\pi}{2}\right)$.

See hint, Problem 25.

By graphical methods, find all elements (approximate values) in each set for $0 \le x \le 2\pi$.

***Example.***            $\{(x, y)\,|\,y = 2 \sin x\} \cap \{(x, y)\,|\,y = 3/2\}.$

*Solution.*   Graph each set and approximate the coordinates of the points of intersection.

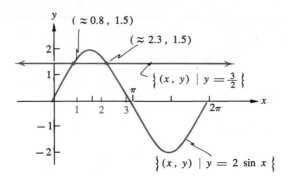

Intersection:   $\{(\approx 0.8, 1.5), (\approx 2.3, 1.5)\}.$

**27.** $\{(x, y)\,|\,y = \cos x\} \cap \{(x, y)\,|\,y = 1\}.$
**28.** $\{(x, y)\,|\,y = \sin 2x\} \cap \{(x, y)\,|\,y = -1\}.$
**•29.** $\{(x, y)\,|\,y = \cos \tfrac{1}{2}x\} \cap \{(x, y)\,|\,y = \tfrac{1}{2}\}.$
**30.** $\{(x, y)\,|\,y = 2 \sin x\} \cap \{(x, y)\,|\,y = -\tfrac{1}{2}\}.$

Find approximate solutions for each equation by graphical methods for $0 \le x \le 2\pi$.

***Example.***   $\sin x = 2 \cos x.$

*Solution.*   Sketch the graph of $y = \sin x$ and of $y = 2 \cos x$ on the same set of axes. Now, $\sin x = 2 \cos x$ for those values of $x$ where the graphs

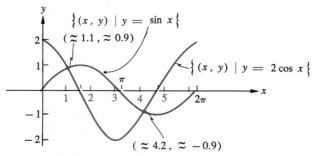

of the two equations intersect. Thus we seek the set of first components for

$$\{(x, y)\,|\,y = \sin x\} \cap \{(x, y)\,|\,y = 2 \cos x\}.$$

The solution set is $\{\approx 1.1, \approx 4.2\}.$

Graphical methods generally lead to approximate results only. In Section 5.4, you shall see how such equations involving circular functions can be solved analytically for exact results.

**31.** $\sin 2x = \frac{1}{2} \sin x$.  **·33.** $2 \cos 2x - \sin \frac{1}{2}x = 0$.

**32.** $\cos 2x = 2 \sin x$.  **34.** $2 \cos \frac{1}{2}x - \cos x = 0$.

### C.

In Problems 35 to 40, find the solution set for each inequality by graphical methods for $0 \le x \le 2\pi$.

*Example.* $\cos (x/2) < 0$.

*Solution.* Sketch the graph of $y = \cos (x/2)$ for $0 \le x \le 2\pi$.
The graph lies below the $x$
axis for those values of $y$
or $\cos (x/2)$ which are less
than 0. The corresponding
values of $x$ are coordinates
of points shown in color on
the $x$ axis. The solution
set of the inequality is
$\{x \mid \pi < x \le 2\pi\}$. Notice that
if the inequality had been $\cos (x/2) > 0$, the solution set would be
$\{x \mid 0 \le x < \pi\}$.

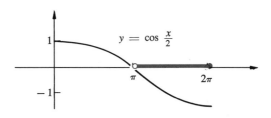

**35.** $\sin x < 0$.  **38.** $\sin \dfrac{x}{2} < 0$.

**36.** $\cos x > 0$.  **39.** $\sin \left(x + \dfrac{\pi}{4}\right) > 0$.

**·37.** $\cos 2x \ge 0$.  **40.** $\cos \left(x - \dfrac{\pi}{4}\right) \le 0$.

**41.** Show analytically that the function defined by $y = \sin x$ is the same function defined by $y = \cos \left(x - \dfrac{\pi}{2}\right)$ or $y = -\cos \left(x + \dfrac{\pi}{2}\right)$.

## 3.3
## Other Circular Functions

The functions defined by $y = \tan x$, $y = \cot x$, $y = \sec x$, and $y = \csc x$ can be graphed in much the same way that we graphed the sine and cosine functions in Section 3.2. Although the graphs of these functions are not sine waves, the graphs do exhibit the periodic nature of the functions and other important features.
First consider the graph of

$$\{(x, y) \mid y = \tan x\}.$$

Recall that the period of this function is $\pi$. Therefore we need consider some special values for $x$ only in the interval 0 to $\pi$. Using Table 2.3, page 71, or from memory, we obtain some ordered pairs in the function which are plotted in the plane in Figure 3.7. Since $\tan x = \sin x/\cos x$, the function defined by

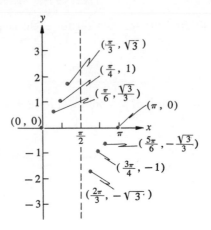

**FIGURE 3.7**

$y = \tan x$ is continuous in the domain $0 \le x < \pi/2$ and $\pi/2 < x \le \pi$ ($\cos x \ne 0$ in these intervals), and we connect the points in Figure 3.7 with a smooth curve to produce one cycle (heavy line) and then we repeat this pattern in both directions to obtain Figure 3.8.

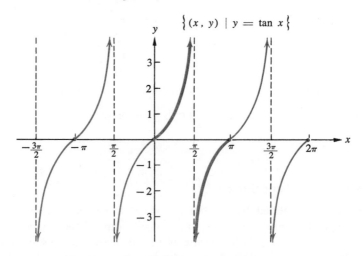

**FIGURE 3.8**

Because $\tan x$ is undefined for $x = \pi/2$, there is no point on the graph of the function when $x = \pi/2$. Notice, however, that $|\tan x|$ *increases indefinitely as x is taken closer and closer to* $\pi/2$. The vertical dashed lines in Figures 3.7 and 3.8, which the curve approaches but does not touch, are called **asymptotes**. The entire set of asymptotes to the curve is the set of lines which are the graphs of $x = (\pi/2) + k\pi, k \in J$. Furthermore, observe that the zeros of the function which are associated with the points where the curve intersects the $x$ axis are $k\pi, k \in J$.

The graph of the function defined by $y = \cot x$ can be similarly obtained by plotting some points over the interval $0 \le x < \pi$, obtained from Table 2.3, page 71, or again from memory. Recall however, that for $x \ne k\pi, k \in J$,

$$\cot x = \frac{1}{\tan x}.$$

Thus, each ordinate of the graph of $y = \cot x$, $(x \ne k\pi)$ is the reciprocal of the corresponding ordinate of the graph of $y = \tan x$. With this information you can quickly sketch the graph of $y = \cot x$ if you first sketch the graph of $y = \tan x$ on the same coordinate system. The graphs appear in Figure 3.9.

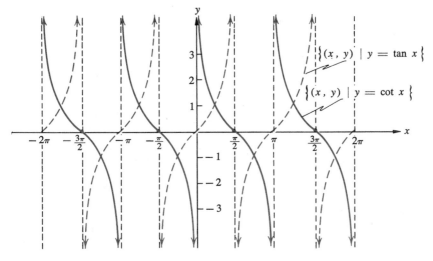

**FIGURE 3.9**

Notice that the zeros of the function defined by $y = \cot x$ are $(\pi/2) + k\pi, k \in J$, and that the asymptotes to the curve are given by $x = k\pi, k \in J$. Also, the range of either the tangent or cotangent function is $\{y \mid y \in R\}$.

The graphs of the functions defined by $y = \sec x$ and $y = \csc x$ can also be obtained by using Table 2.3, page 71, and plotting points. Alternatively, since

$$\sec x = \frac{1}{\cos x}, \qquad x \ne \frac{\pi}{2} + k\pi, k \in J,$$

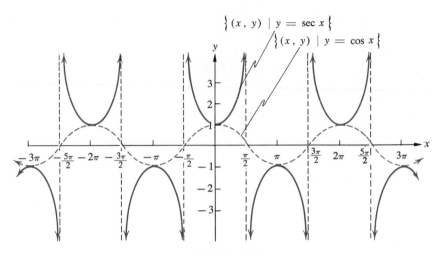

**FIGURE 3.10**

and

$$\csc x = \frac{1}{\sin x}, \qquad x \neq k\pi, k \in J,$$

each ordinate of the graph of $y = \sec x$ is the reciprocal of the corresponding ordinate of the graph of $y = \cos x$, and each ordinate of the graph of $y = \csc x$ is the reciprocal of the corresponding ordinate of the graph of $y = \sin x$, except for the noted restrictions. Thus the graphs of $y = \sec x$ and $y = \csc x$ can be obtained directly from the graphs of $y = \cos x$ and $y = \sin x$ as shown in Figures 3.10 and 3.11. Notice that the asymptotes of the graph of $y = \sec x$

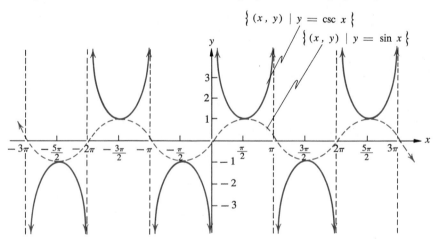

**FIGURE 3.11**

are $x = (\pi/2) + k\pi$, $k \in J$, which are also the zeros of the function defined by $y = \cos x$. The asymptotes of the graph of $y = \csc x$ are $x = k\pi$, $k \in J$, which are also the zeros of the function defined by $y = \sin x$. The range of either the secant or the cosecant function is $\{y \mid |y| \geq 1\}$. Thus there are no zeros of these functions.

Variations in the constants $A$, $B$, and $C$ in equations such as

$$y = A \tan B(x + C)$$

affect their graphs in much the same way that these constants affected the graphs of $y = A \sin B(x + C)$ and $y = A \cos B(x + C)$. Several examples appear in the exercises.

### EXERCISE 3.3

**A.**

Sketch the graph of the function defined by each equation over the interval $0 \leq x \leq 2\pi$ and specify (a) the range; (b) the equations of the asymptotes; (c) the zeros of the function, if they exist.

***Example.*** $y = 3 \tan 2x$.

*Solution.* Since the period of $y = \tan x$ is $\pi$, the period of $y = 3 \tan 2x$ is $\pi/2$ and each ordinate is 3 times the corresponding ordinate of $y = \tan x$. The graph is sketched as shown.

(a) Range: $\{y \mid y \in R\}$.

(b) Equations of the asymptotes:

$$x = \frac{\pi}{4} + \frac{k\pi}{2}, \quad k \in J.$$

(c) Zeros: $\left\{x \mid x = \frac{k\pi}{2}, \, k \in J\right\}$.

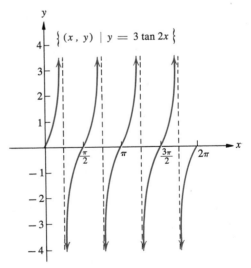

**•1.** $y = 2 \tan x$.

**2.** $y = 3 \csc x$.

**3.** $y = 2 \sec x$.

**4.** $y = \frac{1}{2} \cot x$.

**•5.** $y = \tan 2x$.

**6.** $y = \tan \frac{1}{2}x$.

**7.** $y = \tan \left(x - \frac{\pi}{4}\right)$.

**8.** $y = \cot \left(x + \frac{\pi}{3}\right)$.

**•9.** $y = 2 \tan \left(x + \frac{\pi}{6}\right)$.

**10.** $y = 3 \cot \left(x - \frac{\pi}{6}\right)$.

**B.**

**11.** Show graphically that $\cot x = \tan\left(\dfrac{\pi}{2} - x\right)$.

   *Hint:*  $\tan\left(\dfrac{\pi}{2} - x\right) = \tan\left[-\left(x - \dfrac{\pi}{2}\right)\right]$.

**12.** Show graphically that $\tan x = \cot\left(\dfrac{\pi}{2} - x\right)$.

   *Hint:*  $\cot\left(\dfrac{\pi}{2} - x\right) = \cot\left[-\left(x - \dfrac{\pi}{2}\right)\right]$.

By graphical methods, find elements (approximate values) in each set over the interval $0 \le x \le 2\pi$.

***Example.***  $\{(x, y)\,|\,y = \tan x\} \cap \{(x, y)\,|\,y = 1\}$.

*Solution.*  Sketch the graph of $y = \tan x$ and $y = 1$ as shown. The coordinates of the points of intersection are estimated.

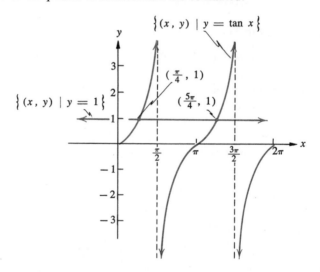

Intersection: $\{(\pi/4, 1), (5\pi/4, 1)\}$.

**▪13.** $\{(x, y)\,|\,y = \tan x\} \cap \{(x, y)\,|\,y = -1\}$.

**14.** $\{(x, y)\,|\,y = \cot x\} \cap \{(x, y)\,|\,y = \sqrt{3}\}$.

**15.** $\{(x, y)\,|\,y = 3 \sec x\} \cap \{(x, y)\,|\,y = 6\}$.

**16.** $\{(x, y)\,|\,y = 4 \csc x\} \cap \{(x, y)\,|\,y = 3\}$.

By graphical methods, approximate values of $x$ in the solution set of the equation over the interval. Check your answers by using Table III and the reduction formulas.

**Example.**   Tan $x = 3$, for $0 \le x \le 2\pi$.

*Solution.*   (1) Graph $y = \tan x$ and $y = 3$ on the same coordinate system.
   (2) Approximate the $x$ coordinate at each point of intersection.

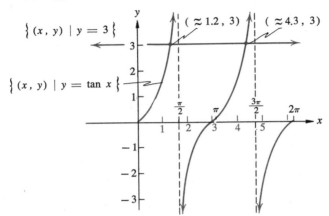

The solution set is $\{\approx 1.2, \approx 4.3\}$.

**•17.** $\cot x = -4$, for $0 \le x \le 2\pi$.      **19.** $\sec x = 2$, for $0 \le x \le 2\pi$.
**18.** $\tan x = \frac{3}{4}$, for $0 \le x \le 2\pi$.      **20.** $\csc x = -3.5$, for $0 \le x \le 2\pi$.

**C.**

In Problems 21 to 26, find the solution set for each inequality by graphical methods for $0 \le x \le 2\pi$. See example on page 101.

**•21.** $\tan x < 0$.                          **24.** $\cot \dfrac{x}{2} < 0$.

**22.** $\sec x \ge 0$.                          **•25.** $\tan \left( x + \dfrac{\pi}{4} \right) \le 0$.

**23.** $\csc \dfrac{x}{2} \ge 0$.                **26.** $\sec \left( x - \dfrac{\pi}{6} \right) < 0$.

**27.** Show analytically that the function defined by $y = \tan x$ is the same function defined by $y = -\cot \left( x + \dfrac{\pi}{2} \right)$.

**28.** Show analytically that the function defined by $y = \sec x$ is the same function defined by $y = \csc \left( x + \dfrac{\pi}{2} \right)$.

You have observed that, in general, approximate values are obtained when you use linear interpolation. State whether the approximation obtained for each of the following is less than or greater than the actual value for $0 < x < \pi/2$.

**29.** $\tan x$.          **30.** $\cot x$.          **31.** $\csc x$.          **32.** $\sec x$.

## 3.4

## Graphical Addition; Parametric Equations

Functions defined by equations such as

$$y = 2 \sin x + \cos x$$

and                                $$y = \cos x + x$$

can be graphed by methods similar to those you used in the preceding sections. You can find some elements in the range for arbitrary elements in the domain by first using available tables to find values for $\sin x$ and $\cos x$ and then performing the indicated operations. Then you can graph the set of ordered pairs obtained and complete the graph by interpolating for other elements in the range as you connect these points.

An alternative, and much simpler method, involves the *graphical addition of ordinates*. We illustrate this method with two examples.

**Example.**   Graph $y = 2 \sin x + \cos x$ over the interval $0 \leq x \leq 2\pi$.

*Solution.*   (1) Sketch the graphs of $y_1 = 2 \sin x$ and $y_2 = \cos x$ on the same coordinate system over the given interval.

(2) The ordinate of the graph of $y = 2 \sin x + \cos x$ for each $x$ is the algebraic sum of the ordinates of the graph of $y_1 = 2 \sin x$ and the graph of $y_2 = \cos x$. For example, for $x = \pi/3$, we have that

$$2 \sin x = 2 \cdot (\sqrt{3}/2) \approx 1.7, \quad \cos x = 0.5,$$

and                    $$y = 2 \sin x + \cos x \approx 1.7 + 0.5 = 2.2.$$

This ordinate, 2.2, can be approximated graphically by "adding" directed vertical distances, $\overrightarrow{AC}$ and $\overrightarrow{AB}$, from the $x$ axis to the respective curves at $x = \pi/3$. Other ordinates can be found graphically in a similar way. The points on the graph of $y = 2 \sin x + \cos x$ for which $2 \sin x$ or $\cos x$ equals zero are particularly easy to locate.

(3) Connect the points with a smooth curve to obtain the graph·

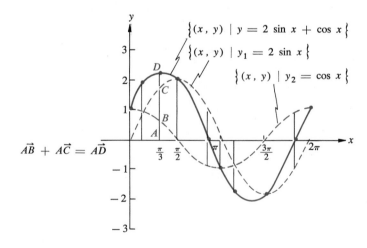

***Example.*** Graph $y = \cos x + x$ over the interval $-\pi \le x \le 2\pi$.

*Solution.* (1) Sketch the graphs of $y_1 = \cos x$ and $y_2 = x$ on the same set of axes over the given interval.

(2) Add ordinates for arbitrary values of $x$. The points on the graph of $y = \cos x + x$ for which $\cos x = 0$ are easy to locate.

(3) Connect the points with a smooth curve, as you interpolate for other points between those already plotted, to obtain the graph:

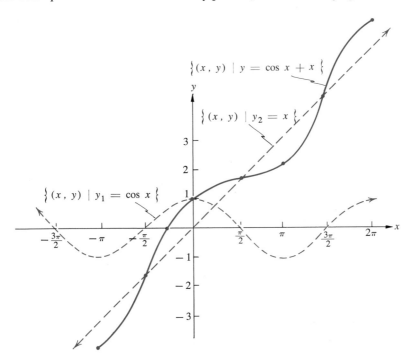

Sometimes it is convenient to define a relation (or function) in which the variable representing an element in the domain and the variable representing an element in the range are each related to a third variable. For example, the pair of equations

$$x = t - 3 \quad \text{and} \quad y = t + 1$$

express $x$ in terms of $t$ and $y$ in terms of $t$. Such equations are called **parametric equations**; $t$ is called the **parameter**. To graph the relation between $x$ and $y$ as shown in the example below, you can arbitrarily choose real number replacement values for $t$, and then obtain the corresponding values for $x$ and $y$. Alternatively, by solving the system of equations, you can eliminate the parameter $t$ to first obtain an equation in the variables $x$ and $y$. Then you can graph the equation in the usual way.

***Example.*** (a) Graph the function defined by the parametric equations

$$x = t - 3, \quad y = t + 1, \tag{1}$$

(*Example continued on the next page.*)

by assigning arbitrary values for $t$ to obtain some ordered pairs $(x, y)$ in the function.

(b) Eliminate $t$ and find the Cartesian equation.

*Solution.* (a) Choosing 0, 1, 2, and 3 as replacements for $t$, the corresponding ordered pairs $(x, y)$ are found to be $(-3, 1)$, $(-2, 2)$, $(-1, 3)$, and $(0, 4)$. The graph is shown in the figure.

(b) Equations (1) on page 109 can be written equivalently as

$t = x + 3,$

$t = y - 1,$

from which

$x + 3 = y - 1,$

or

$x - y = -4,$

the Cartesian equation
of a straight line.

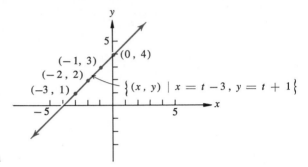

EXERCISE 3.4

A.

In Problems 1 to 12, sketch the graph of each equation over the interval $-\pi \leq x \leq 2\pi$.

**▪1.** $y = 2 \cos x + \sin x.$

**2.** $y = \cos 2x + \frac{1}{2} \sin x.$

**3.** $y = \frac{1}{2} \cos x + \sin \frac{1}{2}x.$

**4.** $y = \cos \dfrac{x}{2} + \frac{1}{3} \sin x.$

**▪5.** $y = \sin x - \cos x.$

**6.** $y = 2 \cos x - \frac{1}{3} \sin x.$

**7.** $y = \cos \dfrac{\pi x}{2} + 2 \sin \dfrac{\pi x}{2}.$

**8.** $y = 2 \cos \dfrac{\pi x}{2} - \cos \dfrac{\pi x}{4}.$

**▪9.** $y = 2 \cos x + 2.$

**10.** $y = 1 - 2 \sin x.$

**11.** $y = \sin x + x.$

**12.** $y = \dfrac{x}{2} - \cos x.$

In Problems 13 to 16,
(a) Graph the curve represented by the parametric equations by assigning values to $t$, where $-5 \leq t \leq 5$, or $\alpha$, where $0 \leq \alpha \leq 2\pi$, to obtain some ordered pairs $(x, y)$.
(b) Eliminate the parameter to find the Cartesian equation.

**•13.** $x = 3t + 2$, $y = 2t - 5$.

**14.** $x = t - 3$, $y = t^2$.

**15.** $x = 2 \cos \alpha$, $y = 2 \sin \alpha$.   *Hint:* $\cos^2 \alpha + \sin^2 \alpha = 1$.

**16.** $x = 2 \cos \alpha$, $y = 5 \sin \alpha$.

**B.**

Sketch the graph of each equation over the interval $0 \le x \le 2\pi$.

**•17.** $y = x - 2 + \cos x$.   *Hint:*  First graph $y_1 = x - 2$ and $y_2 = \cos x$.

**18.** $y = 3 - x + \sin x$.            **•21.** $y = \sin \pi x - x + 4$.

**19.** $y = 2 \cos x + \dfrac{x^2}{12}$.            **22.** $y = \cos \dfrac{\pi}{2} x - \dfrac{x^2}{12}$.

**20.** $y = 2 \sin x + \dfrac{1}{x}$.

**C.**

**23.** Graph the relation between $x$ and $y$ from the parametric equations $x = 2(\theta - \sin \theta)$ and $y = 2(1 - \cos \theta)$ over the interval $0 \le \theta \le 4\pi$.

*Note:*   The graph of these equations is called a cycloid. It is the curve traced by a fixed point on the circumference of a circle which rolls on a fixed straight line.

## 3.5
### Inverse Relations and Functions

Recall from Section 1.6 that the inverse of a function is the relation obtained by interchanging the components of every ordered pair in the function, and that *the defining equation of the inverse relation can be obtained by interchanging the variables in the defining equation of the original function.* Furthermore the graph of the inverse of a function is the *reflection* of the graph of the function about the graph of the equation $y = x$. Consider the graph of the function

$$\text{sine} = \{(x, y) \mid y = \sin x\},$$

with domain $R$ and range, $\{y \mid -1 \le y \le 1\}$, and its inverse,

$$\textbf{sine}^{-1} = \{(x, y) \mid x = \sin y\}$$

with domain $\{x \mid -1 \le x \le 1\}$ and range $R$ in Figure 3.12 on page 112. Notice that the graph of $x = \sin y$ is the reflection of the graph of $y = \sin x$ about the graph of the equation $y = x$. Further, notice that although

$$x = \sin y$$

defines a relation, it does not define a function, because for each element $x$

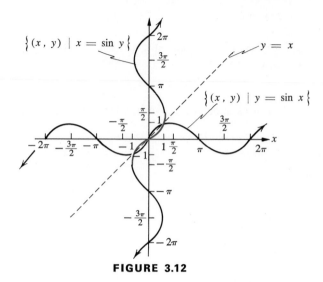

**FIGURE 3.12**

in its domain, there are an unlimited number of elements $y_1$, $y_2$, etc., in its range (see Figure 3.13). The inverse of the sine function is the **arcsine relation**.

Each circular function has an inverse relation. Thus, we have the **arccosine relation**

$$\text{cosine}^{-1} = \{(x, y) \mid x = \cos y\}$$

whose graph is shown in Figure 3.14. The remaining inverse relations of the

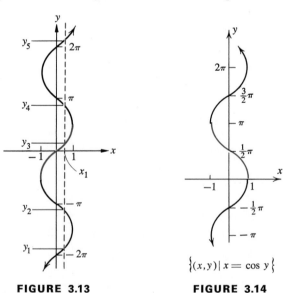

**FIGURE 3.13**                    **FIGURE 3.14**

circular functions are named in a similar way and their graphs are shown in Figure 3.15.

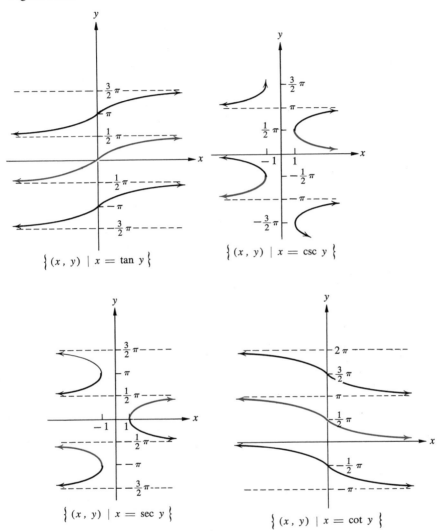

FIGURE 3.15

**Example.** Specify the inverse of each function.

(a) $\{(x, y) \mid y = \frac{1}{2} \tan x\}$.      (b) $\{(x, y) \mid 3y = 2 \sec 5x\}$.

*Solutions.*

(a) $\{(x, y) \mid x = \frac{1}{2} \tan y\}$.      (b) $\{(x, y) \mid 3x = 2 \sec 5y\}$.

By suitably restricting the domains of the circular functions or the ranges of the respective inverse relations, we can define an inverse *function* for each

(*Solutions continued on the next page.*)

circular function. For the sine function, for example, we shall restrict the domain to $\{x \mid -\pi/2 \leq x \leq \pi/2\}$. Thus

$$x = \sin y, \qquad -\frac{\pi}{2} \leq y \leq \frac{\pi}{2} \tag{1}$$

defines the inverse sine function. Its graph over this domain (there is only one $y$ for each $x$ in the domain) is shown in color in Figures 3.12 and 3.13. Generally, we wish to write the defining equation (1) in a form in which $y$ is expressed explicitly in terms of $x$. This is done by using the notation

$$y = \text{Arcsin } x$$

which is read "$y$ equals cap arcsin $x$," or

$$y = \text{Sin}^{-1} x,$$

with the same meaning. It should be understood that $\text{Sin}^{-1} x$ *does not mean* $1/\text{Sin } x$.

Similar notation is used for the other inverse circular functions.

**Definition 3.3.**   *The inverse circular functions\* corresponding to the six circular functions are*

$$\text{Arccosine} = \{(x, y) \mid y = \text{Cos}^{-1} x, \quad 0 \leq y \leq \pi\},$$

$$\text{Arcsine} = \left\{(x, y) \mid y = \text{Sin}^{-1} x, \quad -\frac{\pi}{2} \leq y \leq \frac{\pi}{2}\right\},$$

$$\text{Arctangent} = \left\{(x, y) \mid y = \text{Tan}^{-1} x, \quad -\frac{\pi}{2} < y < \frac{\pi}{2}\right\},$$

$$\text{Arcsecant} = \left\{(x, y) \mid y = \text{Sec}^{-1} x, \quad 0 \leq y < \frac{\pi}{2}, -\pi \leq y < \frac{\pi}{2}\right\},$$

$$\text{Arccosecant} = \left\{(x, y) \mid y = \text{Csc}^{-1} x, \quad 0 < y \leq \frac{\pi}{2}, -\pi < y \leq -\frac{\pi}{2}\right\}.$$

$$\text{Arccotangent} = \{(x, y) \mid y = \text{Cot}^{-1} x, \quad 0 < y < \pi\},$$

Notice that the *ranges* of the inverse functions *are specified in the definition* rather than following the usual procedure of stating the domain. Furthermore, the choices for the ranges are quite *arbitrary*.† Those which have been chosen here involve small values of $y$ and include values between 0 and $\pi/2$, have relatively simple graphs and yield a one-to-one correspondence between elements in the domain and range. They are also the values which are generally specified for use in the calculus.

---

\* In this text we use capital letters to designate inverse functions. However, it should be noted that some authors also use similar notation, with small letters for these functions and function values. For example, Arccosine is written arccosine and $\text{Cos}^{-1} x$ is written $\cos^{-1} x$.

† Some authors use different ranges for the Arcsecant, Arccosecant, and Arccotangent.

The function values in Arcsine, Arccosine, etc. are called the **principal values** in the analogous relation. (We use the word "analogous" to refer to the relation with the same name.) These values can readily be seen from the graphs of their respective functions. The graph of the Arccosine function is shown in color in Figure 3.14, page 112, and the graphs of the remaining inverse circular functions are shown in a similar way in Figure 3.15, page 113.

**Example.** Write $y = \text{Arcsin } \frac{1}{2}$ equivalently without inverse notation and then find the value of $y$ for which the statement is true.

**Solution.** The equivalent form is $\frac{1}{2} = \sin y$, where $-\pi/2 \le y \le \pi/2$, from which we obtain $y = \pi/6$.

## EXERCISE 3.5

A.

Specify the inverse relation of each function by interchanging the variables. In Problems 1 to 4, give the domain and the range of the function $F$ and its inverse $F^{-1}$.

1. $\{(x, y) | y = \tan x\}$.
2. $\{(x, y) | y = \cot x\}$.
3. $\{(v, w) | w = \csc v\}$.
4. $\{(v, w) | w = \sec v\}$.
5. $\{(s, t) | t = \sin 2s\}$.
6. $\{(s, t) | t = \cos 3s\}$.
7. $\left\{(x, y) | y = \cot \dfrac{x}{3}\right\}$.

8. $\left\{(x, y) | y = \tan \dfrac{x}{2}\right\}$.
9. $\{(k, l) | l = \sec k\}$.
10. $\{(k, l) | l = 3 \csc k\}$.
11. $\{(m, n) | n = 3 \cos 2m\}$.
12. $\{(m, n) | n = 5 \sin 4m\}$.
13. $\left\{(u, v) | v = \sec \left(u + \dfrac{\pi}{2}\right)\right\}$.
14. $\{(u, v) | v = \csc (u - \pi)\}$.

Write each equation equivalently without inverse notation and then find the value of $y$ for which the statement is true.

15. $y = \text{Arccos } \dfrac{1}{\sqrt{2}}$.

16. $y = \text{Arctan } \sqrt{3}$.

17. $y = \text{Arccot } 1$.

18. $y = \text{Arccsc } 2$.

19. $y = \text{Arcsec } (-\sqrt{2})$.

20. $y = \text{Arcsin } \left(-\dfrac{\sqrt{3}}{2}\right)$.

21. $y = \text{Arcsin } 2$.

22. $y = \text{Arccot } \dfrac{1}{\sqrt{3}}$.

23. $y = \text{Arctan } (-1)$.

24. $y = \text{Arccsc } \dfrac{2}{\sqrt{3}}$.

25. $y = \text{Arccos } 0.8829$.   Use Table III.
26. $y = \text{Arccot } 2.161$.
27. $y = \text{Arcsin } 0.8542$.

28. $y = \text{Arccsc } (-1.299)$.

Find the value for $y$ for each of the following:

**Example.**   $y = \tan u$ and $u = \text{Arccos } \frac{1}{2}$.

*Solution.*   Since $u = \text{Arccos } \frac{1}{2} = \pi/3$,

$$y = \tan u = \tan \frac{\pi}{3} = \sqrt{3}.$$

**29.** $y = \cos u$ and $u = \text{Arctan } 1$.          **32.** $y = \sin u$ and $u = \text{Arctan } (-\sqrt{3})$.

**30.** $y = \sec u$ and $u = \text{Arcsin } \frac{1}{2}$.          **33.** $y = \csc u$ and $u = \text{Arccos } \frac{1}{2}$.

**31.** $y = \tan u$ and $u = \text{Arccos } (-\frac{1}{2})$.    **34.** $y = \cot u$ and $u = \text{Arctan } \dfrac{1}{\sqrt{3}}$.

Find the value of each of the following.

**Example.**   $\text{Cos } [\text{Arcsin } (\sqrt{3}/2)]$.

*Solution.*   Sketch an arc on the unit circle with the length $s$ such that $\sin s = \sqrt{3}/2$. Since $s = \text{Arcsin } (\sqrt{3}/2) = \pi/3$, then

$$\cos \left( \text{Arcsin } \frac{\sqrt{3}}{2} \right) = \cos \frac{\pi}{3} = \frac{1}{2}.$$

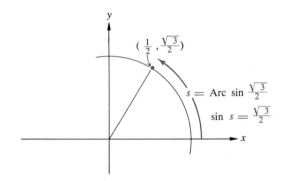

**35.** $\sin (\text{Arcsec } 2)$.                      **41.** $\cos (\text{Arcsin } 0.380)$.

**36.** $\cos (\text{Arccsc } 2)$.                      **42.** $\cos (\text{Arctan } 0.825)$.

**37.** $\sec [\text{Arccos } (-\frac{1}{2})]$.          **43.** $\sec (\text{Arctan } 4.113)$.

**38.** $\cos [\text{Arctan } (-1)]$.                    **44.** $\csc (\text{Arccos } 0.564)$.

**39.** $\cot \left[ \text{Arcsec } \left( -\dfrac{2}{\sqrt{3}} \right) \right]$.          **45.** $\tan [\text{Arccot } (-0.642)]$.

**40.** $\tan \left[ \text{Arcsin } \left( -\dfrac{1}{\sqrt{2}} \right) \right]$.          **46.** $\cot [\text{Arcsec } (-1.129)]$.

**Example.**   $\text{Sin}^{-1} (\tan \pi/4)$.

*Solution.*   Since $\tan \pi/4 = 1$,

$$\text{Sin}^{-1} [\tan (\pi/4)] = \text{Sin}^{-1} 1 = \frac{\pi}{2}.$$

**47.** $\text{Tan}^{-1}\left(\sin\dfrac{\pi}{2}\right)$.

**50.** $\text{Sin}^{-1}\left(\cos\dfrac{\pi}{2}\right)$.

**48.** $\text{Cos}^{-1}(\tan 0)$.

**51.** $\text{Cos}^{-1}(\sin 0.41)$.

**49.** $\text{Sin}^{-1}\left(\cos\dfrac{\pi}{4}\right)$.

**52.** $\text{Tan}^{-1}(\cos 1.32)$.

**B.**

Rewrite each expression without using circular function or inverse notation. Assume values for $x$ for which each function value exists.

**Example.**   $\sin(\text{Arccos } x)$.

*Solution.*   Let $s = \text{Arccos } x$; equivalently $\cos s = x, 0 \le s \le \pi$. Since

$$\sin s = \sqrt{1 - \cos^2 s}, \quad \text{then for } 0 \le s \le \pi,$$

$$\sin(\text{Arccos } x) = \sin s = \sqrt{1 - \cos^2 s} = \sqrt{1 - x^2}.$$

**53.** $\cos(\text{Arcsin } x)$.

**56.** $\csc(\text{Arcsin}\sqrt{1 - x^2})$.

**54.** $\cot(\text{Arcsin } x)$.

**57.** $\sec(\text{Arcsec } x)$.

**55.** $\tan(\text{Arccos}\sqrt{1 - x^2})$.

**58.** $\sin(\text{Arccsc } x)$.

In Problems 59 to 62, evaluate each expression.

**Example.**   $\cos\left(\text{Arcsin}\dfrac{1}{\sqrt{2}} - \text{Arccos}\dfrac{3}{5}\right)$.

*Solution.*   From the formula for $\cos(s_1 - s_2)$, Theorem 2.4 on page 52,

$$\cos\left(\text{Arcsin}\frac{1}{\sqrt{2}} - \text{Arccos}\frac{3}{5}\right)$$

$$= \cos\left(\text{Arcsin}\frac{1}{\sqrt{2}}\right)\cdot\cos\left(\text{Arccos}\frac{3}{5}\right) + \sin\left(\text{Arcsin}\frac{1}{\sqrt{2}}\right)\cdot\sin\left(\text{Arccos}\frac{3}{5}\right)$$

$$= \frac{1}{\sqrt{2}}\cdot\frac{3}{5} + \frac{1}{\sqrt{2}}\cdot\frac{4}{5} = \frac{3}{5\sqrt{2}} + \frac{4}{5\sqrt{2}} = \frac{7}{5\sqrt{2}}.$$

**59.** $\cos\left(\text{Arctan } 1 + \text{Arcsin}\dfrac{\sqrt{3}}{2}\right)$.

**60.** $\sin\left(\text{Arccos}\dfrac{1}{2} - \text{Arcsec } 2\right)$.

**61.** $\sin(\text{Arcsin } 0 + \text{Arccos } 1)$.

**62.** $\cos\left[\text{Arcsin}\left(\dfrac{-\sqrt{3}}{2}\right) - \text{Arctan}(-1)\right]$.

**63.** Show that $\text{Arccos}\dfrac{5}{13} = \text{Arccsc}\dfrac{13}{12}$.

**64.** Show that $\text{Arcsin}\dfrac{1}{\sqrt{10}} = \text{Arcsec}\dfrac{\sqrt{10}}{3}$.

## Chapter Summary

1. Graphs of the form of the graph of the sine function are called **sine waves** or **sinusoidal waves**. The part of the graph over one period is called a **cycle** of the wave. The absolute value of half the difference of the maximum and the minimum ordinates is called the **amplitude** of the wave.

2. The **zeros of the sine function**, that is, values of $x$ for which $\sin x = 0$, are $k\pi$, $k \in J$. These values are associated with the points where the graph of $\{(x, y) \mid y = \sin x\}$ intersects the $x$ axis. **Zeros of the cosine function** are $(\pi/2) + k\pi$, $k \in J$. These values are associated with the points where the graph of $\{(x, y) \mid y = \cos x\}$ intersects the $x$ axis (see Figures 3.2 and 3.4).

3. Functions defined by equations of the form

$$y = A \sin B(x + C)$$

   or

$$y = A \cos B(x + C),$$

   where $A$, $B$, $C$, $x$, $y \in R$ and $A$, $B \neq 0$ always have sine waves for their graphs. The **amplitude** of the wave or function is $|A|$, the **period** is $2\pi/|B|$, and the **phase shift** is $|C|$.

4. The graphs of the functions defined by $y = \tan x$, $y = \cot x$, $y = \sec x$, and $y = \csc x$ exhibit the periodic nature of these functions (see Figures 3.8, 3.9, 3.10 and .311).

5. The **zeros of the function defined by** $y = \tan x$ are $k\pi$, $k \in J$. The **asymptotes** of this function are the graphs of $x = (\pi/2) + k\pi$, $k \in J$.

6. The following table shows the zeros and/or the equations for the asymptotes of the other circular functions.

| *Function* | *Zeros* | *Equations for Asymptotes* |
|---|---|---|
| **cotangent** | $\dfrac{\pi}{2} + k\pi$, $k \in J$ | $x = k\pi$, $k \in J$ |
| **secant** | **none** | $x = \dfrac{\pi}{2} + k\pi$, $k \in J$ |
| **cosecant** | **none** | $x = k\pi$, $k \in J$. |

7. Functions defined by equations such as

$$y = 2 \sin x + \cos x$$

   can be graphed by first graphing $y_1 = 2 \sin x$ and $y_2 = \cos x$, and then "adding" their ordinates graphically for arbitrary values of $x$.

8. A pair of equations of the form

$$x = f(t) \qquad \text{and} \qquad y = g(t),$$

   in which two variables, in this case $x$ and $y$, are each expressed in terms of a third variable, are called **parametric equations**. The relation between $x$ and $y$ can be graphed by arbitrarily choosing real number replacements for $t$ and then obtaining the corresponding values for $x$ and $y$.

9. The inverse of a function is the relation obtained by *interchanging the components of every ordered pair in the function*. The defining equation of the inverse relation can be obtained by *interchanging the variables in the defining equation of the original function*.

10. The graph of the inverse of a function is the *reflection* of the graph of the function about the graph of the equation $y = x$.

11. By *suitably restricting the domain of the circular function, or the ranges of the respective inverse relations,* an inverse *function* is defined for each circular function. (See Definition 3.3.) The ranges have been chosen to *include* values of $y$ such that $0 < y < \pi/2$, the graphs of the functions are relatively simple, and there is a one-to-one correspondence between elements in the domain and the range.

12. The function values in Arcsine, Arccosine, etc. are sometimes called the **principal values** of the function values in the analogous relation.

## Chapter Review

**A.**

Sketch the graph of the function defined by each equation over the interval $-2\pi \le x \le 2\pi$. Specify (a) the amplitude, (b) the period, and (c) the phase (if applicable).

**•1.** $y = 2 \sin x$.

**2.** $y = \frac{1}{4} \cos 2x$.

**3.** $y = -3 \cos \frac{1}{2}x$.

**4.** $y = -\frac{1}{2} \sin \frac{1}{2}x$.

**•5.** $y = \sin\left(x + \dfrac{\pi}{2}\right)$.

**6.** $y = 2 \cos\left(x - \dfrac{\pi}{6}\right)$.

Sketch the graph of the function defined by each equation over the interval $0 \le x \le 2\pi$. Specify (a) the range, (b) the equations of the asymptotes, and (c) the zeros of the function.

**7.** $y = 2 \tan x$.

**8.** $y = \csc 2x$.

**•9.** $y = \frac{1}{2} \sec \frac{1}{2}x$.

**10.** $y = 3 \cot \frac{1}{2}x$.

Using the method of graphical addition of ordinates, graph the equations in Problems 11 and 12 over the interval $0 \le x \le 2\pi$.

**11.** $y = \sin 2x - \frac{1}{2} \cos x$.

**•13.** Graph $x = 4t - 1$, $y = 2t^2$.

**12.** $y = 2 + \frac{1}{3}x - \sin x$.

**14.** Graph $x = 3 \cos \theta$, $y = 4 \sin \theta$.

Specify the inverse function. Give the domain and the range of the function and its inverse.

**15.** $F = \{(x, y) \mid y = 2 \cos 3x\}$.

**16.** $F = \left\{(x, y) \mid y = \cot \dfrac{x}{2}\right\}$.

Write each equation in Problems 17 to 20 equivalently without inverse notation and then find the set of all $y \in R$ for which the statement is true.

**17.** $y = \text{Arccos } \dfrac{1}{\sqrt{2}}$.

**19.** $y = \text{Arccot } (-1)$.

**18.** $y = \text{Arccsc } (-2)$.

**20.** $y = \text{Arcsec } \dfrac{2}{\sqrt{3}}$.

**21.** Find the value for $y$ if $y = \sin u$ and $u = \text{Arccos } \dfrac{1}{\sqrt{2}}$.

**22.** Find the value for $y$ if $y = \csc t$ and $t = \text{Arctan } (-1)$.

Find the value of each of the following.

**23.** $\cos (\text{Arcsec } 1)$.

**26.** $\sin [\text{Arccos } (-0.4331)]$.

**24.** $\tan \left( \text{Arcsin } \dfrac{\sqrt{3}}{2} \right)$.

**27.** $\text{Cot}^{-1} (\cos 0)$.

**25.** $\sec (\text{Arcsin } 0.1908)$.

**28.** $\text{Csc}^{-1} (\tan 1.17)$.

Graph two cycles of each of the following.

**°29.** $y = \cos \dfrac{\pi}{3} x$.

**30.** $y = -2 \sin \dfrac{\pi}{2} x$.

**B.**

Graph each set and find the coordinates of the points of intersection in the interval $0 \le x \le 2\pi$.

**31.** $\{(x, y) \mid y = \sin 2x\} \cap \{(x, y) \mid y = 1\}$.

**32.** $\{(x, y) \mid y = 2 \cos x\} \cap \{(x, y) \mid y = -\frac{1}{2}\}$.

Find the solution set of each equation by graphical methods in the interval $0 \le x \le 2\pi$.

**°33.** $\cos x = 3 \sin x$.

**34.** $\sin 2x = 2 \cos \dfrac{x}{2}$.

Express each of the following without using circular function or inverse notation. Assume values for which each function value exists.

**35.** $\sin (\text{Arccos } x)$.

**36.** $\sec (\text{Arcsin } x)$.

Evaluate each expression.

**37.** $\sin (\text{Arcsec } 2 - \text{Arctan } 1)$.

**38.** $\cos \left[ \text{Arccot } \dfrac{1}{\sqrt{3}} + \text{Arcsin } \left( -\dfrac{\sqrt{3}}{2} \right) \right]$.

## C.

Find the solution set for each inequality by graphical methods for $0 \le x \le 2\pi$.

**‣39.** $\sin 2x \le 0$.

**40.** $\cos\left(x + \dfrac{\pi}{6}\right) > 0$.

**41.** $\cot x > 0$.

**42.** $\tan\dfrac{x}{2} \le 0$.

**‣43.** $\sec\dfrac{x}{2} < 0$.

**44.** $\csc\left(x - \dfrac{\pi}{4}\right) \ge 0$.

# 4

# Trigonometric Functions

ONE TYPE of periodic function introduced in Chapter 2 paired lengths of arcs on a unit circle with elements in the set of real numbers, $R$. Such functions are called circular functions. In this chapter, you will study another type of function in which the set of all angles are paired with elements of $R$. These functions are called **trigonometric functions**, and, as you will see, *they are also periodic and are closely related to the circular functions.*

## 4.1

### Angles

Recall from geometry that an angle is defined as the union of two rays having a common end point and named either by using the names of a point on each ray and the common end point, or by assigning a single symbol to the angle. For example, in Figure 4.1, the angle can be named as $\angle AOB$, or $\angle BOA$, where the symbol $\angle$ means angle, or simply by some Greek letter, in this case alpha, $\alpha$. When the three-letter designation is used, the middle letter is always the common end point of the two rays. This common end point is called the **vertex** of the angle, and the rays are called the **sides** of the angle.

Each angle in the plane can be made to coincide with (or be congruent to) an angle which has one of its sides along the $x$ axis, and vertex at the origin (Figure 4.2). Such an angle is said to be in *standard position*. The side, $OA$, which lies on the $x$ axis, is called the **initial side** of the angle, and the side $OB$

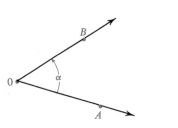

FIGURE 4.1

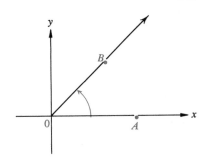

FIGURE 4.2

is called the **terminal side** of the angle. The angle can be visualized as being formed by rotating the initial side $OA$, with $O$ as the pivotal point, to the terminal side, $OB$. Angles for which the rotation is considered are called **directed angles**. The terminal side may lie in any quadrant or on any axis. An angle is said to be in that quadrant in which the terminal side lies. For example in Figure 4.3, $\alpha$ is in Quadrant I, $\beta$ (beta) is in Quadrant II, $\gamma$ (gamma) is in Quadrant III, and $\delta$ (delta) is in Quadrant IV.

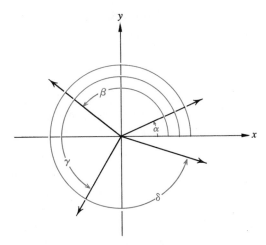

FIGURE 4.3

A measure can be assigned to an angle by using a circle whose center is at the vertex of the angle. An angle in this relation to a circle is called a **central angle**. For convenience, we shall consider angles in standard position, although the process to be described can be applied to angles in any position in the plane. A circle of radius $r$ has a circumference of length $2\pi r$. If the circumference is divided into $p$ arcs of equal length, then each arc would have a length given by $2\pi r/p$. Starting at the point $(r, 0)$, as in Figure 4.4, the circumference of the circle can be scaled in either direction, using the arc length $2\pi r/p$ as a unit measure. Positive numbers are assigned for the measure of angles in

which the rotation from the initial to the terminal side is in a counterclockwise
direction, and negative numbers are assigned for the measure if in a clockwise

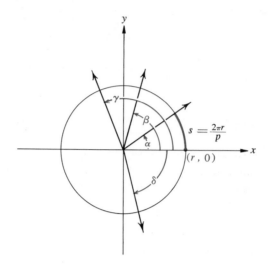

**FIGURE 4.4**

direction. Since the terminal side of any angle in standard position intercepts
the circumference of the circle at some point, *the number which indicates how
many arc lengths are intercepted from $(r, 0)$ to the point, is called the measure
of the angle for the particular unit measure.* For example, in Figure 4.4, if the
circle is considered to be divided into $p$ arcs of equal length where the measure
of $\alpha$ is 1, then the measure of $\beta$ is 2, the measure of $\gamma$ is 3, and the measure of
$\delta$ is $-2$ in $p$-units. We use the word measure here, in conjunction with angles,
to mean a *directed measure*.

The measure assigned to an angle in terms of the same $p$-measure is
independent of the radius of the
measuring circle. This means, for
example, that the measure of $\alpha$ in
Figure 4.5 is the same whether the
measure is determined from either
of the concentric circles shown.
This can be proven by using from
geometry the fact that the lengths
of arcs intercepted by a given
central angle in concentric circles
are proportional to the circum-
ferences of the circles. Thus in
Figure 4.5,

$$\frac{s_1}{2\pi r_1} = \frac{s_2}{2\pi r_2}, \qquad (1)$$

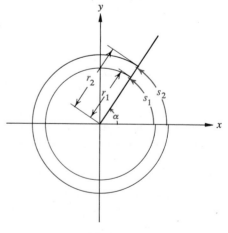

**FIGURE 4.5**

where $s_1$ and $s_2$ are the lengths of the arcs on circles with radii whose lengths are $r_1$ and $r_2$, respectively. If $n_1$ and $n_2$ denote the number of $p$-units contained in the arcs $s_1$ and $s_2$ respectively, then

$$s_1 = n_1 \left(\frac{2\pi r_1}{p}\right) \qquad \text{and} \qquad s_2 = n_2 \left(\frac{2\pi r_2}{p}\right). \tag{2}$$

Substituting the right-hand members of Equation (2) for $s_1$ and $s_2$, respectively, in Equation (1), yields

$$\frac{n_1 \left(\frac{2\pi r_1}{p}\right)}{2\pi r_1} = \frac{n_2 \left(\frac{2\pi r_2}{p}\right)}{2\pi r_2},$$

from which

$$n_1 = n_2.$$

Thus, the measure of $\alpha$ determined from either circle is the same.

Two kinds of angle measure most commonly used are called **degree measure** and **radian measure**. In degree measure, the circumference of the circle is divided into 360 arcs of equal length, so that the unit arc is of length

$$\frac{2\pi r}{360} \qquad \text{or} \qquad \frac{\pi}{180} r,$$

where $r$ is the length of the radius of the circle. To designate the degree-measure of an angle $\alpha$ we use the special notation $m°(\alpha)$. In radian measure, the circumference is divided into $2\pi$ ($\approx 6.28$) arcs of equal length, so that the length of the unit arc becomes

$$\frac{2\pi r}{2\pi} = r,$$

the length of the radius of the circle. For such a measure we use the notation $m^R(\alpha)$. In particular, if the unit circle (radius $= 1$) is used as the basis for radian measure, then the length of the unit arc is 1. Hence, the unit circle is conveniently used as a reference for the radian measure of an angle.

*Notice that the measure of angles are real numbers for any unit of measure.* In Figure 4.6a, $m°(\alpha) = 40$ and in Figure 4.6b, $m^R(\beta) = \pi/3$. We shall use customary conventions concerning symbolism for angles and their measure and write $m°(\alpha) = 40°$ or $m^R(\beta) = (\pi/3)^R$, where the use of the degree or radian symbol in the right-hand member of the equations should be interpreted as a redundancy for emphasis. Furthermore, we shall write $\alpha = 40°$ and $\beta = (\pi/3)^R$, and interpret these statements to mean $m°(\alpha) = 40$ and $m^R(\beta) = \pi/3$, respectively.

The measure of an angle in radians can be related to the measure of the angle in degrees. Since the ratio of the measure of a given angle to the measure

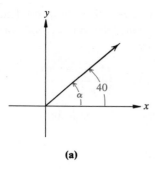

(a)                                    (b)

**FIGURE 4.6**

of the circumference of the circle is the same for any units, we have for degree measure and radian measure in particular that

$$\frac{m^{\circ}(\alpha)}{360} = \frac{m^{R}(\alpha)}{2\pi}.$$

The following theorem, applicable to the elements in the set of angles $A$, is a direct consequence of this statement.

**Theorem 4.1.**   *If $\alpha \in A$, then*

$$\textbf{I.}\quad m^{\circ}(\alpha) = \frac{360}{2\pi}\, m^{R}(\alpha) = \frac{180}{\pi}\, m^{R}(\alpha)$$

and

$$\textbf{II.}\quad m^{R}(\alpha) = \frac{2\pi}{360}\, m^{\circ}(\alpha) = \frac{\pi}{180}\, m^{\circ}(\alpha)$$

Equations in Theorem 4.1 are conversion formulas, relating degree measure and radian measure of an angle.

*Examples.*   Find the degree measure (to the nearest tenth of a degree) of the angle with the given radian measure. Use $\pi \approx 3.14$.
(a) $(3\pi/5)^{R}$.        (b) $0.81^{R}$.

*Solutions.*   From Theorem 4.1-I,

(a) $m^{\circ}(\alpha) = \left(\dfrac{180}{\pi} \cdot \dfrac{3\pi}{5}\right)^{\circ} = 108.0^{\circ}.$        (b) $m^{\circ}(\alpha) = \left(\dfrac{180}{\pi} \cdot 0.81\right)^{\circ} \approx 46.4^{\circ}.$

*Examples.*   Find the radian measure (to the nearest hundredth of a radian) of the angle with the given degree measure. Use $\pi \approx 3.14$.
(a) $60^{\circ}$.        (b) $330^{\circ}$.

*Solutions.*   From Theorem 4.1-II,

(a) $m^{R}(\alpha) = \left(\dfrac{\pi}{180} \cdot 60\right)^{R} = \dfrac{\pi^{R}}{3} \approx 1.05^{R}.$

(b) $m^{R}(\alpha) = \left(\dfrac{\pi}{180} \cdot 330\right)^{R} = \dfrac{11\pi^{R}}{6} \approx 5.76^{R}.$

Notice that in making conversions from one unit of measure to another *the number of units in one measure does not equal the number in another measure* in the usual sense. For example $60 \neq \pi/3$. However, an angle whose measure in degree units is 60 is congruent (has the same amount of rotation) to an angle whose measure in radian units is $\pi/3$. Symbolically, this is written as

$$60° = \frac{\pi}{3}^R,$$

where 60 and $(\pi/3)$ are simply measures in degree units and radian units respectively for the same or congruent angles.

Angles may have the same initial side and the same terminal side, but their measures, may be different. Thus, in Figure 4.7,

$$m^R(\alpha_2) = m^R(\alpha_1 - 2\pi)$$

and

$$m^R(\alpha_3) = m^R(\alpha_1 + 2\pi).$$

Furthermore, $m^R(\alpha_3) = m^R(\alpha_2 + 4\pi)$. Such angles, having the same initial and terminal sides, are called **coterminal angles**.

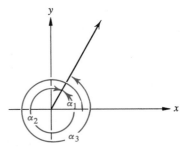

**FIGURE 4.7**

**Examples.** Find the angle of least positive measure which is coterminal with each of the following.
(a) $6.59^R$.      (b) $520°$.

*Solutions.* (a) Sketch the angle, showing its initial and terminal sides. Since the angle generated in one revolution is $2\pi^R \approx 6.28^R$, the angle of least positive measure having the same initial and terminal sides would be $6.59^R - 6.28^R \approx 0.31^R$.

(b) Sketch the angle. In this case, the angle of least positive measure having the same initial and terminal sides is $520° - 360° = 160°$.

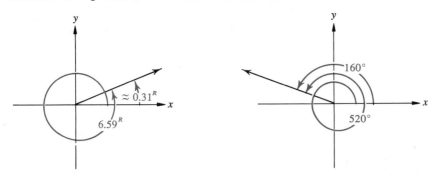

***Example.*** Write a general expression for all angles coterminal with 45°.

***Solution.*** Since one revolution, from a terminal side of any angle brings us back to that terminal side, then all angles coterminal with an angle, 45°, or $(\pi/4)^R$, is given by

$$(45 + 360k)°, \ k \in J \qquad \text{or} \qquad \left(\frac{\pi}{4} + k \cdot 2\pi\right)^R, \ k \in J.$$

## EXERCISE 4.1

A.

1. State the conversion formulas, relating the degree measure and the radian measure of an angle $\alpha$.
2. Complete the following degree-radian conversion wheel for special angles. (It is a handy reference.)

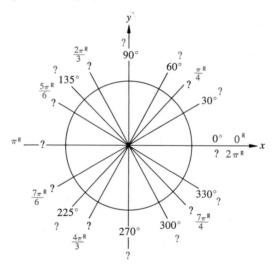

Find the degree measure to the nearest tenth of a degree.

3. $\dfrac{-\pi^R}{3}$.

4. $\dfrac{-2\pi^R}{3}$.

5. $\dfrac{-4\pi^R}{3}$.

6. $\dfrac{-5\pi^R}{4}$.

7. $\dfrac{8\pi^R}{5}$.

8. $\dfrac{11\pi^R}{8}$.

9. $\dfrac{-13\pi^R}{8}$.

10. $\dfrac{-\pi^R}{5}$.

11. $0.18^R$.

12. $0.94^R$.

13. $-1.54^R$.

14. $-7.55^R$.

Find the radian measure to the nearest hundredth of a radian.

15. $-30°$.
16. $-45°$.

17. $75°$.
18. $125°$.

19. $43°$.
20. $137°$.

21. $-202°$.
22. $-342°$.

Find the angle of least positive measure which is coterminal with each of the following. Sketch each angle.

**•23.** $6.72^R$.              **25.** $-379°$.              **•27.** $11.11^R$.              **29.** $792°$.

**24.** $423°$.              **26.** $-8.91^R$.              **28.** $13.48^R$.              **30.** $-863°$.

Write a general expression for all angles coterminal with each of the following.

**31.** $21°$.              **33.** $-\dfrac{5\pi^R}{6}$.              **35.** $-\dfrac{7\pi^R}{6}$.

**32.** $-128°$.              **34.** $\dfrac{5\pi^R}{3}$.              **36.** $310°$.

**37.** Divide the circumference of a circle of radius of length $r$ into 8 equal arcs. (a) What is the length of each arc? (b) What is the measure in radians and in degrees of the central angle which intercepts each of these arcs?

**38.** Divide the circumference of a circle of radius of length $r$ into 12 equal arcs. (a) What is the length of each arc? (b) What is the measure in radians and in degrees of the central angle which intercepts each of these arcs?

# 4.2
## Uniform Circular Motion

If the measure of an angle is given in radians, then the measure of the intercepted arc of a circle with given radius can be found directly. Since $m^R(\alpha)$ indicates how many times the length of the radius $r$ of the circle has been used as a unit length along the circumference of a circle, the length $s$ of an intercepted arc is given by

$$s = r \cdot m^R(\alpha). \tag{1}$$

Notice that the length $s$ is a directed length; it is a positive or negative real number as $m^R(\alpha)$ is a positive or negative real number, respectively. If the measure of an angle is given in degrees, the measure must first be equated to radian measure before Equation (1) can be applied.

**Examples.**   On each circle with the given radius, find the length (to the nearest tenth) of the arc intercepted by the given angle. Use $\pi \approx 3.14$.

(a) $r = 4"$; $\alpha = 2^R$.              (b) $r = 6'$; $\alpha = -150°$.

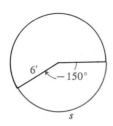

(Solutions on the next page.)

*Solutions.*   (a) From Equation (1), $s = 4 \cdot 2 = 8.0$. Therefore the arc is 8 inches in length.

(b) From Theorem 4.1-II, page 126, the related radian measure of $-150°$ is given by

$$m^R(\alpha) = \left[ \frac{\pi}{180}(-150) \right]^R = -\frac{5\pi^R}{6}.$$

From Equation (1),

$$s = 6\left( -\frac{5\pi}{6} \right) = -5\pi.$$

Since $\pi \approx 3.14$, the arc length to the nearest tenth is $-15.7$ feet.

Now consider a point moving along the circumference of a circle at a constant (uniform) speed. If the point moves from point $P_1$ to $P_2$ in time $t$, the distance traveled in time $t$ equals the length of the arc, equal to $r \cdot m^R(\alpha)$. See Figure 4.8.

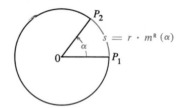

FIGURE 4.8

**Definition 4.1.**   *The **linear velocity** $v$ of a point moving along an arc of a circle at a constant speed is given by*

$$v = \frac{\text{directed length of the arc}}{\text{time}} = \frac{r \cdot m^R(\alpha)}{t}. \qquad (2)$$

**Definition 4.2.**   *The **angular velocity** $\omega$ (the Greek omega) of a point moving along an arc of a circle at a constant speed is given by*

$$\omega = \frac{\text{directed measure of the angle of rotation}}{t}.$$

If the measure of the angle is in radians,

$$\omega = \frac{m^R(\alpha)}{t}, \qquad (3)$$

where $\omega$ is given in radians per unit time.

Notice that the linear velocity and angular velocity are both positive if the motion is in a counterclockwise direction because the length of the arc and the measure of the angle of rotation are both positive. The linear velocity and angular velocity are both negative if the motion is in a clockwise direction.

Substituting $\omega$, the left-hand member of (3), for $m^R(\alpha)/t$ in the right-hand member of (2) leads to the following result.

**Theorem 4.2.**  *The linear velocity v of a point moving along the arc of a circle at a constant speed is given by*

$$v = r\omega, \tag{4}$$

*where r is the length of the radius of the circle and ω is the angular velocity of the moving point in radians per unit time.*

**Example.**  A rocket turbine with a radius of length 4 feet turns at the rate of 100 revolutions per minute. What is the linear velocity of a point on the rim of the turbine?

*Solution.*  Since 1 revolution corresponds to $2\pi$ radians, 100 revolutions per minute corresponds to an angular velocity of $200\pi$ radians per minute. Thus, from Theorem 4.2 with $r = 4$,

$$v = 4 \cdot 200\pi = 800\pi,$$

and the linear velocity is $800\pi$ feet per minute.

### EXERCISE 4.2

A.

On each circle with the given radius, find the length of the arc intercepted by the given angle.

**Example.**  $r = 9''; \alpha = 240°$.

*Solution.*  From Theorem 4.1-II, page 126, the related radian measure of $\alpha$ equal to $240°$ is given by

$$m^R(\alpha) = \left(\frac{\pi}{180} \cdot 240\right)^R = \frac{4\pi^R}{3}.$$

Thus,

$$s = r \cdot m^R(\alpha) = 9 \cdot \frac{4\pi}{3} = 12\pi,$$

and the length of the intercepted arc is $12\pi$ inches.

**1.** $r = 5'; \alpha = 3.1^R$.  
**2.** $r = 11''; \alpha = 1.5^R$.  
**3.** $r = 2''; \alpha = -135°$.  
**4.** $r = 7.5'; \alpha = -240°$.  
**5.** $r = \pi''; \alpha = 150°$.  
**6.** $r = 3.41'; \alpha = 68°$.

In Problems 7 to 10, find the measure (to the nearest tenth of a degree) of the central angle which intercepts the given arc on the circle of the given radius. Use $\pi \approx 3.14$.

**Example.**  $r = 5''; s = 15''$.

*Solution.*  Substituting 5 for $r$ and 15 for $s$ in $s = r \cdot m^R(\alpha)$ yields

$$15 = 5 \cdot m^R(\alpha).$$

*(Solution continued on the next page.)*

Thus,

$$m^R(\alpha) = \frac{15^R}{5} = 3^R,$$

and from Theorem 4.1-I, page 126,

$$m°(\alpha) = \left(\frac{180}{\pi} \cdot 3\right)° = \left(\frac{540}{\pi}\right)°$$

$$\approx \left(\frac{540}{3.14}\right)° \approx 171.9°.$$

**7.** $r = 14'$; $s = 31'$.                    **9.** $r = 1$ yd; $s = 4.9$ yd.
**8.** $r = 2.5''$; $s = 2.5''$.                  **10.** $r = (\pi/2)''$; $s = 6\pi''$.

In Problems 11 to 18, consider a flywheel with uniform circular motion, where $v$ is the linear velocity of a point on the rim and $\omega$ is the angular velocity. In Problems 14 to 18, express each answer in terms of $\pi$. Find:

**11.** $v$ if $r = 3$ feet and $\omega = 12$ radians per second.
**12.** $r$ if $v = 10$ feet per second and $\omega = 10$ radians per second.
**13.** $\omega$ in radians per minute if $v = 3$ feet per second and $r = 6$ inches.
**14.** $\omega$ in degrees per minute if $v = 4$ feet per second and $r = 2$ feet.
**15.** $v$ if $r = 9$ inches and $\omega = -3$ revolutions per second.
**16.** $v$ if $r = 4$ feet and $\omega = -120$ degrees per second.
**17.** $\omega$ in degrees per second if $r = 2.5$ feet and $v = -10$ feet per minute.
**18.** $r$ in inches if $\omega = 1/12$ revolutions per second and $v = 8.2$ feet per minute.

**19.** A pendulum 24 inches long oscillates 4° 10′ on each side of its vertical position. Find the length of arc through which the end of the pendulum swings.

**20.** The center line of a freeway curve is laid out as an arc of a circle of radius 1,000 feet. What is the length of the center line of the curve if it subtends an angle of 1/3 radian at the center of the circle?

**21.** A racing driver moves around a curved track at a speed of 160 miles per hour. If the radius of the circle, of which the driver's path is an arc, is 1,500 feet, what is the driver's angular speed in radians per minute around the curve?

**22.** Find the linear velocity, in inches per minute, of the tip of an hour hand of a clock, if the hour hand has a length of 1 foot.

**23.** The planet Mars moves around the sun in 694 earth-days. If the radius of Mars' orbit (assumed to be circular) around the sun is 142,000,000 miles, find the linear velocity of Mars in its orbit in miles per hour. Assume 1 earth-day equals 24 hours.

**24.** Two pulleys with diameters of length of 5 inches and 10 inches are connected by an open belt, with no sag and not crossing. If the centers of the pulleys are 30 inches apart, find the total length of the belt.

**B.**

25. Show that the area $\mathscr{A}$ of the sector of a circle (see figure below) with central angle $\theta$ is given by

$$\mathscr{A} = \tfrac{1}{2}r^2 m^R(\theta).$$

26. Use the formula in Problem 25 to find the area of the sector of a circle whose radius has a length of 10 inches, and where the sector has a central angle with measure of $(\pi/6)^R$.

27. Find the area of a sector of a circle whose radius has a length of 3 feet, where the sector has a central angle with measure of $148°$.

28. Show that the area $\mathscr{A}$ of the segment of a circle (see figure) with central angle $\theta$ is given by

$$\mathscr{A} = \tfrac{1}{2}r^2[m^R(\theta) - \sin\theta].$$

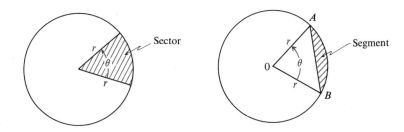

## 4.3
## Trigonometric Functions

Consider the three real numbers, $x$, $y$, and $\sqrt{x^2 + y^2}$ which can be associated with any point on the terminal side of an angle in standard position, as in Figure 4.9. The components of the ordered pair $(x, y)$ are the coordinates of the point and $\sqrt{x^2 + y^2}$ (from the Pythagorean Theorem) is the length of the line segment from the origin to the point. For each angle $\alpha$, there are six possible ratios which may be formed, using the numbers $x$, $y$, and $\sqrt{x^2 + y^2}$.

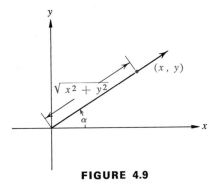

**FIGURE 4.9**

The names given to these ratios are the same as the ones that were used to name elements in the range of the circular functions, because, as you shall see, they are closely related. These ratios are defined as follows.

**Definition 4.3.**   *For all angles $\alpha$ in standard position, if $x$ and $y$ [$(x, y) \neq (0,0)$] are the coordinates of any point on the terminal side of $\alpha$, then*

$$\text{cosine } \alpha = \frac{x}{\sqrt{x^2 + y^2}}, \qquad \text{secant } \alpha = \frac{\sqrt{x^2 + y^2}}{x}, \qquad (x \neq 0),$$

$$\text{sine } \alpha = \frac{y}{\sqrt{x^2 + y^2}}, \qquad \text{cosecant } \alpha = \frac{\sqrt{x^2 + y^2}}{y}, \qquad (y \neq 0),$$

$$\text{tangent } \alpha = \frac{y}{x}, \quad (x \neq 0), \qquad \text{cotangent } \alpha = \frac{x}{y}, \qquad (y \neq 0).$$

Cosine, sine, etc. are abbreviated in the same way as they were for the circular functions. Note that in each ratio where the expression $\sqrt{x^2 + y^2}$ occurs, the *positive* square root of $x^2 + y^2$ is used.

It can be shown by using similar triangles, and this will be left as an exer-

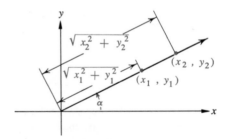

**FIGURE 4.10**

cise, that if $(x_1, y_1)$ and $(x_2, y_2)$ correspond to any two points (excluding the origin) on the terminal side of an angle, then the following equalities of ratios hold (see Figure 4.10):

$$\frac{x_1}{y_1} = \frac{x_2}{y_2}, \tag{1}$$

$$\frac{x_1}{\sqrt{x_1^2 + y_1^2}} = \frac{x_2}{\sqrt{x_2^2 + y_2^2}}, \tag{2}$$

and

$$\frac{y_1}{\sqrt{x_1^2 + y_1^2}} = \frac{y_2}{\sqrt{x_2^2 + y_2^2}}. \tag{3}$$

Similar equalities for the reciprocals of these ratios also hold. Thus, *values of the six ratios in Definition 4.3 are determined only by the position of the terminal side of an angle and not by the point that has been selected on the terminal side.*

Since each angle $\alpha$ in standard position can be paired with a unique number for each of the six ratios of Definition 4.3, each of the equations in the definition defines a function. They are called **trigonometric functions.** The domains of these functions, are determined by their definitions. In each case, for $A = \{\text{all angles}\}$, the domains are as follows:

$$\text{cosine and sine functions, } \{\alpha \mid \alpha \in A\},$$

$$\text{tangent and secant functions, } \left\{\alpha \mid \alpha \in A, \alpha \neq \left(\frac{\pi}{2} + \pi k\right)^{R}, k \in J\right\},$$

and

$$\text{cotangent and cosecant functions, } \{\alpha \mid \alpha \in A, \alpha \neq \pi k^{R}, k \in J\}.$$

## EXERCISE 4.3

**A.**

Determine the quadrant in which the terminal side of each angle lies.

**Example.**   $\sin \alpha < 0$, $\tan \alpha > 0$.

*Solution.*   For $\sin \alpha < 0$, the terminal side of $\alpha$ lies in Quadrant III or IV and for $\tan \alpha > 0$, the terminal side of $\alpha$ lies in Quadrant I or III. Therefore, the terminal side of $\alpha$ must lie in Quadrant III, the intersection of $\{\text{III, IV}\}$ and $\{\text{I, III}\}$.

1. $\sin \alpha < 0$, $\cos \alpha < 0$.
2. $\sin \alpha < 0$, $\tan \alpha < 0$.
3. $\sin \alpha > 0$, $\sec \alpha > 0$.
4. $\sin \alpha > 0$, $\cot \alpha < 0$.

5. $\cos \alpha < 0$, $\tan \alpha > 0$.
6. $\cos \alpha < 0$, $\csc \alpha > 0$.
7. $\cos \alpha > 0$, $\cot \alpha > 0$.
8. $\tan \alpha < 0$, $\csc \alpha < 0$.

In Problems 9 to 18, find the element in the range of each of the six trigonometric functions of $\alpha$, if the terminal side of $\alpha$ (in standard position) contains the given point. Sketch each angle.

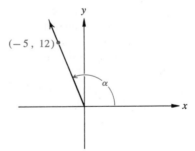

**Example.**   $(-5, 12)$.

*Solution.*   See the figure. By Definition 4.3,

$$\cos \alpha = \frac{x}{\sqrt{x^2 + y^2}} = \frac{-5}{\sqrt{(-5)^2 + (12)^2}} = \frac{-5}{13},$$

$$\sec \alpha = \frac{\sqrt{x^2 + y^2}}{x} = \frac{13}{-5},$$

$$\sin \alpha = \frac{y}{\sqrt{x^2 + y^2}} = \frac{12}{13},$$

$$\csc \alpha = \frac{\sqrt{x^2 + y^2}}{y} = \frac{13}{12},$$

$$\tan \alpha = \frac{y}{x} = \frac{12}{-5},$$

$$\cot \alpha = \frac{x}{y} = \frac{-5}{12}.$$

**⁹9.** $(3, 4)$.                **¹13.** $(2, 3)$.                **16.** $(1, -\sqrt{3})$.

**10.** $(-15, 8)$.          **14.** $(-3, 5)$.           **¹17.** $(3\sqrt{3}, 3)$.

**11.** $(-12, -5)$.        **15.** $(-\sqrt{3}, -1)$.    **18.** $(-5, -5)$.

**12.** $(1, -1)$.

**19.** Which tangent function values are undefined? Which are equal to 0?

**20.** Which cotangent function values are undefined? Which are equal to 0?

**21.** Which secant function values are undefined? Are any secant function values equal to 0?

**22.** Which cosecant function values are undefined? Are any cosecant function values equal to 0?

**B.**

**23.** Prove that if $(x_1, y_1)$ and $(x_2, y_2)$ correspond to any two points (excluding the origin) on the terminal side of an angle $\alpha$ in the first quadrant, then

$$\frac{x_1}{y_1} = \frac{x_2}{y_2}, \qquad \frac{x_1}{\sqrt{x_1^2 + y_1^2}} = \frac{x_2}{\sqrt{x_2^2 + y_2^2}}, \qquad \text{and} \qquad \frac{y_1}{\sqrt{x_1^2 + y_1^2}} = \frac{y_2}{\sqrt{x_2^2 + y_2^2}}.$$

**24.** Show that the statement in Problem 23 is valid for an angle $\alpha$ in any quadrant.

## 4.4
### Relationships Between Circular and Trigonometric Functions

Inasmuch as the circular functions were defined by making use of a unit circle (Sections 2.2, 2.5, and 2.6) and because the unit circle can be used to assign measures to angles, a close relationship exists between the circular and trigonometric functions.

The trigonometric functions of an angle have been defined in terms of the coordinates of any point, except $(0, 0)$, on the terminal side of an angle $\alpha$ in standard position. If a point with coordinates $(x, y)$ is taken such that $\sqrt{x^2 + y^2} = 1$, that is, the point is on the unit circle (Figure 4.11), then

$$\cos \alpha = \frac{x}{\sqrt{x^2 + y^2}} = \frac{x}{1} = x.$$

Equivalently, $x = \cos \alpha$. Similarly,

$$\sin \alpha = \frac{y}{\sqrt{x^2 + y^2}} = \frac{y}{1} = y,$$

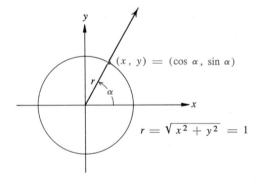

**FIGURE 4.11**

or equivalently, $y = \sin \alpha$. Thus by choosing points on the unit circle, *the elements in the ranges of the trigonometric functions are equal to the corresponding elements in the ranges of the analogous circular functions.* If an element in the range of any one of the six trigonometric functions is denoted by Trig $(\alpha)$, and an element in the range of the analogous circular function by Circ $(x)$, then

$$\textbf{Trig } (\alpha) = \textbf{Circ } (x).$$

where $x$ is the length of the arc intercepted by $\alpha$ on the unit circle. *The elements in the domain of the circular functions are real numbers, while the elements in the domain of the trigonometric functions are angles. The elements in the range of both circular and trigonometric functions are real numbers.* For example, recalling that $(\pi/4)^R$ and $45°$ are measures of the same angle, we write

$$\sin \frac{\pi}{4}^R = \sin \frac{\pi}{4}$$

and

$$\sin 45° = \sin \frac{\pi}{4},$$

where the left-hand member in each equation is an element in the range of the trigonometric function and the right-hand member is an element in the range of the corresponding circular function. Since

$$\textbf{Trig } (\alpha°) = \textbf{Trig } (\alpha^R) = \textbf{Circ } (x),$$

you can find trigonometric function values from the tables used for circular functions. You can use Table 2.3 (page 71) directly for function values of angles with measures $(\pi/6)^R, (\pi/4)^R$, and $(\pi/3)^R$ where these angles are "equated" to the real numbers $\pi/6$, $\pi/4$, and $\pi/3$, respectively. You can obtain these values also, directly from the definition of the trigonometric functions. For example, an angle with measure $(\pi/6)^R$ or $30°$ is shown in Figure 4.12a on page 138 in relation to a unit circle. From Definition 4.3,

$$\cos 30° = \frac{\sqrt{3/2}}{1} = \frac{\sqrt{3}}{2}, \qquad \sec 30° = \frac{1}{\sqrt{3/2}} = \frac{2}{\sqrt{3}},$$

$$\sin 30° = \frac{1/2}{1} = \frac{1}{2}, \qquad \csc 30° = \frac{1}{1/2} = 2,$$

$$\tan 30° = \frac{1/2}{\sqrt{3/2}} = \frac{1}{\sqrt{3}}, \qquad \cot 30° = \frac{\sqrt{3/2}}{1/2} = \sqrt{3}.$$

Similarly, the function values for angles with measures $45°$, $60°$, $0°$, $90°$, $180°$, and $270°$ can be obtained directly from Definition 4.3. The other parts

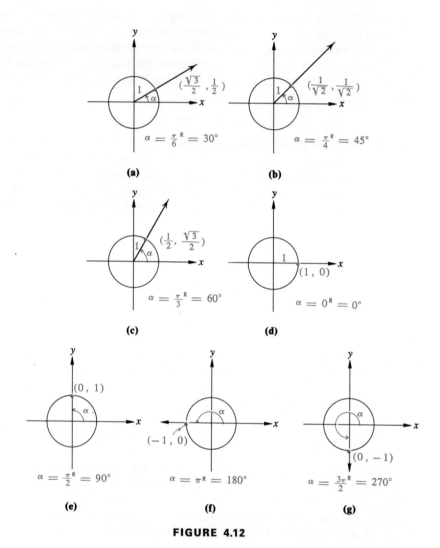

**FIGURE 4.12**

of Figure 4.12 show these angles in relation to a unit circle. The function values are shown as entries in Table 4.1. These entries, which are the same as the entries in Table 2.3, page 71, are shown here again for convenient reference in relation to elements in the domain (angles) of the trigonometric functions. You can verify the entries in Table 4.1 from Definition 4.3 and Figure 4.12.

Since the trigonometric functions are periodic with period $2\pi^R$ or $360°$,

$$\text{Trig } (\alpha) = \text{Trig } (\alpha + k \cdot 2\pi^R) = \text{Trig } (\alpha + k \cdot 360°), \quad k \in J.$$

Thus function values of many other angles can be obtained directly from the entries in Table 4.1.

**TABLE 4.1**

| $m^R(\alpha)$ | $m°(\alpha)$ | $\cos \alpha$ | $\sin \alpha$ | $\tan \alpha$ | $\sec \alpha$ | $\csc \alpha$ | $\cot \alpha$ |
|---|---|---|---|---|---|---|---|
| $0^R$ | $0°$ | $1$ | $0$ | $0$ | $1$ | undef. | undef. |
| $\dfrac{\pi^R}{6}$ | $30°$ | $\dfrac{\sqrt{3}}{2}$ | $\dfrac{1}{2}$ | $\dfrac{1}{\sqrt{3}}$ | $\dfrac{2}{\sqrt{3}}$ | $2$ | $\sqrt{3}$ |
| $\dfrac{\pi^R}{4}$ | $45°$ | $\dfrac{1}{\sqrt{2}}$ | $\dfrac{1}{\sqrt{2}}$ | $1$ | $\sqrt{2}$ | $\sqrt{2}$ | $1$ |
| $\dfrac{\pi^R}{3}$ | $60°$ | $\dfrac{1}{2}$ | $\dfrac{\sqrt{3}}{2}$ | $\sqrt{3}$ | $2$ | $\dfrac{2}{\sqrt{3}}$ | $\dfrac{1}{\sqrt{3}}$ |
| $\dfrac{\pi^R}{2}$ | $90°$ | $0$ | $1$ | undef. | undef. | $1$ | $0$ |
| $\pi^R$ | $180°$ | $-1$ | $0$ | $0$ | $-1$ | undef. | undef. |
| $\dfrac{3\pi^R}{2}$ | $270°$ | $0$ | $-1$ | undef. | undef. | $-1$ | $0$ |

**Examples.** Find: (a) $\cos (7\pi/3)^R$.  (b) $\tan 810°$.

Solutions. (a) $\cos \dfrac{7\pi^R}{3} = \cos \left(\dfrac{\pi}{3} + 2\pi\right)^R = \cos \dfrac{\pi^R}{3} = \dfrac{1}{2}.$

(b) $\tan 810° = \tan (90 + 2 \cdot 360)° = \tan 90°$; undefined.

### EXERCISE 4.4

If the point whose coordinates $x$ and $y$ are on the terminal side of angle $\alpha$ is also on the unit circle, express each of the following in terms of $x$ and/or $y$.

1. $\tan \alpha$.  2. $\sec \alpha$.  3. $\csc \alpha$.  4. $\cot \alpha$.

Using Table 4.1, and the fact that the trigonometric functions are periodic with period $2\pi^R$ or $360°$, find the following.

5. $\sin 390°$.  9. $\csc 480°$.  13. $\sec (-570°)$.

6. $\cos 405°$.  10. $\cot 495°$.  14. $\csc (-585°)$.

7. $\tan \dfrac{7\pi^R}{3}$ .  11. $\cos \dfrac{17\pi^R}{6}$ .  15. $\cot \left(\dfrac{-13\pi^R}{6}\right)$.

8. $\sec \dfrac{5\pi^R}{2}$ .  12. $\tan 3\pi^R$.  16. $\sin \left(\dfrac{-5\pi^R}{4}\right)$.

17. Compare the domains and the ranges of the circular functions and trigonometric functions, respectively.

18. What relationships do you note from Table 4.1 between function values for pairs of complementary angles (sum of their measures equals $90°$)?

## 4.5
### Function Values for $\alpha \in A$

In Figure 4.13, if the terminal side of angle $\alpha$ is taken as the initial side of angle $\beta$, with the same point as vertex for both angles, and then if $\alpha + \beta$ is viewed as the angle with initial side that of $\alpha$'s initial side and terminal side that of $\beta$'s terminal side, we have the relationship

$$m^R(\alpha + \beta) = m^R(\alpha) + m^R(\beta). \qquad (1)$$

Since Equation (1) is true for all such angles $\alpha$ and $\beta$, all of the sum and difference formulas of the circular functions (with domain $R$) which you previously encountered in Chapter 2 are immediately valid for the trigonometric functions (with domain $A$) where the real numbers $x_1$ and $x_2$ are replaced appropriately with angles $\alpha$ and $\beta$. Therefore,

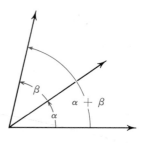

**FIGURE 4.13**

$$\cos(\alpha + \beta) = \cos \alpha \cos \beta - \sin \alpha \sin \beta,$$

$$\cos(\alpha - \beta) = \cos \alpha \cos \beta + \sin \alpha \sin \beta,$$

$$\sin(\alpha + \beta) = \sin \alpha \cos \beta + \cos \alpha \sin \beta,$$

$$\sin(\alpha - \beta) = \sin \alpha \cos \beta - \cos \alpha \sin \beta.$$

It follows further that all of the reduction formulas for the circular functions (Sections 2.3, 2.4, 2.5, and 2.6) are equally applicable to the trigonometric functions including the statement summarizing the reduction formulas (Section 2.8), where real numbers are appropriately "equated" to angles. Furthermore a reference angle $\tilde{\alpha}$ can be used in a way similar to the way a reference arc was used to interpret the reduction formulas. Because

$$\text{Trig}(\alpha°) = \text{Trig}(\alpha^R) = \text{Circ}(x),$$

where the measure of $\alpha$ in radians is "equated" to the real number $x$, we can find values for $\sin \alpha$, $\cos \alpha$, etc., for some additional angles $\alpha \in A$ for which these values are defined from the analogous reduction formulas for the circular functions and from appropriate tables. A figure showing the angle in standard position and the reference angle is helpful to select the appropriate reduction formula as necessary.

Table 4.1, page 139, can be used to find trigonometric function values for special angles in radian or degree measure.

***Examples.***   Find:   (a) $\sin (\pi/3)^R$.        (b) $\cos 135°$.

*Solutions.*  (a) From Table 4.1, or by forming the ratio directly from a sketch of the angle showing the coordinates of an appropriate point on the terminal side,

$$\sin \frac{\pi^R}{3} = \frac{\sqrt{3}}{2}.$$

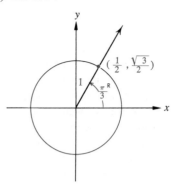

(b) From the appropriate reduction formula, page 54 (note the reference angle $\tilde{\alpha}$ in the figure),

$$\cos 135° = -\cos (180 - 135)° = -\cos 45°.$$

From Table 4.1 or again by forming the appropriate ratio, directly from the sketch,

$$-\cos 45° = -\frac{1}{\sqrt{2}}.$$

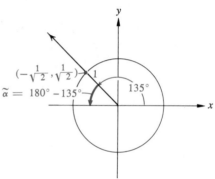

Table III, page 294, which you have previously used for function values for real numbers, also gives function values for angles $\alpha$ in radian measure in the interval $0^R \le \alpha \le 1.57^R$.

***Examples.***   Find:   (a) $\tan 0.47^R$.        (b) $\sin 3.74^R$.

*Solutions.*   (a) From Table III, $\tan 0.47^R \approx 0.5080$.
(b) From the appropriate reduction formula, page 60,

$$\sin 3.74^R \approx -\sin (3.74 - 3.14)^R = -\sin 0.60^R.*$$

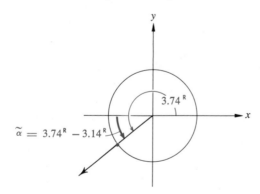

*(Solution continued on the next page.)*

* See the footnote on page 77 for an interpretation of a chain of statements involving both the symbols " $=$ " and " $\approx$ ."

From Table III,

$$-\sin 0.60^R \approx -0.5646.$$

Table IV, page 298, has been prepared from Table III and gives rational number approximations for function values of selected angles over the interval $0° \le \alpha \le 90°$. The table is graduated in intervals of ten minutes (10′) where one minute is equal to one-sixtieth of a degree. Observe that the table reads from top to bottom in the left margin, with function values identified at the tops of the columns. It reads from bottom to top in the right margin with function values identified at the bottoms of the columns. The equivalent measure in radians is given in the second column from the left and the second column from the right.

**Examples.** Find:
(a) sin 23° 10′.        (b) cot 57° 40′.        (c) cos 123° 50′.

*Solutions.*   (a) From Table IV, sin 23° 10′ $\approx$ 0.3934.
           (b) From Table IV, cot 57° 40′ $\approx$ 0.6330.
           (c) From the appropriate reduction formula, page 54,

$$\cos 123° 50′ = -\cos (180° - 123° 50′) = -\cos 56° 10′.$$

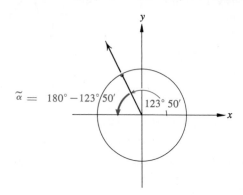

From Table IV,

$$-\cos 56° 10′ \approx -0.5568.$$

Function values for angles between those listed in Tables III and IV can be approximated by using linear interpolation as illustrated in an example in the exercise.

The inverse circular relations and functions were defined in Section 3.5. The analogous inverse trigonometric relations and functions are defined in a similar way. Thus an equation such as

$$\sin \alpha = \frac{1}{2}, \qquad -\frac{\pi}{2} \le \alpha \le \frac{\pi}{2}$$

can be written using inverse notation as

$$\alpha = \text{Arcsine } \tfrac{1}{2} \quad \text{or} \quad \alpha = \text{Sin}^{-1}\,\tfrac{1}{2}.$$

You can interpret the right-hand member of either equation as another name (symbol) for the angle $\alpha$, $-\pi/2 \le \alpha \le \pi/2$.

### EXERCISE 4.5

A.

Find each of the following. In each case, draw a sketch and show the reference angle.

**⋅1.** $\tan \dfrac{4\pi^{R}}{3}$ .

**2.** $\cot \dfrac{7\pi^{R}}{6}$ .

**3.** $\csc \dfrac{7\pi^{R}}{4}$ .

**4.** $\sin 135°$.

**⋅5.** $\cos 120°$.

**6.** $\sec 150°$.

**7.** $\cos (-240°)$.

**8.** $\sin (-150°)$.

Find each of the following. Use Table III, page 294. If a reduction formula is necessary, draw a sketch and show the reference angle.

**9.** $\sin 0.83^{R}$.

**10.** $\tan (-1.24^{R})$.

**11.** $\cos (-0.16^{R})$.

**12.** $\csc 1.34^{R}$.

**13.** $\cot 0.91^{R}$.

**14.** $\sec 1.11^{R}$.

Find each of the following. Use Table IV, page 298. If a reduction formula is necessary, draw a sketch and show the reference angle.

**15.** $\sin 2° 20'$.

**16.** $\cos 9° 40'$.

**17.** $\cos 46° 40'$.

**18.** $\cot 76° 10'$.

**19.** $\csc 62° 50'$.

**20.** $\tan (-18° 10')$.

**⋅21.** $\csc (-83° 30')$.

**22.** $\sec 231°$.

**23.** $\cot 339°$.

**24.** $\tan 153° 10'$.

**⋅25.** $\sec 266° 30'$.

**26.** $\sin 189° 20'$.

Use linear interpolation to find the following.

*Example.* $\sin 9° 14'$.

*Solution.* Table IV gives values of 0.1593 and 0.1622 for $\sin 9° 10'$ and $9° 20'$ respectively. The data arranged for interpolation appears as

$$10' \left\{ 4' \begin{cases} \sin 9° 10' \approx 0.1593 \\ \sin 9° 14' \approx \quad ? \end{cases} d \right\} 0.0029.$$
$$\sin 9° 20' \approx 0.1622$$

Therefore,

$$\frac{4}{10} \approx \frac{d}{0.0029},$$

from which

$$d \approx \frac{4}{10}(0.0029) \approx 0.0012.$$

*(Solution continued on the next page.)*

Thus,

$$\sin 9° \, 14' \approx 0.1593 + 0.0012 = 0.1605.$$

**27.** $\sin 3° \, 38'$.        **▪29.** $\cos 110° \, 51'$.        **31.** $\tan 219° \, 22'$.
**28.** $\csc 45° \, 59'$.        **30.** $\cot 163° \, 03'$.        **32.** $\sec 354° \, 24'$.

For each of the following, find the measure of $\alpha$ in degrees.

***Examples.***    (a) $\alpha = \text{Arcsin} \, (\sqrt{3}/2)$.        (b) $\alpha = \text{Arcsin} \, (-\sqrt{3}/2)$.

*Solutions.*    (a)   Find an $\alpha$ such that $\sin \alpha = \sqrt{3}/2$, $-\pi/2 \le \alpha \le \pi/2$. Since the function value is positive, $\alpha$ is in the first quadrant. From Table 4.1, page 139 (or memory), $\alpha = 60°$.

           (b) Find an $\alpha$ such that $\sin \alpha = -\sqrt{3}/2$, $-\pi/2 \le \alpha \le \pi/2$. Since the function value is negative, $\alpha$ is the fourth quadrant. From (a) above the reference angle $\tilde{\alpha}$ equals $60°$. Thus, $\alpha = -60°$.

**33.** $\alpha = \text{Arctan} \, 1$.       **36.** $\alpha = \text{Arccot} \, \sqrt{3}$.       **39.** $\alpha = \text{Arccos} \left( -\dfrac{1}{\sqrt{2}} \right)$.

**34.** $\alpha = \text{Arccos} \, \dfrac{1}{2}$.       **37.** $\alpha = \text{Arcsin} \left( -\dfrac{1}{2} \right)$.       **40.** $\alpha = \text{Arcsin} \left( -\dfrac{\sqrt{3}}{2} \right)$.

**35.** $\alpha = \text{Arccsc} \, 2$.       **38.** $\alpha = \text{Arctan} \, (-1)$.

**41.** $\alpha = \text{Arcsin} \, 0.2784$.          **46.** $\alpha = \text{Arccot} \, 0.3314$.
**42.** $\alpha = \text{Arccos} \, 0.9367$.          **47.** $\alpha = \text{Arccos} \, (-0.9936)$.
**43.** $\alpha = \text{Arctan} \, 0.5930$.          **48.** $\alpha = \text{Arctan} \, (-0.3739)$.
**44.** $\alpha = \text{Arccsc} \, 1.309$.          **49.** $\alpha = \text{Arccsc} \, (-1.106)$.
**45.** $\alpha = \text{Arcsec} \, 1.476$.          **50.** $\alpha = \text{Arcsec} \, (-1.133)$.

**B.**

Use the formulas for sums and differences to prove each of the following.

**51.** $\cos (90° - \alpha) = \sin \alpha$.        **54.** $\csc (90° - \alpha) = \sec \alpha$.

**52.** $\sin (90° - \alpha) = \cos \alpha$.        **55.** $\tan (45° - \alpha) = \dfrac{1 - \tan \alpha}{1 + \tan \alpha}$.

**53.** $\sec (90° - \alpha) = \csc \alpha$.        **56.** $\cot (45° - \alpha) = \dfrac{\cot \alpha + 1}{\cot \alpha - 1}$.

## Chapter Summary

1. The following symbols have been introduced: $\alpha$, $\beta$, $\gamma$, etc., $\measuredangle$, $m°(\alpha)$, $m^R(\alpha)$, $x°$, $x^R$, $v$, $\omega$, Trig $(\alpha)$, Circ $(x)$, and $\tilde{\alpha}$.

2. An angle is the union of two rays which have a common end point, called the **vertex**. The rays are called the **sides of the angle**.

3. An angle which has its initial side along the $x$ axis and its vertex at the origin is said to be in **standard position**. Such an angle is said to be in that quadrant in which its terminal side lies.

4. If the circumference of a circle of radius $r$ is divided into $p$ arcs of equal length, then each arc would have a length of $2\pi r/p$. The measure assigned to an angle with vertex at the center is equal to the number of such arcs which it intercepts. If the rotation of the initial side to the terminal side is in a counter-clockwise direction the measure is positive; if clockwise, the measure is negative.

5. The measure assigned to an angle in terms of the same unit of measure is independent of the radius of the circle used to assign the measure.

6. Two kinds of angle measure most commonly used are called **degree measure** and **radian measure**, where the unit arc is of length $\pi r/180$ and $r$, respectively. If the unit circle (radius $= 1$) is used as the basis for radian measure, then the length of the unit arc is 1.

7. The measure of an angle in radians is related to the measure of the angle in degrees by the equations

$$m^\circ(\alpha) = \frac{180}{\pi}\, m^R(\alpha)$$

and

$$m^R(\alpha) = \frac{\pi}{180}\, m^\circ(\alpha).$$

8. Angles having the same initial side and the same terminal side, but whose measures differ by integral multiples of $2\pi$, are called **coterminal angles.**

9. The length $s$ of the arc of a circle of radius of length $r$ intercepted by an angle $\alpha$ whose vertex is at the center of the circle is given by

$$s = r \cdot m^R(\alpha).$$

10. The **linear velocity** $v$ of a point moving along the circumference of a circle at a constant speed is given by

$$v = \frac{s}{t} = \frac{r \cdot m^R(\alpha)}{t},$$

where $t$ is the time to move through the length of arc $s = r \cdot m^R(\alpha)$. The **angular velocity** $\omega$ of the moving point is given by

$$\omega = \frac{m^R(\alpha)}{t},$$

where $\omega$ is given in radians per unit time.

11. The linear velocity $v$ and the angular velocity $\omega$ of a point on a circle with radius of length $r$ are related by the equation

$$v = r \cdot \omega,$$

where $\omega$ is given in radians per unit time.

**12.** If $\alpha$ is an angle in standard position, and $x$ and $y$ [$(x, y) \neq (0, 0)$] are the coordinates of any point on the terminal side of $\alpha$, then

$$\text{cosine } \alpha = \frac{x}{\sqrt{x^2 + y^2}}, \qquad \text{secant } \alpha = \frac{\sqrt{x^2 + y^2}}{x}, \qquad (x \neq 0),$$

$$\text{sine } \alpha = \frac{y}{\sqrt{x^2 + y^2}}, \qquad \text{cosecant } \alpha = \frac{\sqrt{x^2 + y^2}}{y}, \qquad (y \neq 0),$$

$$\text{tangent } \alpha = \frac{y}{x}, \qquad (x \neq 0), \qquad \text{cotangent } \alpha = \frac{x}{y}, \qquad (y \neq 0).$$

**13.** The values of the six ratios in 12 above are determined only by the position of the terminal side of an angle in standard position and not by the point selected on the terminal side.

**14.** The elements in the ranges of the trigonometric functions are equal to the corresponding elements in the ranges of the analogous circular functions. That is,

$$\textbf{Trig } (\alpha^\circ) = \textbf{Trig } (\alpha^R) = \textbf{Circ } (x),$$

where $\alpha$ intercepts the arc with length $x$ on a unit circle.

**15.** Many theorems applicable for circular function values are, with appropriate rewording, also valid for trigonometric function values. These include the sum and difference formulas as well as all of the reduction formulas.

**16.** Inverse trigonometric relations and functions are defined in a manner similar to the definitions for the inverse circular relations and functions.

## Chapter Review

Find the degree measure of each angle $\alpha$ to the nearest tenth of a degree.

**1.** $\alpha = \dfrac{2\pi^R}{5}$.

**2.** $\alpha = -2.41^R$.

Find the radian measure of each angle $\alpha$ to the nearest hundredth of a radian.

**3.** $\alpha = 75^\circ$.

**4.** $\alpha = -257^\circ$.

Find the angle of least positive measure coterminal with each angle $\alpha$.

**5.** $\alpha = 7.23^R$.

**6.** $\alpha = -938^\circ$.

Write a general expression for the measure of all angles coterminal with each angle $\alpha$.

**7.** $m^\circ(\alpha) = 131^\circ$.

**8.** $m^R(\alpha) = -\dfrac{7\pi^R}{4}$.

9. Find the length of the arc intercepted on a circle by a central angle $\alpha$ with measure $2.2^R$ if the radius of the circle is 3.1 feet in length.
10. Find the measure of the central angle which intercepts an arc with length 2.1 feet on a circle whose radius is 9.1 feet in length.
11. Find the linear velocity $v$ of a point on a circle moving with constant speed if $r = 2$ inches and $\omega = 6$ radians per minute.
12. Find the angular velocity $\omega$ of a point on a circle moving with constant speed if $r = 1.5$ feet and $v = 19.7$ feet per second.

Determine the quadrant in which the terminal side of each angle $\alpha$ lies.

13. $\tan \alpha > 0$, $\sec < 0$.                    14. $\csc \alpha < 0$, $\cot \alpha < 0$.

In Problems 15 and 16, find the six trigonometric function values of $\alpha$, if the terminal side of $\alpha$ (in standard position) contains the point corresponding to the given ordered pair.

15. $(-4, 3)$.                                16. $(-15, -8)$.

17. Find $\sin(-135°)$.                       18. Find $\tan 0.79^R$.

Use linear interpolation to find each value.

19. $\cos 52° \, 27'$.                         20. $\tan 301° \, 24'$.

Find each measure of $\alpha$ in degrees.

21. $\alpha = \operatorname{Arcsec} 2$.                     22. $\alpha = \operatorname{Arccos}(-0.5348)$.

# 5

# Identities and Conditional Equations

ECALL from Section 1.3 that equations such as

$$x + y = y + x \quad \text{and} \quad (x + y)^2 = x^2 + 2xy + y^2$$

which are true for all real numbers $x$ and $y$ are called **identities** and that equations such as

$$x + 1 = 7 \quad \text{and} \quad x^2 - 3x + 2 = 0$$

which are true for some real number replacements of the variable and false for others are called **conditional equations**. We shall now consider identities that involve circular or trigonometric function values. A knowledge of these identities will be useful in later work to simplify expressions and solve equations. In our work with such identities we shall not explicitly specify the replacement set of the variable in each case. The variables $\alpha$, $\beta$, $\gamma$, and $\theta$ will be elements of the set of all angles, and the variables $x$ and $y$ will be considered either as elements of the set of real numbers or of the set of all angles. Furthermore, it should be understood that *an equation which is specified to be an identity without noting the restrictions is in fact an identity only for all angles and for all real number replacements of the variable for which both members are defined.*

## 5.1
### Basic Identities

Some of the identities in this section are not new to you. They are repeated here as a convenient reference. First consider the identities

$$\tan x = \frac{\sin x}{\cos x}, \qquad \cot x = \frac{\cos x}{\sin x},$$

$$\csc x = \frac{1}{\sin x}, \qquad \text{and} \qquad \sec x = \frac{1}{\cos x}.$$

These identities are simply definitions. Other identities such as

$$\sin^2 x + \cos^2 x = 1,$$

are logical consequences of definitions and other identities (Theorem 2.1). This identity was established in Section 2.2 by showing that the equation is implied from a statement known to be true. Such identities, that are referred to frequently, shall be specified as theorems. Two such identities are given by the following.

**Theorem 5.1.** *If* $x \in R$ *or* $x \in A$, *then*

    I.    $\tan^2 x + 1 = \sec^2 x$,     $x \neq \dfrac{\pi}{2} + k\pi$ *or* $\left( \dfrac{\pi}{2} + k\pi \right)^R$, $k \in J$,

    II.   $\cot^2 x + 1 = \csc^2 x$,     $x \neq k\pi$ *or* $k\pi^R$, $k \in J$.

*Proof of I.* Since $\sin^2 x + \cos^2 x = 1$ and because $\cos^2 x \neq 0$ for any $x$ for which $\tan x$ and $\sec x$ are defined, we can multiply each member by $1/\cos^2 x$ to obtain the equivalent equation

$$\frac{\sin^2 x}{\cos^2 x} + \frac{\cos^2 x}{\cos^2 x} = \frac{1}{\cos^2 x},$$

from which we obtain

$$\tan^2 x + 1 = \sec^2 x.$$

The proof of Theorem 5.1-II is similar and is left as an exercise.

There is no general method for proving identities. Sometimes, as in the proof above, it is convenient to multiply each member of an equation by a nonzero real number. At other times, it is more convenient to generate equivalent equations by making substitutions in only one member of the equation. In this case if one member appears to be the more complicated, it should be transformed so that it reduces to the simpler member. At each step you should look for any application of the field properties, page 288, that will enable you to generate an equivalent equation which is clearly an identity —by substituting one expression for another. The following example illustrates such a procedure. The proof also illustrates a basic approach in proving identities. Each term in both members of the equation is first expressed in terms of $\sin x$ and $\cos x$ and then the left-hand and/or right-hand member is simplified.

**Example.** Show that the equation

$$\frac{\cos x - \sin x}{\cos x} = 1 - \tan x$$

is an identity.                                        *(Solution on the next page.)*

*Solution.*   By substituting $(\sin x)/(\cos x)$ for $\tan x$ in the right-hand member we obtain

$$\frac{\cos x - \sin x}{\cos x} = 1 - \frac{\sin x}{\cos x}.$$

Writing the right-hand member as a single term yields

$$\frac{\cos x - \sin x}{\cos x} = \frac{\cos x - \sin x}{\cos x},$$

which is clearly an identity.

Often it is helpful if you recognize the basic identities in the following equivalent forms. For example, showing equivalent equations by using the symbol $\Leftrightarrow$, we have that

$$\sec x = \frac{1}{\cos x} \Leftrightarrow \cos x \cdot \sec x = 1 \Leftrightarrow \cos x = \frac{1}{\sec x};$$

$$\csc x = \frac{1}{\sin x} \Leftrightarrow \sin x \cdot \csc x = 1 \Leftrightarrow \sin x = \frac{1}{\csc x};$$

$$\cot x = \frac{\cos x}{\sin x} \Leftrightarrow \cot x = \frac{1}{\tan x} \Leftrightarrow \tan x \cot x = 1 \Leftrightarrow \tan x = \frac{1}{\cot x};$$

$$\sin^2 x + \cos^2 x = 1 \Leftrightarrow \sin^2 x = 1 - \cos^2 x \Leftrightarrow \cos^2 x = 1 - \sin^2 x;$$

$$\tan^2 x + 1 = \sec^2 x \Leftrightarrow \tan^2 x = \sec^2 x - 1;$$

$$\cot^2 x + 1 = \csc^2 x \Leftrightarrow \cot^2 x = \csc^2 x - 1.$$

Recall from Section 1.3 that, in general, it is a relatively simple matter to show that an equation is not an identity, if such is the case, by finding a counterexample. For example, substituting 0 for $x$ in the equation

$$\sin x = \cos x$$

yields

$$\sin 0 = \cos 0,$$

$$0 = 1,$$

which is a false statement. Thus, the equation $\sin x = \cos x$ is not an identity.

### EXERCISE 5.1

A.

Transform each expression on the left and show that it is identical to the expression on the right for all values of the variable for which each expression is defined.

**Example.**   $(1 - \cos^2 x)(1 + \cot^2 x)$; 1.

*Solution.*   Substituting $\sin^2 x$ for $1 - \cos^2 x$ and $\csc^2 x$ for $1 + \cot^2 x$ yields

$$(1 - \cos^2 x)(1 + \cot^2 x) = \sin^2 x \cdot \csc^2 x.$$

Since $\csc^2 x = 1/(\sin^2 x)$,

$$(1 - \cos^2 x)(1 + \cot^2 x) = \sin^2 x \cdot \frac{1}{\sin^2 x} = 1.$$

Although we have assumed appropriate restrictions, notice that the expression $(1 - \cos^2 x)(1 + \cot^2 x)$ does not equal 1 for values of $x = k\pi$, $k \in J$. $\cot^2 x$ is undefined for these values.

**1.** $\sec x \sin x$; $\tan x$.

**2.** $\tan x \cos x$; $\sin x$.

**3.** $\sin x \cot x$; $\cos x$.

**4.** $\csc x \tan x$; $\sec x$.

**5.** $\dfrac{\sin x}{\tan x}$; $\cos x$.

**6.** $\dfrac{\cos^2 x}{\cot^2 x}$; $\sin^2 x$.

**7.** $\cot^2 x \sin^2 x$; $\cos^2 x$.

**8.** $\tan^2 x \cos^2 x$; $\sin^2 x$.

**9.** $\cos^2 x\,(1 + \tan^2 x)$; 1.

**10.** $(\csc^2 x - 1)(\sin^2 x)$; $\cos^2 x$.

**11.** $\dfrac{(1 + \sin x)(1 - \sin x)}{\cos x}$; $\cos x$.

**12.** $\dfrac{(1 + \cos x)(1 - \cos x)}{\sin x}$; $\sin x$.

**13.** $\dfrac{\sin x \sec x}{\tan x}$; 1.

**14.** $\dfrac{1 + \tan^2 x}{\csc^2 x}$; $\tan^2 x$.

Verify that each equation in Problems 15 to 38 is an identity for those values of $x$ for which each member is defined.

**15.** $\cos x \tan x \csc x = 1$.

**16.** $\csc x = \cot x \sec x$.

**17.** $\sec x - \cos x = \tan x \sin x$.

**18.** $\cos x = \sec x - \tan x \sin x$.

**19.** $\cos^2 x - \sin^2 x = 2 \cos^2 x - 1$.

**20.** $(\sec^2 x - 1)(\csc^2 x - 1) = 1$.

**21.** $\tan^2 x + \sec^2 x = 2 \sec^2 x - 1$.

**22.** $(\tan x + \cot x) \sin x \cos x = 1$.

**23.** $\dfrac{\cos x - \sin x}{\cos x} = 1 - \tan x$.

**24.** $\dfrac{1 - \tan^2 x}{\tan x} = \cot x - \tan x$.

**25.** $\tan^2 x - \sin^2 x = \tan^2 x \sin^2 x$.

**26.** $\dfrac{1}{1 + \sin x} + \dfrac{1}{1 - \sin x} = 2 \sec^2 x$.

**27.** $\tan x + \cot x = \sec x \csc x$.

**28.** $\dfrac{1 + \tan^2 x}{\tan^2 x} = \csc^2 x$.

**29.** $\cos^4 x - \sin^4 x = \cos^2 x - \sin^2 x$.   *Hint:* Factor the left-hand member.

**30.** $\dfrac{1 + \sec x}{\sec x} = \dfrac{\sin^2 x}{1 - \cos x}$.   *Hint:* $\sin^2 x = (1 - \cos x)(1 + \cos x)$.

**31.** $\dfrac{1 - \cos x}{\sin x} = \dfrac{\sin x}{1 + \cos x}$.

*Hint*: Multiply the right-hand member by $\dfrac{1 - \cos x}{1 - \cos x}$.

**32.** $\dfrac{1 + \sin x}{\cos x} = \dfrac{\cos x}{1 - \sin x}$.

**33.** $\csc x = \dfrac{\sec x + \csc x}{1 + \tan x}$.

**34.** $\dfrac{1}{\sec x - \tan x} = \sec x + \tan x$.

**35.** $\dfrac{\sec x - 1}{\sec x + 1} = \dfrac{1 - \cos x}{1 + \cos x}$.

**36.** $\dfrac{\cos^2 x}{1 - \sin x} = \dfrac{\cos x}{\sec x - \tan x}$.

**37.** $\sec^2 x + \csc^2 x = \sec^2 x \csc^2 x$.

**38.** $\tan^4 x + \tan^2 x = \sec^4 x - \sec^2 x$.

Show by using counterexamples that each equation in Problems 39 to 42 is not an identity.

**39.** $\sin x + \cos x \tan x = 2$.

**40.** $\sin^2 x + 2 \cos x - \cos^2 x = 1$.

**41.** $2 \sin x \cos x + \sin x = 0$.

**42.** $\tan^2 x - 2 \tan x = 0$.

**43.** If $x = a \cos \alpha$ and $y = a \sin \alpha$, show that $x^2 + y^2 = a^2$.

**44.** If $x = a \cos \alpha$ and $y = b \sin \alpha$, show that $\dfrac{x^2}{a^2} + \dfrac{y^2}{b^2} = 1$.

**45.** Prove Theorem 5.1-II.

## 5.2
### Identities Involving cos 2x, sin 2x, and tan 2x

Additional identities follow from the sum and difference formulas which were previously developed. For example, if $x$ is substituted for $x_1$ and $x_2$ in

$$\cos (x_1 + x_2) = \cos x_1 \cos x_2 - \sin x_1 \sin x_2,$$

we obtain

$$\cos (x + x) = \cos x \cos x - \sin x \sin x,$$

from which

$$\cos 2x = \cos^2 x - \sin^2 x. \tag{1}$$

The right member can be written in a variety of convenient forms. Thus, if $1 - \sin^2 x$ is substituted for $\cos^2 x$, we have

$$\cos 2x = (1 - \sin^2 x) - \sin^2 x$$
$$= 1 - 2 \sin^2 x.$$

If $1 - \cos^2 x$ is substituted for $\sin^2 x$ in (1), we have

$$\cos 2x = \cos^2 x - (1 - \cos^2 x)$$
$$= 2 \cos^2 x - 1.$$

The result of this argument is summarized in Theorem 5.2.

**Theorem 5.2.** *If $x \in R$ or $x \in A$, then*

$$\cos 2x = \cos^2 x - \sin^2 x,$$
$$\cos 2x = 1 - 2\sin^2 x,$$
$$\cos 2x = 2\cos^2 x - 1.$$

If $x$ is substituted for $x_1$ and $x_2$ in

$$\sin(x_1 + x_2) = \sin x_1 \cos x_2 + \cos x_1 \sin x_2,$$

we obtain

$$\sin(x + x) = \sin x \cos x + \cos x \sin x,$$

from which

$$\sin 2x = 2\sin x \cos x.$$

The result of this argument is stated in Theorem 5.3.

**Theorem 5.3.** *If $x \in R$ or $x \in A$, then*

$$\sin 2x = 2\sin x \cos x.$$

The following theorem can be demonstrated in a similar way and is left as an exercise.

**Theorem 5.4.** *If $x \in R$ or $x \in A$, then*

$$\tan 2x = \frac{2\tan x}{1 - \tan^2 x}, \qquad x \neq \frac{\pi}{4} + k\frac{\pi}{2}, \text{ or } x \neq \left(\frac{\pi}{4} + k\frac{\pi}{2}\right)^R, k \in J.$$

The sum and difference formulas and the above identities are often helpful to rewrite expressions in more useful forms. You will prove additional identities in this section to become familiar with the important relationships above.

**Example.** Show that

$$\frac{2}{\sin 2x} = \tan x + \cot x$$

is an identity for all values of $x$ for which both members are defined.

*Solution.* Using the approach that is frequently effective, we first rewrite each term of each member of the equation in terms of $\sin x$ or $\cos x$. Substituting $2\sin x \cos x$ for $\sin 2x$, $\sin x/\cos x$ for $\tan x$, and $\cos x/\sin x$ for $\cot x$, we obtain

$$\frac{2}{2\sin x \cos x} = \frac{\sin x}{\cos x} + \frac{\cos x}{\sin x}.$$

Rewriting the right-hand member as a single term yields

$$\frac{2}{2\sin x \cos x} = \frac{\sin^2 x + \cos^2 x}{\sin x \cos x}.$$

*(Solution continued on the next page.)*

Substituting 1 for $\sin^2 x + \cos^2 x$ and simplifying the left-hand member yields

$$\frac{1}{\sin x \cos x} = \frac{1}{\sin x \cos x},$$

which is clearly an identity for those values of $x$ for which the expression is defined. (For what values of $x$ is the expression $1/\sin x \cos x$ undefined?)

Relationships similar to those developed in this section for sine, cosine, and tangent can be derived for cotangent, secant, and cosecant. However they are of less importance and will not be considered.

Identities, such as

$$\sin 2x = 2 \sin x \cos x,$$

are sometimes called **double-angle formulas**. This name derives from the fact that historically such identities were studied with domains, $x \in A$, before they were considered for $x \in R$.

### EXERCISE 5.2

**A.**

Express each of the following as a single function of $kx$, where $k$ is a positive integer. Assume that $x$ takes on no value for which the expression is undefined.

*Examples.*  (a) $1 - 2 \sin^2 x$.     (b) $\dfrac{2 \tan 3x}{1 - \tan^2 3x}$.

*Solutions.*  (a) From Theorem 5.2,
$$1 - 2 \sin^2 x = \cos 2x.$$
(b) From Theorem 5.4,
$$\frac{2 \tan 3x}{1 - \tan^2 3x} = \tan 2(3x) = \tan 6x.$$

**1.** $2 \sin x \cos x$.

**2.** $\cos^2 x - \sin^2 x$.

**3.** $2 \cos^2 x - 1$.

**4.** $\dfrac{2 \tan x}{1 - \tan^2 x}$.

**5.** $\cos^2 (3x) - \sin^2 (3x)$.

**6.** $2 \sin (5x) \cos (5x)$.

**7.** $1 - 2 \sin^2 (4x)$.

**8.** $2 \cos^2 (2x) - 1$.

**9.** $\sin x \cos x$.

**10.** $\dfrac{4 \tan x}{1 - \tan^2 x}$.

**11.** $4 \cos^2 (5x) - 2$.

**12.** $4 \sin 3x \cos 3x$.

**13.** $\sin 2x \cos x - \cos 2x \sin x$.

**14.** $\cos 3x \cos 5x - \sin 3x \sin 5x$.

Verify that each equation in Problems 15 to 24 is an identity for those values of $x$ for which each member is defined.

**15.** $\cos 2x + 2 \sin^2 x = 1$.

**19.** $\sec^2 x = \dfrac{2}{1 + \cos 2x}$.

**16.** $\sin 2x \cdot \csc x = 2 \cos x$.

**20.** $\cot x = \dfrac{1 + \cos 2x}{\sin 2x}$.

**17.** $\dfrac{\sin 2x}{1 + \cos 2x} = \tan x$.

**21.** $(\cos^2 x - \sin^2 x)^2 + \sin^2 2x = 1$.

**18.** $\sin 2x = \dfrac{2 \tan x}{1 + \tan^2 x}$.

**22.** $\cos^4 x - \sin^4 x = \cos 2x$.

**23.** $\sin 3x = 3 \sin x - 4 \sin^3 x$.   *Hint:* First rewrite $\sin 3x$ as $\sin (2x + x)$.
**24.** $\cos 3x = 4 \cos^3 x - 3 \cos x$.

**B.**

**25.** If $\cos x = \frac{-4}{5}$ and $\cot x > 0$, find $\cos 2x$.
**26.** If $\sin x = \frac{12}{13}$ and $\tan x < 0$, find $\sin 2x$.
**27.** If $\cos 2x = \frac{-8}{17}$, find $\sin x$.
**28.** If $\tan 2x = -\frac{5}{12}$, find $\tan x$.
**29.** Prove Theorem 5.4.
**30.** Show that $\cos^2 x = \dfrac{1 + \cos 2x}{2}$.   **31.** Show that $\sin^2 x = \dfrac{1 - \cos 2x}{2}$.

**32.** Show that $\cot (x_1 + x_2) = \dfrac{\cot x_1 \cot x_2 - 1}{\cot x_1 + \cot x_2}$.   *Hint:* Use a similar argument to the one shown on page 65 for $\tan (x_1 + x_2)$.

**33.** Show that $\cot 2x = \dfrac{\cot^2 x - 1}{2 \cot x}$.

**34.** Show that $\sin x_1 \cos x_2 = \frac{1}{2}[\sin (x_1 + x_2) + \sin (x_1 - x_2)]$
$= \frac{1}{2}[\sin (x_1 + x_2) - \sin (x_2 - x_1)]$.
**35.** Show that $\cos x_1 \cos x_2 = \frac{1}{2}[\cos (x_1 + x_2) + \cos (x_1 - x_2)]$.
**36.** Show that $\sin x_1 \sin x_2 = \frac{1}{2}[\cos (x_1 - x_2) - \cos (x_1 + x_2)]$.

Using the results from Problems 34 to 36, express each of the following products as a sum or a difference.

**37.** $\sin 4x \cos 3x$.

**39.** $3 \cos \dfrac{\pi}{2} \cos \dfrac{\pi}{3}$.

**38.** $\cos 6x \sin 3x$.

**40.** $4 \sin \dfrac{2\pi}{3} \sin \dfrac{\pi}{3}$.

Using the results from Problems 34 to 36, express each of the following sums as a product.

**41.** $\sin \dfrac{5\pi}{6} + \sin \dfrac{\pi}{6}$.

**42.** $\cos \dfrac{7\pi}{9} + \cos \dfrac{\pi}{9}$.

C.

**43.** Show that if $x_1 + x_2 + x_3 = \pi$, then

$$\tan x_1 + \tan x_2 + \tan x_3 = \tan x_1 \tan x_2 \tan x_3.$$

## 5.3
### Identities Involving cos (x/2), sin (x/2), and tan (x/2)

In Problems 30 and 31 of Section 5.2, you were asked to prove the following.

**Theorem 5.5.**   *If $x \in R$ or $x \in A$, then*

$$\text{I.} \quad \cos^2 x = \frac{1 + \cos 2x}{2},$$

$$\text{II.} \quad \sin^2 x = \frac{1 - \cos 2x}{2}.$$

These identities enable you to write the second degree terms, $\cos^2 x$ and $\sin^2 x$, in terms of first-degree terms. They also enable you to write additional useful identities.

**Theorem 5.6.**   *If $x \in R$ or $x \in A$, then*

$$\text{I.} \quad \cos \frac{x}{2} = \pm\sqrt{\frac{1 + \cos x}{2}},$$

$$\text{II.} \quad \sin \frac{x}{2} = \pm\sqrt{\frac{1 - \cos x}{2}}.$$

*Proof of I.*   Substituting $\dfrac{x}{2}$ for $x$ in Theorem 5.5-I yields

$$\cos^2 \frac{x}{2} = \frac{1 + \cos 2\left(\frac{x}{2}\right)}{2}$$

$$= \frac{1 + \cos x}{2},$$

from which

$$\cos \frac{x}{2} = \pm\sqrt{\frac{1 + \cos x}{2}}.$$

The proof of Theorem 5.6-II is left as an exercise.

Values for cos $(x/2)$ and sin $(x/2)$ are positive or negative, depending on the value of $x/2$. Figure 2.11, page 53, can be used to help you select the appropriate sign for the expression in the right-hand member.

Formulas for $\tan x/2$ also follow directly from identities we have already developed. From Definition 2.3,

$$\tan \frac{x}{2} = \frac{\sin \dfrac{x}{2}}{\cos \dfrac{x}{2}}.$$

The numerator and the denominator of the right-hand member each multiplied by $2 \sin (x/2)$ yields

$$\tan \frac{x}{2} = \frac{2 \sin^2 \dfrac{x}{2}}{2 \sin \dfrac{x}{2} \cos \dfrac{x}{2}}$$

from which, by Theorems 5.3 and 5.5-II,

$$\tan \frac{x}{2} = \frac{2 \left[ \dfrac{1 - \cos 2\left(\dfrac{x}{2}\right)}{2} \right]}{\sin 2\left(\dfrac{x}{2}\right)}$$

$$\tan \frac{x}{2} = \frac{1 - \cos x}{\sin x}. \tag{1}$$

By multiplying the numerator and the denominator of the right-hand member of Equation (1) by $1 + \cos x$, we obtain

$$\tan \frac{x}{2} = \frac{(1 - \cos x)(1 + \cos x)}{\sin x \, (1 + \cos x)}$$

$$= \frac{\sin^2 x}{\sin x \, (1 + \cos x)}$$

$$= \frac{\sin x}{1 + \cos x}.$$

The above arguments justify the following.

**Theorem 5.7.** *If* $x \in R$ *or* $x \in A$, *then*

I.   $\tan \dfrac{x}{2} = \dfrac{1 - \cos x}{\sin x}$,     $x \neq k\pi$ *or* $x \neq k\pi^R$, $k \in J$,

II.   $\tan \dfrac{x}{2} = \dfrac{\sin x}{1 + \cos x}$,     $x \neq \pi + k \cdot 2\pi$ *or* $x \neq (\pi + k \cdot 2\pi)^R$, $k \in J$.

Just as identities such as $\sin 2x = 2 \sin x \cos x$ are sometimes called double-angle formulas, the identities in this section such as $\sin (x/2) = \sqrt{(1 - \cos x)/2}$ are sometimes called **half-angle formulas.**

Relationships similar to those developed in this section for sine, cosine, and tangent can be derived for cotangent, secant, and cosecant. They are of less importance and will not be considered.

### EXERCISE 5.3

**A.**

Verify that each identity is true for the specified replacements.

**1.** $\cos^2 x = \dfrac{1 + \cos 2x}{2}$, for $x = \dfrac{\pi^R}{6}$; for $x = 45°$.

**2.** $\sin^2 x = \dfrac{1 - \cos 2x}{2}$, for $x = \dfrac{\pi^R}{6}$; for $x = 45°$.

**3.** $\cos \dfrac{x}{2} = -\sqrt{\dfrac{1 + \cos x}{2}}$, for $x = \dfrac{3\pi^R}{2}$; for $x = -240°$.

**4.** $\sin \dfrac{x}{2} = \sqrt{\dfrac{1 - \cos x}{2}}$, for $x = \pi^R$; for $x = 270°$.

**5.** $\tan \dfrac{x}{2} = \dfrac{1 - \cos x}{\sin x}$, for $x = -\dfrac{2\pi^R}{3}$; for $x = -240°$.

**6.** $\tan \dfrac{x}{2} = \dfrac{\sin x}{1 + \cos x}$, for $x = -2\pi^R$; for $x = -450°$.

Verify that each equation in Problems 7 to 10 is an identity for those values of $x$ for which each member is defined.

**7.** $\tan \dfrac{x}{2} = \csc x - \cot x.$

**9.** $\left(\cos \dfrac{x}{2} - \sin \dfrac{x}{2}\right)^2 = 1 - 2 \sin x.$

**8.** $\sin \dfrac{x}{2} \cos \dfrac{x}{2} = \dfrac{\sin x}{2}.$

**10.** $\sin x - \cos x \tan \dfrac{x}{2} = \tan \dfrac{x}{2}.$

**B.**

**11.** If $\sin x = \dfrac{\sqrt{3}}{2}$ and $\cos x < 0$, find $\tan \dfrac{x}{2}$.

**12.** If $\sin x = \dfrac{1}{2}$ and $\sec x > 0$, find $\sin \dfrac{x}{2}$.

**13.** Find $\cos x$, if $\sin \dfrac{x}{2} = -\dfrac{1}{\sqrt{2}}$.

**14.** Find $\sin x$, if $\cos \dfrac{x}{2} = \dfrac{\sqrt{3}}{2}$.

**15.** Prove Theorem 5.6-II.

**16.** Show that $\cos \left( \dfrac{\pi}{2} - \dfrac{x}{2} \right) = \sin \dfrac{x}{2}$.

**C.**

Rewrite each expression as an algebraic expression in $x$.

**17.** $\sin (2 \operatorname{Cos}^{-1} x)$.

**18.** $\sin (\tfrac{1}{2} \operatorname{Cos}^{-1} x)$.

**19.** $\sin (\tfrac{1}{2} \operatorname{Sin}^{-1} x)$.

**20.** $\sin (2 \operatorname{Sin}^{-1} x)$.

# 5.4

## Conditional Equations

You have observed that certain equations are *identities*; that is, *they are true for all real number or angle replacements of the variable for which both members are defined.* You will now find solutions for certain equations, such as

$$\sin x = \tfrac{1}{2}, \tag{1}$$

$$\sin x = \cos x, \tag{2}$$

and

$$\cos^2 x + \cos x - 2 = 0, \tag{3}$$

which are *conditional equations* because *they are true for a limited number of replacements of the variable.*

Recall that in Section 1.5 you found the solution set of simple conditional algebraic equations by inspection. You have already solved many simple equations involving circular or trigonometric function values such as

$$\sin x = \tfrac{1}{2} \tag{1}$$

in a similar way. For example, some solution sets of Equation (1) over various domains are

$$\left\{ x \mid \sin x = \frac{1}{2},\, x \in R,\, -\frac{\pi}{2} \le x \le \frac{\pi}{2} \right\} = \left\{ \frac{\pi}{6} \right\},$$

$$\left\{ x \mid \sin x = \frac{1}{2},\, x \in A,\, 0^{R} \le x \le 2\pi^{R} \right\} = \left\{ \frac{\pi^{R}}{6}, \frac{5\pi^{R}}{6} \right\},$$

and for $k \in J$,

$$\{ x \mid \sin x = \tfrac{1}{2},\, x \in A \} = \{ x \mid x = (30 + k \cdot 360)^{\circ} \} \cup \{ x \mid x = (150 + k \cdot 360)^{\circ} \}.$$

These solution sets can be obtained simply by inspection and your knowledge of the ordered pairs in the sine function or Table III and Table IV.

Furthermore, you have seen that the solution set of (1) over the interval $-\pi/2 \le x \le \pi/2$ can be written using inverse notation, as

$$\{x \mid x = \text{Arcsin } \tfrac{1}{2}\}.$$

Perhaps you can determine the solution set of

$$\sin x = \cos x, \qquad 0 < x < \frac{\pi}{2}, \tag{2}$$

by inspection—perhaps not. However, you can generate equivalent equations and hopefully you will generate one from which the solution(s) will be evident by inspection. Generally, solutions can be determined most easily from equations which involve *values of one function only*. The identities developed in the preceding sections are useful in obtaining equivalent equations involving such values from equations involving values of more than one function. For example, in (2), for $0 < x < \pi/2$, $\cos x \ne 0$. Thus each member can be multiplied by $1/\cos x$ to obtain

$$\frac{\sin x}{\cos x} = 1.$$

Substituting $\tan x$ for $\sin x/\cos x$ yields

$$\tan x = 1.$$

Since, in this case, values of $x$ are restricted in the interval $0 < x < \pi/2$, the solution set is found by inspection to be

$$\{x \mid x = \text{Arctan } 1\} = \left(\frac{\pi}{4}\right).$$

Because function values ($\cos x$, $\sin x$, etc) are real numbers for $x \in R$ or $x \in A$, equations, quadratic in a given circular or trigonometric function value, such as

$$\cos^2 x + \cos x - 2 = 0, \qquad x \in R, \tag{3}$$

can be solved by first solving the quadratic equation for the function value involved. Since the left member of Equation (3) is factorable, the equation can be written equivalently as

$$(\cos x + 2)(\cos x - 1) = 0.$$

Since $\cos x + 2$ cannot equal zero (there is no real number $x$ such that $\cos x = -2$), we seek values of $x$ for which

$$\cos x - 1 = 0$$

or

$$\cos x = 1.$$

By inspection the solution set of this latter equation and therefore of Equation (3) is

$$\{x \mid x = k \cdot 2\pi, k \in J\}.$$

## EXERCISE 5.4

### A.

Solve the equations for $0^R \leq x \leq (\pi/2)^R$.

1. $2 \sin x - 1 = 0$.
2. $2 \cos x - 1 = 0$.
3. $2 \cos^2 x = 1$.
4. $4 \sin^2 x = 3$.
5. $3 \csc^2 x - 3 = 0$.
6. $\sec^2 x = 4$.
7. $(2 \cos^2 x - 1)(2 \sin x - 2) = 0$.
8. $(4 \sin^2 x - 1)(\sec x - 2) = 0$.
9. $\cos x \sin x = 0$.
10. $3 \tan x \sec x = 0$.

Solve the equations for $0° \leq x < 360°$.

11. $\cos x = -\sin x$.
12. $\sec x - 2 \cos x = 0$.
13. $\csc x + \sin x = -2$.
14. $2 \cos x + \sec x = 3$.
15. $2 \sin^2 x - \cos x - 1 = 0$.
16. $2 \cos^2 x - 3 \sin x - 3 = 0$.
17. $\cot^2 x + 5 = 2 \csc^2 x$.
18. $\sec^2 x - 4 \cos^2 x = 0$.
19. $2 \sec^2 x - \cos^2 x = \sin^2 x$.
20. $\tan x + \cot x = -2$.
21. $2 \cos^3 x - \cos x = 0$.
22. $2 \sin^3 x - \sin x = 0$.

### B.

Solve the equations for $x \in R$. Use Table III, page 294, as necessary.

23. $2 \cos^2 x - \cos x - 1 = 0$.
24. $2 \cos^2 x + 3 \cos x - 2 = 0$.
25. $25 \sin^2 x - 20 \sin x = -4$.
26. $9 \cos^2 x - 8 = 21 \cos x$.
27. $\tan^4 x - 12 \tan^2 x = -27$.
28. $3 \sin^4 x + 2 = -7 \sin^2 x$.
29. $4 \sin x + 3 \cos x = 5$.
30. $\sin^2 x + \sin x = 1$. *Hint*: Use the quadratic formula to obtain an approximation for $\sin x$.

### C.

It can be shown that for any given real numbers $B$, $D$, and $E$ a positive real number $K$ and a real number $C$ exist such that

$$D \sin Bx + E \cos Bx = K \sin (Bx + C), \qquad (1)$$

for all values of $x \in R$, where $K = \sqrt{D^2 + E^2}$, $\sin C = E/K$, and $\cos C = D/K$. Equation (1) can be used to good advantage in solving a certain type of trigonometric equation. For $x \in A$, Equation (1) is also valid and in this case, $C \in A$.

***Example.*** Solve

$$\sqrt{3} \sin x + \cos x = 1, \qquad x \in R. \qquad (2)$$

*Solution.* Compare the left-hand member of (2) with the left-hand member of Equation (1) and observe that

$$B = 1, \qquad D = \sqrt{3}, \qquad \text{and } E = 1.$$

*(Solution continued on the next page.)*

Thus

$$K = \sqrt{D^2 + E^2} = \sqrt{(\sqrt{3})^2 + (1)^2} = \sqrt{3+1} = 2,$$

$\sin C = 1/2$, and $\cos C = \sqrt{3}/2$. Since both $\sin C$ and $\cos C$ are positive, $0 < C < \pi/2$, and $C = \pi/6$. Hence, $\sqrt{3} \sin x + \cos x = 1$ can be rewritten, by Equation (1),

$$2 \sin \left( x + \frac{\pi}{6} \right) = 1,$$

or

$$\sin \left( x + \frac{\pi}{6} \right) = \frac{1}{2},$$

from which

$$x + \frac{\pi}{6} = \frac{\pi}{6} + k \cdot 2\pi, \quad k \in J \qquad \text{or} \qquad x + \frac{\pi}{6} = \frac{5\pi}{6} + k \cdot 2\pi, \quad k \in J.$$

The solution set is

$$\{x \mid x = k \cdot 2\pi\} \cup \left\{ x \mid x = \frac{2\pi}{3} + k \cdot 2\pi \right\}, \qquad k \in J.$$

Solve each equation for $x \in R$.

**31.** $\sin x + \cos x = 1$.

**32.** $\sin x - \cos x = 1$.

**33.** $\sqrt{3} \cos x - \sin x = -1$.

**34.** $3 \sin x + 4 \cos x = 5$.

## 5.5
## Conditional Equations for Multiples

Conditional equations such as

$$\tan 3\alpha = 1, \tag{1}$$

$$2 \sin \alpha \cos \alpha = \frac{\sqrt{3}}{2}, \tag{2}$$

and

$$\cos 2x = \cos^2 x - 1 \tag{3}$$

need further discussion.

First consider the solution set of

$$\tan 3\alpha = 1 \tag{1}$$

for $\alpha \in A$ over the interval $0° \leq \alpha < 360°$. By inspection we have that

$$3\alpha = (45 + k \cdot 360)° \qquad \text{or} \qquad 3\alpha = (225 + k \cdot 360)°,$$

from which

$$\alpha = (15 + k \cdot 120)° \qquad \text{or} \qquad \alpha = (75 + k \cdot 120)°, \qquad k \in J.$$

To find the solution set over the interval $0° \leq \alpha \leq 360°$, we need only take 0, 1, and 2 as replacements for $k$ and obtain

$$\{15°, 135°, 255°, 75°, 195°, 315°\}.$$

The double angle and half-angle identities are often useful to help you solve certain kinds of equations. For example, the solution set of

$$2 \sin \alpha \cos \alpha = \frac{\sqrt{3}}{2}, \tag{2}$$

for $\alpha \in A$ over the interval $0^R \leq \alpha \leq (\pi/2)^R$, can readily be found if $\sin 2\alpha$ is substituted for the left-hand member, $2 \sin \alpha \cos \alpha$, to obtain the equivalent equation

$$\sin 2\alpha = \frac{\sqrt{3}}{2}.$$

By inspection we have that

$$2\alpha = \frac{\pi}{3}^R \quad \text{or} \quad 2\alpha = \frac{2\pi}{3}^R,$$

from which

$$\alpha = \frac{\pi}{6}^R \quad \text{or} \quad \alpha = \frac{\pi}{3}^R,$$

and the solution set of Equation (2) is $\left(\frac{\pi}{6}^R, \frac{\pi}{3}^R\right)$.

One method of solving

$$\cos 2x = \cos^2 x - 1, \tag{3}$$

for $x \in R$ consists of substituting $2 \cos^2 x - 1$ for $\cos 2x$ to obtain the equivalent equation

$$2 \cos^2 x - 1 = \cos^2 x - 1,$$

from which

$$\cos^2 x = 0.$$

Then, we have as the solution set,

$$\left\{x \mid x = \frac{\pi}{2} + k\pi, k \in J\right\}.$$

Alternatively, we could have substituted $(1 + \cos 2x)/2$ for $\cos^2 x$ in the right-hand member of Equation (3) to obtain the equivalent equations

$$\cos 2x = \frac{1 + \cos 2x}{2} - 1,$$

$$2 \cos 2x = 1 + \cos 2x - 2,$$

and

$$\cos 2x = -1.$$

(Solution continued on the next page.)

By inspection, we have that

$$2x = \pi + k \cdot 2\pi, \qquad k \in J,$$

from which

$$x = \frac{\pi}{2} + k\pi, \qquad k \in J,$$

and the solution set is given by

$$\left\{ x \mid x = \frac{\pi}{2} + k\pi, k \in J \right\},$$

as we found above using the other method.

### EXERCISE 5.5

**A.**

Solve the equations for $0^R \leq \alpha \leq (\pi/2)^R$.

**1.** $\cos 2\alpha - \sin \alpha = 0.$                **5.** $\sin 2\alpha = \cos 2\alpha.$

**2.** $\cos 2\alpha + \cos \alpha = 0.$                **6.** $\sin \dfrac{\alpha}{2} = \tan \dfrac{\alpha}{2}.$

**3.** $\sin 2\alpha + \sin \alpha = 0.$                **7.** $\tan \dfrac{\alpha}{2} - \cos \alpha = -1.$

**4.** $\sin 2\alpha - \cos \alpha = 0.$                **8.** $\cos \alpha = \sin \dfrac{\alpha}{2}.$

Solve the equations for $0° \leq \beta < 360°$.

**9.** $\sin 3\beta = \dfrac{1}{2}.$                **13.** $\left(1 + \cos \dfrac{\beta}{2}\right) \cos \dfrac{\beta}{2} = 0.$

**10.** $\cos 4\beta = -1.$                **14.** $\sin^2 \left(\dfrac{\beta}{4}\right) - \dfrac{1}{2} \sin \left(\dfrac{\beta}{4}\right) = 0.$

**11.** $\sin 3\beta = \cos 3\beta.$                **15.** $5 \sec 3\beta = 2 \cos 3\beta - 3.$
**12.** $\tan 2\beta = \cot 2\beta.$                **16.** $\tan \beta \cdot \tan 2\beta = 1.$

**B.**

Solve the equations for $x \in R$. Use Table III, page 294, as necessary.

**17.** $\sin 2x = \cos x.$                **20.** $(\sin 2x - 1)(\cos 3x + 1) = 0.$
**18.** $2 \cos 2x - 1 = -3 \sin^2 x.$                **21.** $\tan 2x - 3 \sec^2 2x = -5.$
**19.** $\sin x \cos 3x = 0.$                **22.** $\cos^2 2x + 3 \sin 2x - 3 = 0.$

**23.** $\sin 2x \cos x + \cos 2x \sin x = \dfrac{1}{\sqrt{2}}.$

**24.** $\sin 3x \cos x - \cos 3x \sin x = \dfrac{\sqrt{3}}{2}.$

## C.

**25.** Solve the system

$$\rho = \tfrac{1}{2} \cos \theta,$$
$$\rho = \tfrac{1}{2} \sin 2\theta,$$

analytically for $0° \leq \theta < 360°$. Write solutions in the form $(\rho, \theta)$.

**26.** Solve the system

$$\rho = 4 \cos \theta,$$
$$\rho = 4 \sin \theta \tan \theta,$$

analytically for $0° \leq \theta < 360°$. Write solutions in the form $(\rho, \theta)$.

**•27.** Use graphical methods to approximate a positive solution for

$$\cos 2x = x, \quad x \in R.$$

**28.** Use graphical methods to approximate a positive solution for

$$x^2 - 3 \sin \frac{x}{2} = 0, \quad x \in R.$$

## Chapter Summary

**1.** Equations involving circular or trigonometric function values are **identities** if they are satisfied for all real number replacements of the variable (or for all angles) for which both members of the identity are defined.

**2.** An equation can be shown not to be an identity by replacing the variable with a real number (or an angle) such that the resulting statement is false.

**3.** For convenience the basic identities presented in this chapter as well as some presented in previous chapters are listed on page 166. Although no restrictions are given for variables here, they should be kept in mind when using any identity. You can refer to the appropriate definition or theorem, as necessary, to recall these restrictions. Furthermore, while $x \in R$ or $x \in A$, are used below, other variables $\alpha$, $\beta$, $\gamma$, and $\theta$ are also used.

**4.** **Conditional equations** involving circular or trigonometric function values can be solved by inspection or by generating equivalent equations from which the solutions are evident by inspection. The identities introduced in this chapter are useful in obtaining such equivalent equations.

## Summary of Identities

**1.** $\cos(-x) = \cos x$.

**6.** $\csc x = \dfrac{1}{\sin x}$.

**2.** $\sin(-x) = -\sin x$.

**7.** $\cot x = \dfrac{\cos x}{\sin x}$.

**3.** $\tan x = \dfrac{\sin x}{\cos x}$.

**8.** $\sin^2 x + \cos^2 x = 1$.

**4.** $\tan(-x) = -\tan x$.

**9.** $\tan^2 x + 1 = \sec^2 x$.

**5.** $\sec x = \dfrac{1}{\cos x}$.

**10.** $\cot^2 x + 1 = \csc^2 x$.

**11.** $\cos(x_1 + x_2) = \cos x_1 \cos x_2 - \sin x_1 \sin x_2$.

**12.** $\cos(x_1 - x_2) = \cos x_1 \cos x_2 + \sin x_1 \sin x_2$.

**13.** $\sin(x_1 + x_2) = \sin x_1 \cos x_2 + \cos x_1 \sin x_2$.

**14.** $\sin(x_1 - x_2) = \sin x_1 \cos x_2 - \cos x_1 \sin x_2$.

**15.** $\tan(x_1 + x_2) = \dfrac{\tan x_1 + \tan x_2}{1 - \tan x_1 \tan x_2}$.

**16.** $\tan(x_1 - x_2) = \dfrac{\tan x_1 - \tan x_2}{1 + \tan x_1 \tan x_2}$.

**17.** $\cos 2x = \cos^2 x - \sin^2 x$
$$= 2\cos^2 x - 1 = 1 - 2\sin^2 x.$$

**18.** $\sin 2x = 2 \sin x \cos x$.

**22.** $\cos \dfrac{x}{2} = \pm\sqrt{\dfrac{1 + \cos x}{2}}$.

**19.** $\tan 2x = \dfrac{2 \tan x}{1 - \tan^2 x}$.

**23.** $\sin \dfrac{x}{2} = \pm\sqrt{\dfrac{1 - \cos x}{2}}$.

**20.** $\cos^2 x = \dfrac{1 + \cos 2x}{2}$.

**24.** $\tan \dfrac{x}{2} = \dfrac{1 - \cos x}{\sin x}$.

**21.** $\sin^2 x = \dfrac{1 - \cos 2x}{2}$.

**25.** $\sin x_1 \cos x_2 = \frac{1}{2}[\sin(x_1 + x_2) + \sin(x_1 - x_2)]^*$
$$= \tfrac{1}{2}[\sin(x_1 + x_2) - \sin(x_2 - x_1)].^*$$

**26.** $\cos x_1 \cos x_2 = \frac{1}{2}[\cos(x_1 + x_2) + \cos(x_1 - x_2)].^*$

**27.** $\sin x_1 \sin x_2 = \frac{1}{2}[\cos(x_1 - x_2) - \cos(x_1 + x_2)].^*$

* These identities appear in Exercise 5.2, Problems 34 to 36.

## Chapter Review

### A.

Prove that the equation is an identity or show by a counterexample that it is not an identity.

**1.** $\tan x = \dfrac{1}{\cos x \csc x}$.

**4.** $\dfrac{1 - \cos x}{\sin x} = \dfrac{1 + \sin x}{\cos x}$.

**2.** $1 + \tan^2 x = \csc^2 x \tan^2 x$.

**5.** $\dfrac{1 + \cos 2x}{\cos^2 x} = 2$.

**3.** $\sin x \sec x = 2 \tan x$.

**6.** $2 \cos 2x = \sin 2x \cot x - 2 \sin^2 x$.

Verify that each identity is true for the specified replacement.

**7.** $\tan \dfrac{x}{2} = \dfrac{1 - \cos x}{\sin x}$, for $x = \dfrac{\pi}{3}$.

**8.** $\cos 2x = 1 - 2 \sin^2 x$, for $x = -135°$.

In Problems 9 and 10 solve each equation for $0^R \le x \le (\pi/2)^R$.

**9.** $2 \sin x - \sqrt{3} = 0$.

**10.** $\csc^2 x = 4$.

**11.** Solve:     $\tan 2x + \cot 2x = 2$, for $0° \le x < 360°$.

### B.

**12.** Solve:     $\cos 2x + \cos^2 x = 1$, for $x \in R$.

**13.** If $\sin x = \tfrac{1}{3}$, find $\cos 2x$.

**14.** If $\sec x = \tfrac{5}{4}$ and $\cot x < 0$, find $\sin x/2$.

### C.

**15.** Solve the system

$$\rho = \tfrac{1}{4} \sin 2\theta,$$

$$\rho = -\tfrac{1}{4} \sin \theta,$$

analytically for $0° \le \theta < 360°$. Write solutions in the form $(\rho, \theta)$.

**16.** Use graphical methods to approximate a positive solution for

$$2 \sin x = x - 1, \quad x \in R.$$

# 6

# Solutions of Triangles; Geometric Vectors

**T**HE PERIODIC properties of the circular or trigonometric functions have become of primary importance in trigonometry. However, the ability to "solve" triangles, which motivated earlier development of the subject, is also important. When we use given information about the lengths of one, two, or three sides of a triangle and the measure of one or two angles to find the remaining, or unknown, lengths and measures, we are said to be solving a triangle. We are, as in common practice, referring to the angles formed by the rays containing the sides of a triangle as the angles of the triangle.

In this chapter you will first solve different kinds of triangles. Then, you will be introduced to the notion of a geometric vector, an important prerequisite for the study of advanced topics in mathematics.

## 6.1
### Right Triangles

Triangles, in which the measure of one angle equals 90°, or $(\pi/2)^R$, are called **right triangles**. Such triangles can be solved by using the ratios of pairs of sides, which are actually function values of the acute angles (angles whose measures are between 0° and 90°). In fact, historically, such ratios were used many years before functions of the general angle were considered. Figure 6.1 shows that, in the first quadrant, any point $(x, y)$ on the terminal side of an angle $\alpha$ in standard position determines a right triangle with sides of length

$x$ and $y$ and with hypotenuse of length $\sqrt{x^2 + y^2}$. Therefore, in the special case where $\alpha$ is an acute angle in a right triangle,

$$\cos \alpha = \frac{\text{length of side adjacent to } \alpha}{\text{length of hypotenuse}}, \qquad \sec \alpha = \frac{\text{length of hypotenuse}}{\text{length of side adjacent to } \alpha},$$

$$\sin \alpha = \frac{\text{length of side opposite } \alpha}{\text{length of hypotenuse}}, \qquad \csc \alpha = \frac{\text{length of hypotenuse}}{\text{length of side opposite } \alpha},$$

$$\tan \alpha = \frac{\text{length of side opposite } \alpha}{\text{length of side adjacent to } \alpha}, \qquad \cot \alpha = \frac{\text{length of side adjacent to } \alpha}{\text{length of side opposite } \alpha}.$$

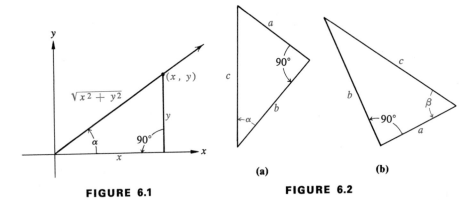

FIGURE 6.1 FIGURE 6.2

It is not necessary that $\alpha$ be in standard position. The above relations involving the ratios of the lengths of the sides of a right triangle are equally applicable to a right triangle in any position. For example, in Figure 6.2a, the six trigonometric ratios of the angle $\alpha$ are:

$$\cos \alpha = \frac{b}{c}, \qquad \sec \alpha = \frac{c}{b},$$

$$\sin \alpha = \frac{a}{c}, \qquad \csc \alpha = \frac{c}{a},$$

$$\tan \alpha = \frac{a}{b}, \qquad \cot \alpha = \frac{b}{a}.$$

In Figure 6.2b, the trigonometric ratios of the angle $\beta$ are

$$\cos \beta = \frac{a}{c}, \qquad \sec \beta = \frac{c}{a},$$

$$\sin \beta = \frac{b}{c}, \qquad \csc \beta = \frac{c}{b},$$

$$\tan \beta = \frac{b}{a}, \qquad \cot \beta = \frac{a}{b}.$$

Generally, the sides of a triangle are labeled with their lengths, $a$, $b$, and $c$, and the angles opposite $a$, $b$, and $c$ are labeled $\alpha$, $\beta$, and $\gamma$, or $A$, $B$, and $C$, respectively. The measure of angles in triangles is usually given in degree units.

Right triangles can be solved directly by using Table IV, page 298.

**Example.** If one angle of a right triangle measures 34° and the side opposite this angle has a length of 12 inches, solve the triangle and find its area.

*Solution.* Sketch the right triangle, showing the given information. In this case

$$\beta = 90° - \alpha = 90° - 34° = 56°.$$

By definition,

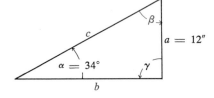

$$\cot 34° = \frac{b}{a} = \frac{b}{12} \quad \text{or} \quad b = 12 \cot 34°,$$

$$\csc 34° = \frac{c}{a} = \frac{c}{12} \quad \text{or} \quad c = 12 \csc 34°.$$

From Table IV, $\cot 34° \approx 1.483$ and $\csc 34° \approx 1.788$, thus,

$$b \approx 12(1.483) \approx 17.8$$

and

$$c \approx 12(1.788) \approx 21.5.$$

The area $\mathscr{A}$ of a triangle equals one-half of the product of the length of the altitude and the length of the base. Thus,

$$\mathscr{A} \approx \tfrac{1}{2}(12)(17.8) \approx 107.$$

Note that the ratio selected in each case in the above example, $\cot 34°$ and $\csc 34°$, was one in which the desired length, $b$ and $c$, respectively, appeared in the numerator of the ratio. With these choices, $b$ and $c$ were found by multiplication. If the ratios

$$\tan 34° = \frac{12}{b} \quad \text{and} \quad \sin 34° = \frac{12}{c}$$

had been used, then equivalently,

$$b = \frac{12}{\tan 34°} \quad \text{and} \quad c = \frac{12}{\sin 34°}$$

and the operation of division would have to be performed to find $b$ and $c$.*

---

* Some tables do not include function values for cotangent, secant, and cosecant, in which case the desired length cannot be placed in the numerator and division is necessary.

In the foregoing example the lengths $b$ and $c$ and the area $\mathscr{A}$ were expressed by an approximation using three *significant digits*.* Unless otherwise stated, we shall continue to specify lengths of line segments and values for areas to this degree of accuracy when solutions are given in decimal notation and we shall specify measures of angles to the *nearest ten minutes in degree measure* $(0° \, 10')$ *or nearest hundredth in radian measure* $(0.01^R)$. Furthermore, we shall assume that all given data have similar accuracy. Thus, $a = 12$ in the above example implies that $a = 12.0$ and $m°(\alpha) = 34°$ implies that $m°(\alpha) = 34° \, 00'$.

Certain angles are given special names in problems pertaining to surveying and navigation. For example, the angle formed by a horizontal ray and an observer's "line of sight" to any object above the horizontal is called an **angle of elevation** (Figure 6.3a). The angle formed by a horizontal ray and an observer's "line of sight" to any object below the horizontal is called an **angle of depression** (Figure 6.3b).

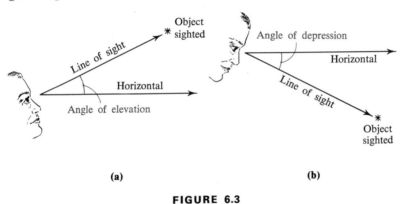

(a)                                    (b)

**FIGURE 6.3**

***Example.*** On level ground at point $A$ the measure of the angle of elevation of the top of a tower is $57° \, 20'$. The distance from $A$ to the base of the tower is 100 feet. How high is the tower?

*Solution.* By definition,

$$\tan 57° \, 20' = \frac{h}{100},$$

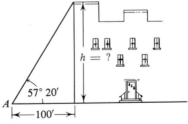

where $h$ represents the height of the tower. Equivalently,

$$h = 100 \tan 57° \, 20' \approx 100 \, (1.560) = 156.$$

Thus, the height of the tower is approximately 156 feet.

---

\* Recall that significant digits are the digits required for specifying $n$ when a number is written in scientific notation as $n \cdot 10^k$, $k \in J$ and $1 \leq n < 10$. For example, since $0.030552 = 3.0552 \times 10^{-2}$, an approximation for $0.030552$ to the *first three* significant digits is given by $0.0306$, where the third significant digit was arbitrarily changed from 5 to 6 because the following digit was 5.

## EXERCISE 6.1

**A.**

In Problems 1 to 8, solve each right triangle and find its area. In each case, $C = 90°$.

1. $a = 2$, $b = 3$.
2. $c = 41$, $b = 9$.
3. $c = 17$, $a = 15$.
4. $c = 14$, $A = 17°$.

5. $c = 23.5$, $B = 51°$.
6. $a = 6.1$, $A = 73°$.
7. $a = 11$, $B = 22° 10'$.
8. $b = 5$, $A = 64° 40'$.

9. A circle has a radius of length of 8.5 feet. Find the measure of the central angle $\alpha$ that subtends a chord with a length of 3.3 feet.

10. A rectangle has dimensions 7.5 inches by 4 inches. Find the length of the diagonal and the measure of the angles that the diagonal forms with the sides.

11. The measure of an angle of elevation from a surveyor's position (ignoring the surveyor's height) to the top of a hill is $37° 40'$. The top of the hill is 275 feet above a level line through the surveyor's position. How far is the surveyor from a point directly below the top of the hill?

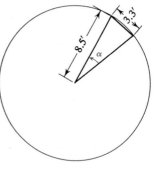

*Problem 9*

12. From the top of a cliff, 80 feet above the surface of a lake, the measure of the angle of depression to an aquanaut's sea chamber is $18° 10'$. How far is the sea chamber from the foot of the cliff?

13. A 40-foot ladder is used to reach the top of a 24-foot wall. If the ladder extends 10 feet past the top of the wall, find the measure of the angle that the ladder forms with the horizontal ground.

14. The lengths of the sides of an isosceles triangle are 17, 17, and 30 inches. Find the measures of the angles in the triangle. *Hint*: Use the altitude of the triangle.

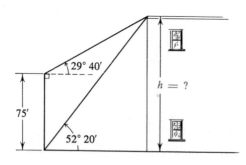

15. From a window in a house, 75 feet above the ground, the measure of the angle of elevation of the top of a nearby building is $29° 40'$. From a point on the ground directly below the window, the measure of the angle of elevation of the top of the same building is $52° 20'$. Find the height of the building.

**B.**

16. Show that the length of the perpendicular $h$ from $C$ to the side $AB$ of any triangle $ABC$ can be found by

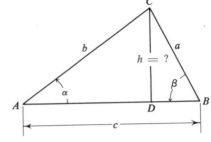

$$h = \frac{c}{\cot \alpha + \cot \beta}.$$

*Hint:* $AD = h \cot \alpha$.

17. Show that if $l$ is the length of the side of a regular polygon of $n$ sides and $r$ is the length of the radius of the inscribed circle, then $r = \frac{1}{2}l \cot (\pi/n)$.

18. Show that the length of the radius $r$ of the circumscribed circle of a regular polygon of $n$ sides, each of length $l$, is given by $r = (l/2) \csc (\pi/n)$.

19. Show that the perimeter $P$ of a regular polygon of $n$ sides, inscribed in a circle of radius with length $r$, is given by $P = 2nr \sin (\pi/n)$.

20. Show that the area $\mathcal{A}$ of a regular polygon of $n$ sides, each of length $l$, is given by $\mathcal{A} = \frac{1}{4}nl^2 \cot (\pi/n)$.

21. Show that the area $\mathcal{A}$ of a regular polygon of $n$ sides inscribed in a circle of radius with length $r$, is given by $\mathcal{A} = \frac{1}{2}nr^2 \sin (2\pi/n)$.

22. If $ABC$ is a right triangle with right angle at $C$ and if $CD$ is perpendicular to $AB$, show that the following formulas for the area $\mathcal{A}$ of triangle $ABC$ are true.
    (a) $\mathcal{A} = \frac{1}{2}bc \sin \alpha = \frac{1}{2}ac \sin \beta = \frac{1}{2}ab \sin \gamma$,
    (b) $\mathcal{A} = \frac{1}{2}c^2 \sin \alpha \cos \alpha$,
    (c) $\mathcal{A} = \frac{1}{2}b^2 \tan \alpha$,
    (d) $\mathcal{A} = \frac{1}{2}a^2 \cot \alpha$.

23. Prove that when a simple pendulum of length $l$ is inclined at an angle $\alpha$ with the vertical, the bob is at a height $h$ above its lowest position, where $h = 2l \sin^2 (\alpha/2)$.

## 6.2
## The Law of Sines

In Section 6.1 you solved right triangles using the trigonometric ratios and the Pythagorean theorem. In this section and in the next, special formulas will be derived by which any triangle can be solved. One such formula can be derived directly from expressions relating the area of a triangle to the lengths of two sides and the measure of the angle formed by these two sides.

Consider Figure 6.4 on page 174. Since the area $\mathcal{A}$ of either triangle equals one half of the product of the length of its base, $c$, and the length of its altitude, $h$, ($b \sin \alpha$), it follows that

$$\mathcal{A} = \frac{1}{2}(c)(b \sin \alpha) = \frac{1}{2}bc \sin \alpha.$$

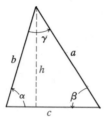
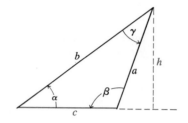

**FIGURE 6.4**

Using perpendiculars from the other vertices, it can be shown in a similar way that

$$\mathscr{A} = \tfrac{1}{2}ab \sin \gamma \quad \text{and} \quad \mathscr{A} = \tfrac{1}{2}ac \sin \beta.$$

Since each triangle has only one area, the right-hand members of these equations can be equated to obtain

$$\tfrac{1}{2}bc \sin \alpha = \tfrac{1}{2}ac \sin \beta = \tfrac{1}{2}ab \sin \gamma.$$

If each member here is multiplied by $2/abc$, the following result, called the **Law of Sines**, is readily established.

**Theorem 6.1.**  *If $\alpha$, $\beta$, and $\gamma$ are the angles of any triangle, and if $a$, $b$, and $c$ are the lengths of the sides opposite $\alpha$, $\beta$, and $\gamma$, respectively, then*

$$\frac{a}{\sin \alpha} = \frac{b}{\sin \beta} = \frac{c}{\sin \gamma}.$$

In many applications the Law of Sines is more conveniently used in the form

$$\frac{\sin \alpha}{a} = \frac{\sin \beta}{b} = \frac{\sin \gamma}{c}.$$

The Law of Sines can be used to solve certain triangles. Consider first the case in which the measure of two angles and the length of one side of a triangle are given.

**Example.**  Solve the triangle for which $\alpha = 30°$, $\beta = 45°$, and $a = 20$.

*Solution.*  A sketch of the triangle showing the given information is helpful. First, using the fact that $\alpha + \beta + \gamma = 180°$, it follows that

$$\gamma = 180° - \alpha - \beta = 180° - 30° - 45° = 105°.$$

From the Law of Sines,

$$\frac{20}{\sin 30°} = \frac{b}{\sin 45°},$$

or equivalently

$$b = \frac{20 \sin 45°}{\sin 30°}.$$

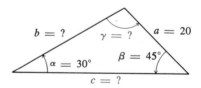

From Table 4.1, page 139, (or memory),

$$b = \frac{20\,(1/\sqrt{2})}{1/2} = 20\,\sqrt{2} \approx 20(1.414) \approx 28.3.$$

Then again from the Law of Sines,

$$\frac{20}{\sin 30°} = \frac{c}{\sin 105°},$$

or equivalently, using Table IV, we have

$$c = \frac{20\,\sin 105°}{\sin 30°} \approx \frac{20(0.9659)}{1/2} = 40(0.9659) \approx 38.6.$$

Hence,

$$b \approx 28.3, \qquad c \approx 38.6, \qquad \text{and } \gamma = 105°.$$

A second case in which the Law of Sines is used is that in which the lengths of two sides of any triangle, say $a$ and $b$, and the measure of an angle opposite one of them, say $\alpha$, are given. Ambiguity may arise here, depending on the length $a$, in relation to the length $b$ and the measure of $\alpha$. Figure 6.5, in which $\alpha$ is an *acute* angle, shows the various possibilities that may occur. If $a < h$ (Figure 6.5a), the side with length $a$ cannot intersect the side opposite $C$ and no triangle is possible. Since $h = b \sin \alpha$, in this case, $a < b \sin \alpha$.

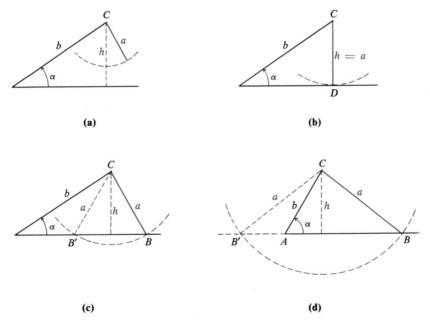

(a)                              (b)

(c)                              (d)

FIGURE 6.5

If $a = h = b \sin \alpha$ (Figure 6.5b), the side with length $a$ will touch the side opposite $C$ at one point $D$. In this case, one right triangle is possible.

If $a > h$, $(b \sin \alpha)$, and $a < b$, or $b \sin \alpha < a < b$ (Figure 6.5c), the side with length $a$ can intersect the side opposite $C$ in two distinct points, $B$ and $B'$, and two triangles are possible.

If $a \geq b$ (Figure 6.5d), the side with length $a$ intersects the side opposite $C$ at the points $B$ and $B'$. The triangle $AB'C$ does not contain the given angle $\alpha$, and therefore does not lead to a possible solution. Hence, in this case, only one triangle is possible. In particular, the triangle is isosceles if $a = b$. The following table summarizes these possibilities.

Given $a$, $b$, and $\alpha$ an *acute* angle;

| If: | Then: |
|---|---|
| I. $a < b \sin \alpha$ | no triangle possible |
| II. $a = b \sin \alpha$ | one right triangle |
| III. $b \sin \alpha < a < b$ | two triangles |
| IV. $a \geq b$ | one triangle |

Given $a$, $b$, and $\alpha$ an *obtuse* angle, there are only two possibilities (Figure 6.6).

| If: | Then: |
|---|---|
| I. $a > b$ | one triangle |
| II. $a \leq b$ | no triangle possible |

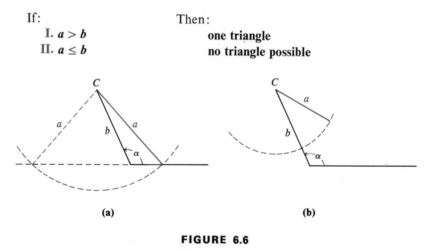

(a)                                              (b)

**FIGURE 6.6**

While we have concerned ourselves only with the variables $a$, $b$, and $\alpha$ in the above discussion, the results are valid when the other variables, $c$, $\beta$, and $\gamma$ are involved in the given information. A sketch of the triangle using the given information will help to indicate which possibility exists.

***Example.***  Solve the triangle(s) for which $a = 7$, $b = 5$, and $\beta = 30°$.

*Solution.*  Sketch the triangle using the given information. Since the lengths of two sides and the measure of an angle $\beta$, opposite the side with measure $b$, are given, a check must be made to see how many triangles are possible. Since

$$h = a \sin \beta = 5 \sin 30°$$

$$= 5 \cdot \tfrac{1}{2} = 2.5,$$

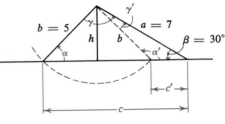

we have that    $h < b < a$,

and two triangles are possible, as indicated in the figure. From the Law of Sines

$$\frac{\sin \alpha}{7} = \frac{\sin 30°}{5},$$

from which

$$\sin \alpha = \frac{7 \sin 30°}{5} = \frac{7 \cdot 1/2}{5} = 0.700.$$

Since $\sin \alpha > 0$ in Quadrants I and II, then from Table IV, to the nearest $10'$,

$$\alpha \approx 44° \; 30' \quad \text{or} \quad \alpha' \approx 180° - 44° \; 30' = 135° \; 30'.$$

Taking $\alpha \approx 44° \; 30'$, the solution of one of the two possible triangles is completed as follows. Since

$$\gamma = 180° - \alpha - \beta \approx 180° - 44° \; 30' - 30° = 105° \; 30',$$

then from the Law of Sines

$$\frac{5}{\sin 30°} = \frac{c}{\sin 105° \; 30'},$$

from which

$$c = \frac{5 \sin 74° \; 30'}{\sin 30°} \approx \frac{5(0.9636)}{1/2} = 2(5)(0.9636) \approx 9.64.$$

Taking $\alpha' \approx 135° \; 30'$, the solution of the second possible triangle can be completed. Since

$$\gamma' \approx 180° - 135° \; 30' - 30° = 14° \; 30',$$

by the Law of Sines

$$\frac{5}{\sin 30°} = \frac{c'}{\sin 14° \; 30'},$$

from which

$$c' \approx (2)(5)(0.2504) \approx 2.50.$$

(Solution continued on the next page.)

Thus, the two solutions are

$$\alpha \approx 44° \ 30', \qquad \gamma = 105° \ 30', \qquad c \approx 9.64,$$

and

$$\alpha' \approx 135° \ 30', \qquad \gamma' \approx 14° \ 30', \qquad c' \approx 2.50.$$

## EXERCISE 6.2

A.

Solve the following triangles.

1. $a = 5$, $\alpha = 20°$, $\beta = 75°$.
2. $a = 32$, $\alpha = 26° \ 20'$, $\gamma = 81° \ 50'$.
3. $a = 58.4$, $\beta = 37° \ 10'$, $\gamma = 100°$.
4. $b = 1.9$, $\alpha = 111° \ 40'$, $\beta = 5° \ 10'$.
5. $b = 0.42$, $\alpha = 35° \ 40'$, $\gamma = 91° \ 30'$.
6. $b = 0.88$, $\beta = 63° \ 50'$, $\gamma = 34° \ 10'$.
7. $c = 13.6$, $\alpha = 30° \ 20'$, $\beta = 72° \ 10'$.
8. $c = 1.4$, $\alpha = 135° \ 10'$, $\gamma = 34° \ 50'$.

Determine the number of triangles that satisfy the conditions in each problem. Sketch a figure in each case.

*9. $a = 7$, $b = 5$, $A = 30°$.     13. $b = 16$, $c = 32$, $B = 30°$.
10. $a = 10$, $b = 9$, $B = 60°$.     14. $b = 0.3$, $c = 0.4$, $C = 61° \ 50'$.
*11. $a = 4.2$, $c = 6.1$, $A = 32° \ 10'$.  *15. $b = 3.9$, $a = 5$, $B = 42° \ 40'$.
12. $a = 38$, $c = 45$, $C = 35° \ 20'$.   16. $b = 37$, $a = 51$, $A = 135° \ 30'$.
17–24. Solve the triangles in Problems 9 to 16 above.

In Problems 25 to 30, find the area of each triangle.

***Example.***  $a = 2.3$, $\beta = 115°$, $\gamma = 36°$.

*Solution.*   Make a sketch showing the given information. As shown in this section,

$$\mathscr{A} = \tfrac{1}{2}ab \sin \gamma. \qquad (1)$$

It is first necessary to find $b$. Since

$$\alpha = 180° - \beta - \gamma$$

$$= 180° - 115° - 36° = 29°,$$

from the Law of Sines

$$\frac{2.3}{\sin 29°} = \frac{b}{\sin 115°},$$

from which

$$b = \frac{2.3 \sin 115°}{\sin 29°} = \frac{2.3 \sin 65°}{\sin 29°} \approx \frac{2.3(0.9063)}{0.4848} \approx 4.30.$$

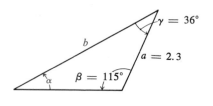

Substituting appropriately in Equation (1),

$$\mathcal{A} \approx \tfrac{1}{2}(2.3)(4.30) \sin 36° \approx 4.945(0.5878) \approx 2.91.$$

**25.** $a = 3.4$,  $\beta = 13°$,  $\alpha = 50°$.      **28.** $c = 5.2$,  $a = 3.2$,  $\gamma = 31° \ 20'$.
**26.** $c = 3.8$,  $\beta = 15° \ 10'$,  $\gamma = 50°$.    **29.** $c = 2.2$,  $a = 3.2$,  $\alpha = 60° \ 30'$.
**27.** $b = 5.3$,  $c = 7.1$,  $\gamma = 29°$.      **30.** $b = 1.9$,  $c = 7.2$,  $\gamma = 18°$.

**31.** To find the distance from an observer's point $A$ on one side of a river
(see figure) to a point $P$ on the other
side, a line $AP'$ measuring 2,540 feet
was laid out on one side and the
angles $P'AP$ and $AP'P$ were measured and found to be 47° 20′ and
78° 10′, respectively. Find the distance from $A$ to $P$.

**32.** Two men, 400 feet apart, observe a
balloon between them that is in the
same vertical plane with the men.
The respective angles of elevation
of the balloon are observed by the
men to be 75° 20′ and 49° 30′. Find
the height of the balloon above the ground.

**B.**

Show that, in any triangle $ABC$, each of the statements in Problems 33 to 38,
is true.   *Hint*: In Problems 33 and 34, start with the Law of Sines.

**33.** $\dfrac{a+b}{b} = \dfrac{\sin \alpha + \sin \beta}{\sin \beta}.$

**36.** $\dfrac{a-b}{a+b} = \dfrac{\tan \frac{1}{2}(\alpha - \beta)}{\tan \frac{1}{2}(\alpha + \beta)}.$

**34.** $\dfrac{a-b}{b} = \dfrac{\sin \alpha - \sin \beta}{\sin \beta}.$

**37.** $\dfrac{a+b}{c} = \dfrac{\cos \frac{1}{2}(\alpha - \beta)}{\sin \frac{1}{2}\gamma}.$

**35.** $\dfrac{a-b}{a+b} = \dfrac{\sin \alpha - \sin \beta}{\sin \alpha + \sin \beta}.$

**38.** $\dfrac{a-b}{c} = \dfrac{\sin \frac{1}{2}(\alpha - \beta)}{\cos \frac{1}{2}\gamma}.$

**C.**

The formula in Problem 36, called the **Law of Tangents**, is sometimes used
in engineering and surveying for the solution of triangles, given the lengths
of two sides and the measure of the angle formed by the rays which include the
given sides.

*Example.*   Solve the triangle in which $a = 27$, $b = 23$, and $\gamma = 30°$.

*Solution.*   From the Law of Tangents,

$$\tan \tfrac{1}{2}(\alpha - \beta) = \frac{(a - b) \tan \frac{1}{2}(\alpha + \beta)}{a + b}. \tag{1}$$

(Solution continued on the next page.)

Because $\gamma = 30°$, then $\alpha + \beta = 180° - \gamma = 150°$. Substituting appropriate values for the variables in Equation (1),

$$\tan \tfrac{1}{2}(\alpha - \beta) = \frac{(27 - 23) \tan \tfrac{1}{2}(150°)}{27 + 23} = \frac{4 \tan 75°}{50}.$$

Substituting appropriate approximations from Table IV yields

$$\tan \tfrac{1}{2}(\alpha - \beta) \approx \frac{4(3.732)}{50} \approx 0.2986,$$

from which,

$$\tfrac{1}{2}(\alpha - \beta) \approx 16° \, 40',$$

$$\alpha - \beta \approx 33° \, 20'.$$

The measures of the angles $\alpha$ and $\beta$ can now be found by solving the system

$$\alpha + \beta = 150°,$$

$$\alpha - \beta = 33° \, 20',$$

which gives

$$\alpha = 91° \, 40', \qquad \beta = 58° \, 20'.$$

The length of side $c$ can now be found by using the Law of Sines.

In Problems 39 to 42, solve each triangle by first using the Law of Tangents.

**39.** $b = 1$,   $c = 4$,   $\alpha = 45°$.          **41.** $a = 0.13$,   $b = 2.5$,   $\gamma = 18° \, 20'$.
**40.** $c = 3$,   $a = 7.5$,   $\beta = 100°$.          **42.** $b = 1.06$,   $c = 0.97$,   $\alpha = 132° \, 40'$.

## 6.3

### The Law of Cosines

The following theorem called the **Law of Cosines** specifies formulas which are convenient for solving triangles in which the lengths of two sides and the measure of the angle formed by their respective rays are specified or in which the lengths of the three sides are specified.

**Theorem 6.2.**   *If $\alpha$, $\beta$, and $\gamma$ are the angles of any triangle, and $a$, $b$, and $c$ are the lengths of the sides opposite $\alpha$, $\beta$, and $\gamma$, respectively, then*

$$c^2 = a^2 + b^2 - 2ab \cos \gamma, \tag{1}$$

$$b^2 = a^2 + c^2 - 2ac \cos \beta, \tag{2}$$

and

$$a^2 = b^2 + c^2 - 2bc \cos \alpha. \tag{3}$$

*Proof of* (1).   Figure 6.7 shows points $P(x_1, y_1)$ and $Q(x_2, y_2)$ on the terminal sides of angles $\alpha$ and $\beta$, respectively. Now, if $P$ is located $a$ units from

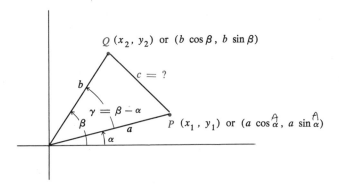

**FIGURE 6.7**

the origin and $Q$ is located $b$ units from the origin, then the coordinates of $P$ are $(a \cos \alpha, a \sin \alpha)$ and the coordinates of $Q$ are $(b \cos \beta, b \sin \beta)$. Using the distance formula,

$$d^2 = (x_2 - x_1)^2 + (y_2 - y_1)^2,$$

we have

$$c^2 = (b \cos \beta - a \cos \alpha)^2 + (b \sin \beta - a \sin \alpha)^2$$
$$= b^2 \cos^2 \beta - 2ab \cos \beta \cos \alpha + a^2 \cos^2 \alpha$$
$$\qquad + b^2 \sin^2 \beta - 2ab \sin \alpha \sin \beta + a^2 \sin^2 \alpha$$
$$= b^2(\cos^2 \beta + \sin^2 \beta) + a^2(\cos^2 \alpha + \sin^2 \alpha)$$
$$\qquad - 2ab (\cos \beta \cos \alpha + \sin \beta \sin \alpha).$$

Substituting 1 for $\cos^2 \beta + \sin^2 \beta$, 1 for $\cos^2 \alpha + \sin^2 \alpha$, and $\cos (\beta - \alpha)$ for $\cos \beta \cos \alpha + \sin \beta \sin \alpha$ in the right-hand member of the preceding equation yields

$$c^2 = b^2 + a^2 - 2ab \cos (\beta - \alpha).$$

Because $\beta - \alpha = \gamma$, the angle between the terminal rays of the angles $\beta$ and $\alpha$, this equation can be written equivalently as

$$c^2 = a^2 + b^2 - 2ab \cos \gamma. \tag{1}$$

By substituting appropriate variables, the validity of equations (2) and (3) can also be shown.

Note that when $\gamma$ is $90°$, $\cos \gamma = 0$ and the Law of Cosines reduces to

$$c^2 = a^2 + b^2 - 2ab(0),$$
$$c^2 = a^2 + b^2,$$

the familiar formula applicable to right triangles.

Now, consider how the Law of Cosines can be used to solve a triangle when the measure of an angle and the lengths of the adjacent sides are known.

**Example.**  Solve the triangle for which $\beta = 75°$, $a = 7$, and $c = 5$.

*Solution.*  Sketch a triangle showing the given information. The length $b$ can be found by using the Law of Cosines expressed in the form $b^2 = a^2 + c^2 - 2ac \cos \beta$. Substituting the known values gives

$$b^2 = 7^2 + 5^2 - 2(7)(5) \cos 75°,$$

$$b^2 \approx 49 + 25 - 70(0.2588) \approx 55.9,$$

and by linear interpolation from Table I,

$$b \approx 7.48.$$

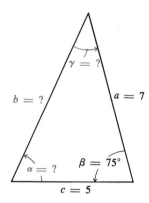

The Law of Sines can now be used to complete the solution since the length $b$, opposite the given angle $\beta$, is known. However, we will use the Law of Cosines here again in order to show how it can be used to determine the measure of an angle if the lengths of three sides of a triangle are known. To find the measure of $\alpha$, the equation

$$a^2 = b^2 + c^2 - 2bc \cos \alpha,$$

is first written in the form

$$\cos \alpha = \frac{b^2 + c^2 - a^2}{2bc}.$$

From the previous calculations, $b^2 \approx 55.9$ and $b \approx 7.48$. Thus

$$\cos \alpha \approx \frac{55.9 + 25 - 49}{2(7.48)(5)} \approx 0.426.$$

Because $\alpha$, an angle in a triangle, must have a measure less than $180°$, and $\cos \alpha > 0$, we have from Table IV,

$$\alpha \approx 64° \, 50'.$$

Because $\alpha + \beta + \gamma = 180°$,

$$\gamma \approx 180° - 64° \, 50' - 75° \, 00' = 40° \, 10'.$$

Thus, the solution to the triangle is

$$b \approx 7.48, \qquad \alpha \approx 64° \, 50', \qquad \text{and} \qquad \gamma \approx 40° \, 10'.$$

### EXERCISE 6.3

**A.**

Solve the following triangles.

**1.** $b = 4$, $\quad c = 3.5$, $\quad \alpha = 71°$.
**2.** $c = 0.3$, $\quad a = 0.1$, $\quad \beta = 29° \, 40'$.
**3.** $a = 3.2$, $\quad b = 2.2$, $\quad \gamma = 35° \, 20'$.

**4.** $b = 6.0$, $\quad a = 5.1$, $\quad \gamma = 43° \, 30'$.
**5.** $a = 0.7$, $\quad c = 0.8$, $\quad \beta = 141° \, 30'$.
**6.** $b = 3.4$, $\quad a = 2.1$, $\quad \gamma = 122° \, 10'$.

In Problems 7 to 10, solve the triangles for which the lengths of the three sides are given. Find the measures of *all three* angles by the Law of Cosines or the Law of Sines. Check your results by noting if $\alpha + \beta + \gamma \approx 180°$.

***Example.*** $a = 3.4, \quad b = 2.7, \quad c = 1.3.$

*Solution.* Draw a sketch showing the given information. From the Law of Cosines,

$$\cos \alpha = \frac{(2.7)^2 + (1.3)^2 - (3.4)^2}{2(2.7)(1.3)}$$

$$= \frac{7.29 + 1.69 - 11.56}{7.02} \approx -0.3675.$$

Because $\cos \alpha < 0$, $90° < \alpha < 180°$; thus, from Table IV, and by using the proper reduction formula,

$$\alpha \approx 180° - 68° \ 30' \approx 111° \ 30'.$$

Either the Law of Cosines or the Law of Sines can now be used to find $\beta \approx 47° \ 40'$ and $\gamma \approx 20° \ 50'$. Does

$$\alpha + \beta + \gamma = 111° \ 30' + 47° \ 40' + 20° \ 50' = 180°? \quad \text{Yes.}$$

7. $a = 9, \quad b = 7, \quad c = 5.$  9. $a = 1.2, \quad b = 9, \quad c = 10.$
8. $a = 2.7, \quad b = 5.1, \quad c = 4.4.$  10. $a = 4.5, \quad b = 7.5, \quad c = 5.8.$

11. Find the measure of the smallest angle of the triangle whose sides have lengths 4.3, 5.1, and 6.3. *Hint*: The angle with least measure is opposite the side with least length.
12. Find the measure of the largest angle of the triangle whose sides have lengths 2.9, 3.3, and 4.1.
13. Find the area of the triangle in Problem 11 above.
14. Find the area of the triangle in Problem 12 above.

**B.**

The formulas in Problems 15 to 21, where $s = (a + b + c)/2$, are sometimes used in the solution of triangles. Show that each statement is true for all $\alpha \in A$. *Hint*: Use the fact that $\cos \alpha = (b^2 + c^2 - a^2)/2bc$.

15. $1 + \cos \alpha = \dfrac{(b + c + a)(b + c - a)}{2bc}.$

16. $1 - \cos \alpha = \dfrac{(a - b + c)(a + b - c)}{2bc}.$

17. $\cos \tfrac{1}{2}\alpha = \sqrt{\dfrac{s(s - a)}{bc}}.$ *Hint*: $\cos \tfrac{1}{2}\alpha = \sqrt{\dfrac{1 + \cos \alpha}{2}}.$

**18.** $\sin \frac{1}{2}\alpha = \sqrt{\dfrac{(s-b)(s-c)}{bc}}$.   *Hint*: $\sin \frac{1}{2}\alpha = \sqrt{\dfrac{1-\cos\alpha}{2}}$.

**19.** Use the results of Problems 17 and 18 to prove Hero's (or Heron's) formula,

$$\mathscr{A} = \sqrt{s(s-a)(s-b)(s-c)},$$

which can be used to find the area of any triangle directly from the lengths of its three sides.

**20.** Use Hero's formula to find the area of the triangle in Problem 7 above.

**21.** Use Hero's formula to find the area of the triangle in Problem 8 above.

**22.** Two sides of a parallelogram are of lengths 6.8 and 8.3 inches, and one of the diagonals is of length 4.2 inches. Find the area of the parallelogram.

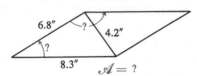

**C.**

The half-angle formulas in Problems 17 and 18 are sometimes used in engineering and surveying in the solution of triangles when the lengths of the three sides are known.*

*Example.*   Solve the triangle in which $a = 5$, $b = 6$, and $c = 7$.

*Solution.*

$$s = \frac{a+b+c}{2} = \frac{5+6+7}{2} = 9.$$

Thus,

$$\cos\frac{1}{2}\alpha = \sqrt{\frac{9(9-5)}{6\cdot 7}} = \sqrt{\frac{36}{42}} = \sqrt{\frac{6}{7}} \approx 0.9258,$$

from which from Table IV,

$$\tfrac{1}{2}\alpha \approx 22°\ 10',$$

$$\alpha \approx 44°\ 20'.$$

The Law of Sines can now be used to find the measures for $\beta$ and $\gamma$.

In Problems 23 to 26, solve each triangle by first using a half-angle formula.

**23.** $b = 8$,   $c = 10$,   $a = 6$.          **25.** $a = 1.12$,   $b = 0.54$,   $c = 1.3$.
**24.** $c = 4.5$,   $a = 11.0$,   $b = 12.3$.          **26.** $b = 0.07$,   $c = 0.042$,   $a = 0.108$.

---

\* A half-angle formula or a form of the Law of Tangents discussed on page 179 is more convenient to use than the Law of Cosines when logarithms (or the slide rule) are to be used. See Section A.4 in Appendix A.

## 6.4
### Geometric Vectors

Some physical quantities, such as length and mass, are characterized by *magnitude alone*. If units of measure are specified, the quantities can be represented by real numbers and associated with points on a number line. Thus, for an appropriate unit, the real numbers 10, 15, and 20 are lengths of line segments and they can be represented on a number line.

In the description of some physical quantities such as force, velocity, and acceleration, *direction as well as magnitude* is required. Such a quantity can be represented by a pair of real numbers and associated with a line segment, called a **geometric vector**, in the form of an arrow with length equal to the magnitude and pointing in an assigned direction.

In this section you will study geometric vectors in a plane (2-dimensional) and in Chapter 7 you will look at the analytic notion of 2-dimensional vectors as ordered pairs of real numbers. First, we will make some agreements on terminology and symbolism.

**Definition 6.1.**  *A 2-dimensional **geometric vector** designated by $\vec{v}$ is a line segment with a specified direction* (see Figure 6.8a). *The set of all such geometric vectors is designated by $\vec{V}$.*

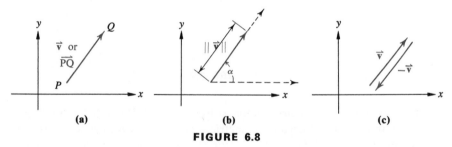

**FIGURE 6.8**

If the end points of a geometric vector are named, it is sometimes convenient to specify the vector by using these names. For example, in Figure 6.8a the geometric vector can also be designated by $\overrightarrow{PQ}$, where the point named $P$ is called the **initial point** and the point named $Q$ is called the **terminal point**.

We shall assume that each geometric vector in the plane has a length which can be associated with some real number depending upon the unit of measurement. Notice that line segments associated with directed distances parallel to the $x$ axis and $y$ axis that we discussed in Section 1.4 are special cases of geometric vectors. They have both magnitude and direction.

**Definition 6.2.**  *For all geometric vectors $\vec{v}$, the **norm** (or **magnitude**) of $\vec{v}$ is the length of $\vec{v}$ and designated by $\|\vec{v}\|$* (see Figure 6.8b). *From the vector viewpoint, the real number $\|\vec{v}\|$ is called a **scalar**.*

**Definition 6.3.** *The **direction angle** of a geometric vector* $\vec{v}$ *is an angle* $\alpha$, $-180° < \alpha \leq 180°$, *such that:*

> *The initial side of* $\alpha$ *is the ray from the initial point of* $\vec{v}$ *parallel to the x axis and directed in the positive x direction; the terminal side of* $\alpha$ *is the ray from the initial point of* $\vec{v}$ *and containing* $\vec{v}$ *(see Figure 6.8b).*

**Definition 6.4.** *For all geometric vectors* $\vec{v}$, *the geometric vector* $-\vec{v}$ *has the same magnitude as* $\vec{v}$ *but is of the opposite direction (see Figure 6.8c).*

**Definition 6.5.** *Two geometric vectors* $\vec{v}_1$ *and* $\vec{v}_2$ *are **equivalent** if they have the same magnitude and the same direction (see Figure 6.9a). The symbol* " $=$ " *will be used to indicate this relationship.*

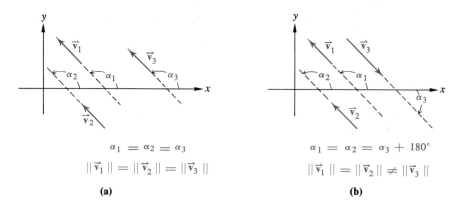

**FIGURE 6.9**

If we imagine a geometric vector as free to "slide" parallel to its initial position in the plane, then any geometric vector can always "slide" onto any vector equivalent to it. Geometric vectors which are viewed as free to "slide" are called **free vectors** as distinguished from bound geometric vectors whose initial points are always tied to a fixed position in the plane. In this section and the next you will be working with free geometric vectors; in Chapter 7, you will be working with bound vectors with initial points at the origin.

Two geometric vectors are said to be **collinear** if they have the *same direction or opposite directions* regardless of whether or not they have the same magnitude. Thus, in Figure 6.9b, $\vec{v}_1$, $\vec{v}_2$, and $\vec{v}_3$ are all collinear, and $\vec{v}_1$ is equivalent to $\vec{v}_2$; however, neither $\vec{v}_1$ nor $\vec{v}_2$ is equivalent to $\vec{v}_3$.

**Definition 6.6.** *For all geometric vectors* $\vec{v}_1$ *and* $\vec{v}_2$, $\vec{v}_1 + \vec{v}_2$ *is the geometric vector having as its initial point the initial point of* $\vec{v}_1$ *(or an equivalent geometric vector) and as its terminal point the terminal point of* $\vec{v}_2$ *(or an equivalent geometric vector), where the terminal point of* $\vec{v}_1$ *is the initial point of* $\vec{v}_2$.

*The geometric vector* $\vec{v}_1 + \vec{v}_2$ *is called the **sum** (or **resultant**) of* $\vec{v}_1$ *and* $\vec{v}_2$.

From Definitions 6.5 and 6.6 it follows that for $\vec{v}_2'$ equivalent to $\vec{v}_2$, the sum $\vec{v}_1 + \vec{v}_2 (= \vec{v}_1 + \vec{v}_2')$ is the geometric vector shown in Figure 6.10a. Since it is customary to use the same symbol for all equivalent geometric vectors, the sum is simply designated by $\vec{v}_1 + \vec{v}_2$. Because a triangle is formed by $\vec{v}_1$, $\vec{v}_2$, and $\vec{v}_1 + \vec{v}_2$, we sometimes say that geometric vectors are added

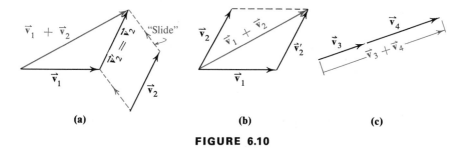

(a)          (b)          (c)

**FIGURE 6.10**

according to "the triangle law." From the properties of a parallelogram we have that if $\vec{v}_1$ and $\vec{v}_2$ have the same initial point (Figure 6.10b), the sum $\vec{v}_1 + \vec{v}_2$ is a diagonal of the parallelogram with adjacent sides $\vec{v}_1$ and $\vec{v}_2$. Therefore, we sometimes also say that geometric vectors are added according to "the parallelogram law." Figure 6.10c shows the sum $\vec{v}_3 + \vec{v}_4$, if $\vec{v}_3$ and $\vec{v}_4$ are collinear.

Notice that the symbol $+$ in a vector sum is being used in a different sense than it is ordinarily used to pair two real numbers in a sum, although the operation is commonly re- ferred to as the addition of geometric vectors.

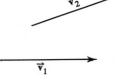

***Examples.*** Consider geometric vectors $\vec{v}_1$ and $\vec{v}_2$ in the figure. Construct the sum $\vec{v}_1 + \vec{v}_2$ in two ways.

*Solution.* (a) "Slide" the initial point of $\vec{v}_2$ onto the terminal point of $\vec{v}_1$. The sum $\vec{v}_1 + \vec{v}_2$ is the geometric vector from the initial point of $\vec{v}_1$ to the terminal point of $\vec{v}_2$.

    (b) "Slide" the initial point of $\vec{v}_2$ onto the initial point of $\vec{v}_1$ and form a parallelogram. The sum $\vec{v}_1 + \vec{v}_2$ is the diagonal shown.

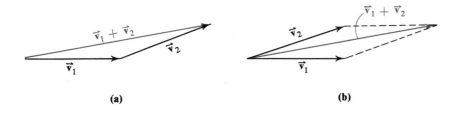

(a)                             (b)

*For all geometric vectors* $\vec{v}$ *and* $c$ *any real number,* $c\vec{v}$ *is a geometric vector collinear to* $\vec{v}$ *with magnitude multiplied by a factor of* $c$.

For $c > 0$, $c\vec{v}$ is in the same direction as $\vec{v}$. Note that if $0 < c < 1$, then $\|c\vec{v}\| < \|\vec{v}\|$; if $c < 0$, then $c\vec{v}$ is in the opposite direction to $\vec{v}$ (Figure 6.11).

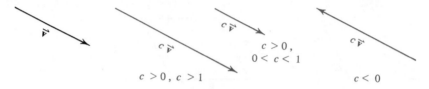

**FIGURE 6.11**

**Definition 6.8.** *The* **zero geometric vector** $\vec{0}$ *has magnitude equal to zero and any direction may be assigned to it.*

The above definitions determine the following properties for geometric vectors.

**Theorem 6.3.** *If* $\vec{v}_1, \vec{v}_2, \vec{v}_3$ *are geometric vectors and* $c, d \in R$, *then*

    I. $\vec{v}_1 + \vec{v}_2$ *is a geometric vector,*
    II. $\vec{v}_1 + \vec{v}_2 = \vec{v}_2 + \vec{v}_1$,
    III. $(\vec{v}_1 + \vec{v}_2) + \vec{v}_3 = \vec{v}_1 + (\vec{v}_2 + \vec{v}_3)$,
    IV. $\vec{v}_1 + \vec{0} = \vec{v}_1$,
    V. $\vec{v}_1 + (-\vec{v}_1) = \vec{0}$,
    VI. $(c + d)\vec{v}_1 = c\vec{v}_1 + d\vec{v}_1$,
    VII. $c(\vec{v}_1 + \vec{v}_2) = c\vec{v}_1 + c\vec{v}_2$.

We shall prove Parts I, II, and III of Theorem 6.3 and leave the proof for parts IV to VII as exercises.

*Proof of I.*   Since any two points in a plane can be joined by a line segment, it follows from Definition 6.1 that for any $\vec{v}_1$ and $\vec{v}_2$, $\vec{v}_1 + \vec{v}_2$ is a geometric vector (see Figure 6.12).

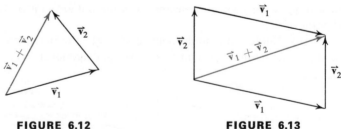

**FIGURE 6.12**                    **FIGURE 6.13**

*Proof of II.* Since the opposite sides of a parallelogram have the same magnitude and are parallel, it follows from Definitions 6.5 and 6.6 (see Figure 6.13) that $\vec{v}_1 + \vec{v}_2 = \vec{v}_2 + \vec{v}_1$.

*Proof of III.* From Definition 6.6 (see Figure 6.14) it follows directly that $\vec{v}_1 + (\vec{v}_2 + \vec{v}_3) = (\vec{v}_1 + \vec{v}_2) + \vec{v}_3$.

The difference of two geometric vectors $\vec{v}_2 - \vec{v}_1$ is defined somewhat analogously to the difference of two real numbers. Recall (see page 7) that if $a, b \in R$, then the difference $a - b$ can be viewed in two ways. First, $a - b$ is defined to be a number such that

$$b + (a - b) = a.$$

For example, $7 - 4$ (also written as 3) is a number such that

$$4 + (7 - 4) = 7.$$

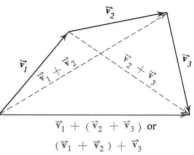

$\vec{v}_1 + (\vec{v}_2 + \vec{v}_3)$ or
$(\vec{v}_1 + \vec{v}_2) + \vec{v}_3$

**FIGURE 6.14**

Secondly, for $a, b \in R$, it can be shown that

$$a - b = a + (-b).$$

Thus, in the preceding example,

$$7 - 4 = 7 + (-4) = 3.$$

Similarly, the difference of two geometric vectors can be defined in either of the following two ways, which are shown to be consistent with each other.

**Definition 6.9.** *For all geometric vectors $\vec{v}_1$ and $\vec{v}_2$, $\vec{v}_2 - \vec{v}_1$ is a vector such that*

    **I.** $\vec{v}_1 + (\vec{v}_2 - \vec{v}_1) = \vec{v}_2$.

    **II.** $\vec{v}_2 - \vec{v}_1 = \vec{v}_2 + (-\vec{v}_1)$.

*The vector, $\vec{v}_2 - \vec{v}_1$, is called a **difference**.*

The construction of $\vec{v}_2 - \vec{v}_1$ is shown in two ways in Figure 6.15. Observe that the geometric vector $\vec{v}_2 - \vec{v}_1$ in Figure 6.15a is equivalent to $\vec{v}_2 - \vec{v}_1$ in Figure 6.15b. We could "slide" one onto the other.

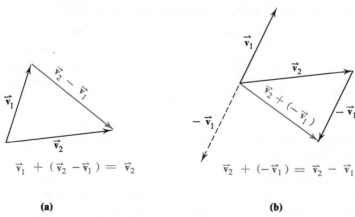

$\vec{v}_1 + (\vec{v}_2 - \vec{v}_1) = \vec{v}_2$         $\vec{v}_2 + (-\vec{v}_1) = \vec{v}_2 - \vec{v}_1$

    **(a)**                                 **(b)**

**FIGURE 6.15**

## EXERCISE 6.4

**A.**

In Problems 1 to 40, use the following geometric vectors:

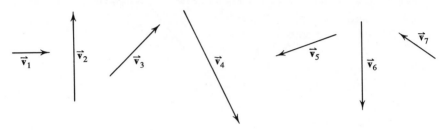

If $\|\vec{v}_1\|$ is 1, estimate the magnitude and the measure of the direction angle for each of the following geometric vectors where $\vec{v}_1$ is in the direction of the positive $x$ axis.

| | | |
|---|---|---|
| **1.** $\vec{v}_2$. | **3.** $\vec{v}_4$. | **5.** $\vec{v}_6$. |
| **2.** $\vec{v}_3$. | **4.** $\vec{v}_5$. | **6.** $\vec{v}_7$. |

Using "free-hand" methods, construct a single vector representing each of the following sums.

**Example.** $\vec{v}_3 + \vec{v}_5$.

**Solution.** $\vec{v}_3 + \vec{v}_5$

| | | |
|---|---|---|
| **■7.** $\vec{v}_1 + \vec{v}_2$. | **■11.** $\vec{v}_1 + \vec{v}_6$. | **■15.** $\vec{v}_2 + \vec{v}_5 + \vec{v}_6$. |
| **8.** $\vec{v}_1 + \vec{v}_3$. | **12.** $\vec{v}_1 + \vec{v}_7$. | **16.** $\vec{v}_2 + \vec{v}_6 + \vec{v}_7$. |
| **9.** $\vec{v}_1 + \vec{v}_4$. | **13.** $\vec{v}_2 + \vec{v}_3$. | **17.** $\vec{v}_2 + \vec{v}_3 + \vec{v}_5$. |
| **10.** $\vec{v}_1 + \vec{v}_5$. | **14.** $\vec{v}_2 + \vec{v}_4$. | **18.** $\vec{v}_3 + \vec{v}_4 + \vec{v}_7$. |

Using "free-hand" methods, construct each of the following.

| | | |
|---|---|---|
| **■19.** $-\vec{v}_1$. | **24.** $-\vec{v}_6$. | **29.** $\frac{1}{2}\vec{v}_4 + \vec{v}_7$. |
| **20.** $-\vec{v}_3$. | **25.** $\frac{1}{2}\vec{v}_3$. | **30.** $\vec{v}_2 + \frac{3}{4}\vec{v}_3$. |
| **21.** $2\vec{v}_7$. | **26.** $-\frac{1}{3}\vec{v}_4$. | **■31.** $2\vec{v}_3 + (-3\vec{v}_1)$. |
| **22.** $4\vec{v}_1$. | **■27.** $2\vec{v}_1 + 3\vec{v}_5$. | **32.** $\vec{v}_7 + (-2\vec{v}_2)$. |
| **■23.** $-2\vec{v}_5$. | **28.** $4\vec{v}_7 + 2\vec{v}_6$. | |

**B.**

By appropriate "free-hand" sketches show that each statement in Problems 33 to 40 is true.

| | |
|---|---|
| **33.** $\vec{v}_2 + \vec{v}_3 = \vec{v}_3 + \vec{v}_2$. | **37.** $2(\vec{v}_3 + \vec{v}_5) = 2\vec{v}_3 + 2\vec{v}_5$. |
| **34.** $\vec{v}_5 + \vec{v}_4 = \vec{v}_4 + \vec{v}_5$. | **38.** $(2 + 3)\vec{v}_7 = 2\vec{v}_7 + 3\vec{v}_7$. |
| **■35.** $(\vec{v}_6 + \vec{v}_7) + \vec{v}_2 = \vec{v}_6 + (\vec{v}_7 + \vec{v}_2)$. | **■39.** $\frac{1}{3}\vec{v}_1 + 2\vec{v}_1 = \frac{7}{3}\vec{v}_1$. |
| **36.** $\vec{v}_3 + (\vec{v}_7 + \vec{v}_5) = (\vec{v}_3 + \vec{v}_7) + \vec{v}_5$. | **40.** $\vec{v}_2 + (-2\vec{v}_2) = -\vec{v}_2$. |

## C.

In Problems 41 to 44, copy the diagram and sketch the geometric vectors $c_1\vec{v}_1$ and $c_2\vec{v}_2$, so that $c_1\vec{v}_1 + c_2\vec{v}_2 = \vec{v}$. By inspection, approximate the values of $c_1$, $c_2$, and $\|\vec{v}\|$ if $\|\vec{v}_1\| = \|\vec{v}_2\| = 1$.

*Example.*

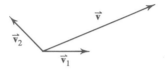

*Solution.* Extend $\vec{v}_1$ and $\vec{v}_2$, and form the parallelogram in the figure for which $\vec{v}$ is the diagonal.

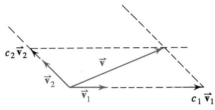

Thus,

$$c_1\vec{v}_1 + c_2\vec{v}_2 = \vec{v}.$$

By inspection,

$$c_1 \approx 4, \qquad c_2 \approx \tfrac{3}{2}, \qquad \text{and} \qquad \|\vec{v}\| \approx 3.$$

**41.**

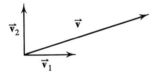

**43.**

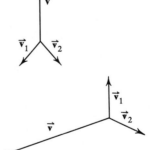

**42.**

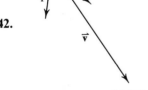

**44.**

**45.** Prove Theorem 6.3-IV.
**46.** Prove Theorem 6.3-V.

**47.** Prove Theorem 6.3-VI.
**48.** Prove Theorem 6.3-VII.

# 6.5
## Elementary Applications of Vectors

Many kinds of practical problems involving such quantities as velocities, accelerations, forces, and directed line segments can be solved by the use of geometric vectors.

First, consider the *projections* of a geometric vector on the $x$ and $y$ axes of a Cartesian coordinate system. Figure 6.16 depicts a geometric vector $\vec{v}$ and perpendicular lines drawn to the $x$ and $y$ axes from the end points $P_1$ and $P_2$ of $\vec{v}$. The geometric vector projections, $\vec{v}_x$ and $\vec{v}_y$ of $\vec{v}$ on the $x$ and $y$ axes,

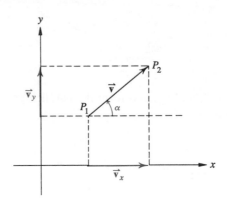

**FIGURE 6.16**

respectively, are called the **rectangular components** of $\vec{v}$; $\vec{v}_x$ is called the **horizontal component** of $\vec{v}$ and $\vec{v}_y$ is called the **vertical component.** These components are determined by the magnitude of $\vec{v}$ and its direction. Recall that the direction of $\vec{v}$ can be specified by the direction angle $\alpha$, such that $-180° < \alpha \le 180°$ (see Definition 6.3).

**Theorem 6.4.**   *If $\vec{v}_x$ and $\vec{v}_y$ are the geometric vector projections of $\vec{v}$ on the $x$ and $y$ axes, respectively, and $\alpha$ is the direction angle of $\vec{v}$, then*

$$\|\vec{v}_x\| = \|\vec{v}\| \, |\cos \alpha|$$

*and*

$$\|\vec{v}_y\| = \|\vec{v}\| \, |\sin \alpha|.$$

*Proof.*   In Figure 6.17, $P_1P_3$ is drawn parallel to the $x$ axis and $P_2P_3$ is drawn parallel to the $y$ axis.
Because

$$|\cos \alpha| = \frac{\|\vec{v}_x\|}{\|\vec{v}\|},$$

$$\|\vec{v}_x\| = \|\vec{v}\| \, |\cos \alpha|. \quad \textbf{(1)}$$

Also because

$$|\sin \alpha| = \frac{\|\vec{v}_y\|}{\|\vec{v}\|},$$

$$\|\vec{v}_y\| = \|\vec{v}\| \, |\sin \alpha|. \quad \textbf{(2)}$$

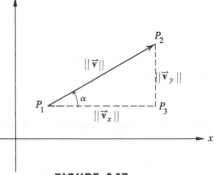

**FIGURE 6.17**

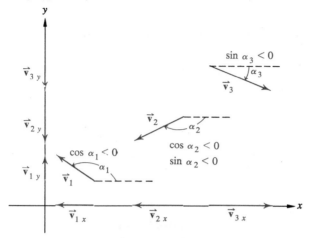

**FIGURE 6.18**

The possible directions of a vector for which $\sin \alpha < 0$ or $\cos \alpha < 0$ as shown in Figure 6.18, illustrates why the absolute values of $\cos \alpha$ and $\sin \alpha$ were specified in Equations (1) and (2), respectively, if $\|\vec{v}_x\|$ and $\|\vec{v}_y\|$ are to be positive real numbers.

If the geometric vectors $\vec{v}_x$ and $\vec{v}_y$ are known, the magnitude $\|\vec{v}\|$ and direction angle $\alpha$ of $\vec{v}$ can be determined from the relationships in a right triangle (see Figure 6.17):

$$\|\vec{v}\| = \sqrt{\|\vec{v}_x\|^2 + \|\vec{v}_y\|^2},$$

and

$$|\cos \alpha| = \frac{\|\vec{v}_x\|}{\|\vec{v}\|} \quad \text{or} \quad |\sin \alpha| = \frac{\|\vec{v}_y\|}{\|\vec{v}\|}.$$

*Example.* The horizontal component, $\vec{v}_x$, has its initial point at $(4, 0)$ with $\|\vec{v}_x\| = 6$. The vertical component $\vec{v}_y$ has its initial point at $(0, 3)$ with $\|\vec{v}_y\| = 8$. Find $\alpha$ and $\|\vec{v}\|$ and sketch $\vec{v}$.

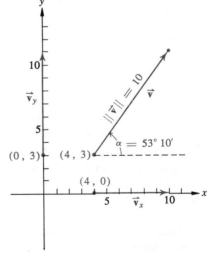

*Solution.* The magnitude,

$$\|\vec{v}\| = \sqrt{\|\vec{v}_x\|^2 + \|\vec{v}_y\|^2}$$

$$= \sqrt{6^2 + 8^2} = \sqrt{100} = 10.$$

Because

$$\cos \alpha = \frac{6}{10} = 0.6000,$$

from Table IV, $\alpha \approx 53° 10'$ to the nearest ten minutes. Thus, the geometric

*(Solution continued on the next page.)*

vector $\vec{v}$ has its initial point at (4, 3), has a direction angle whose measure is approximately 53° 10′, and has a magnitude of 10 units.

Consider two forces $F_1$ and $F_2$ acting on a particle at the point $O$, as in Figure 6.19. Since a force has magnitude and direction, we can represent it

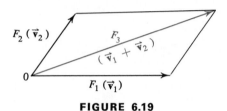

**FIGURE 6.19**

with a geometric vector, with magnitude given in terms of the units of force and with direction angle given by the direction of the force. It is shown in physics that the two forces represented by $\vec{v}_1$ and $\vec{v}_2$ can be replaced by a single equivalent force $F_3$ associated with the vector $\vec{v}_1 + \vec{v}_2$, called the resultant of the forces represented by $\vec{v}_1$ and $\vec{v}_2$. The resultant $\vec{v}_1 + \vec{v}_2$ can be obtained, as discussed in Section 6.4, by constructing the parallelogram with adjacent sides, $\vec{v}_1$ and $\vec{v}_2$, and drawing the diagonal $\vec{v}_1 + \vec{v}_2$ which is then associated with the resultant force $F_3$.

*Example.* Two forces, the first one of 5 pounds and the second of 12 pounds, act on a body at right angles to each other. Find the magnitude of the resultant force and the measure of the angle which the resultant force makes with the 5 pound force.

*Solution.* Because

$$\|\vec{v}_1\| = 5 \qquad \text{and} \qquad \|\vec{v}_2\| = 12,$$

$$\|\vec{v}\| = \sqrt{5^2 + 12^2} = \sqrt{169} = 13$$

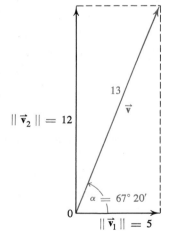

and $\cos \alpha = 5/13 \approx 0.3845$. From Table IV, $\alpha = 67° 20′$ to the nearest 10 minutes. Thus, the magnitude of the resultant force is 13 pounds, and the angle it makes with the 5 pound force has a measure of approximately 67° 20′.

In cases where the measure of an angle between two forces is not 90°, it is sometimes convenient to use the Law of Cosines and the Law of Sines in order to find the resultant force.

*Example.*   A force of 5 pounds and another force of 8 pounds act on an object at an angle of 60° with respect to each other. Find the magnitude of the resultant force and the angle it forms with respect to the 8-pound force.

*Solution.*   The figure shows the forces of 5 pounds and 8 pounds represented by the geometric vectors $\vec{v}_1$ and $\vec{v}_2$, respectively, and the resultant $\vec{v}_3 = \vec{v}_1 + \vec{v}_2$. Since $\vec{v}_2'$ is equivalent to $\vec{v}_2$, and since the adjacent angles of a parallelogram are supplementary, $m°\angle ABC = 180° - 60° = 120°$ and it follows from the Law of Cosines that

$$\|\vec{v}_3\|^2 = \|\vec{v}_1\|^2 + \|\vec{v}_2\|^2 - 2\|\vec{v}_1\| \cdot \|\vec{v}_2\| \cos 120°$$

$$= 5^2 + 8^2 - 2(5)(8)(-\cos 60°)$$

$$= 25 + 64 + 80(0.5000) = 129.$$

Therefore, $\|\vec{v}_3\| = \sqrt{129} \approx 11.4$. From the Law of Sines,

$$\frac{\|\vec{v}_1\|}{\sin \alpha} = \frac{\|\vec{v}_3\|}{\sin 120°},$$

or equivalently,

$$\sin \alpha = \frac{\|\vec{v}_1\| \sin 120°}{\|\vec{v}_3\|}$$

$$\approx \frac{5(0.8660)}{11.4} \approx 0.3798.$$

From Table IV, correct to the nearest ten minutes, $\alpha \approx 22° \, 20'$. Thus, the two forces of 5 and 8 pounds, acting at an angle of 60° with respect to each other can be represented by the single resultant force of 11.4 pounds. The angle it makes with the 8-pound force has a measure of approximately $22° \, 20'$.

Geometric vectors also can facilitate the solution of problems involving velocities and accelerations. As examples, we shall look at several air navigation problems. First, however we consider some terminology that is commonly used. The **heading** of an airplane is the direction in which it is pointed and the **course** is the direction in which it is actually flying over the ground. Its **air speed** is the speed relative to the air and its **ground speed** is the speed relative to the ground. The two directions and two speeds may differ due to wind effect.

If we use one geometric vector, say $\vec{v}_1$, to represent the heading and air speed of an airplane (see Figure 6.20, page 196) and another, say $\vec{v}_2$, to represent the wind direction and speed, then $\vec{v}_3 = \vec{v}_1 + \vec{v}_2$ represents the course and ground speed. The angle $\alpha$ between these geometric vectors $\vec{v}_1$ and $\vec{v}_3$, is called the **drift angle.** The measure of the angles for the heading, wind, and course directions in air navigation are generally given in relation to

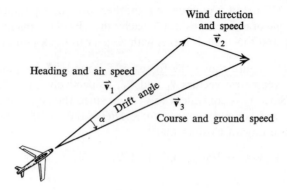

**FIGURE 6.20**

the Earth's meridian measured clockwise from the ray directed toward true
north. This angle is called a **bearing**. In this case, a positive number is
assigned as the measure of an angle formed by a rotation in a clockwise direction. In graphical representations, we shall use the positive $y$ axis as the true
north direction, and the positive $x$ axis as the true east direction. For example,
in Figure 6.21, the bearing of the geometric vector $\vec{v}_1$ is 120°, that of $\vec{v}_2$ is
285°, and that of $\vec{v}_3$ is 354°.*

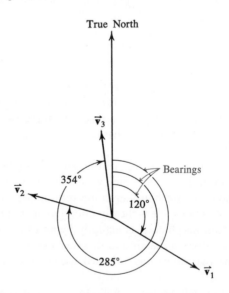

**FIGURE 6.21**

* Other types of bearings, each of which has its own reference, or initial, ray from which
the angle is measured, are also defined. A bearing used in surveying is designated by an
acute angle measured to the east or west of a ray pointing towards true north or true south
as N 35° E, N 38° W, S 22° E, and S 47° W.

***Example.*** An airplane is headed northeast (bearing 45°) with an air speed of 500 miles per hour, with a wind blowing from the southeast (bearing 315°) at a speed of 75 miles per hour. Find the drift angle, the ground speed, and the course of the plane.

*Solution.* The figure shows the geometric vector $\vec{v}_1$ representing the heading and air speed, $\vec{v}_2$ representing the wind direction and speed, and $\vec{v}_3$ representing the course and ground speed. From geometry we have that $m°\angle ABD = 180° - 45° = 135°$ and $m°\angle CBD = 360° - 315° = 45°$. Therefore $m°\angle ABC = 135° - 45° = 90°$.

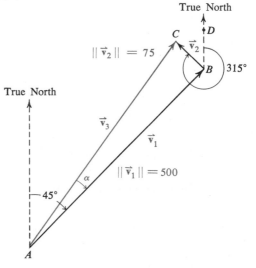

Because $\angle ABC$ is a right angle,

$$\tan \alpha = \frac{75}{500} = 0.1500,$$

and, from Table IV, the drift angle $\alpha \approx 8° 30'$ to the nearest ten minutes. The bearing, or direction, of the course is

$$45° - \alpha = 45° - 8° 30'$$
$$= 36° 30'.$$

Now, because $\sec \alpha = \dfrac{\|\vec{v}_3\|}{500}$,

$$\|\vec{v}_3\| = 500 \sec 8° 30'$$
$$\approx 500 \,(1.011) \approx 506.$$

Thus, the ground speed (magnitude of the course vector) is approximately 506 miles per hour.

### EXERCISE 6.5

**A.**

In Problems 1 to 4, the direction angle $\alpha$ and the magnitude of $\vec{v}$ are given. Find $\|\vec{v}_x\|$ and $\|\vec{v}_y\|$.

***Example.*** $\alpha = 60°$, $\|\vec{v}\| = 8$.

*Solution.* The figure shows the geometric vector $\vec{v}$ and its rectangular components, $\vec{v}_x$ and $\vec{v}_y$. From Theorem 6.4,

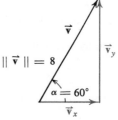

$$\|\vec{v}_x\| = \|\vec{v}\| \,|\cos 60°| = 8(\tfrac{1}{2}) = 4,$$

and

$$\|\vec{v}_y\| = \|\vec{v}\| \,|\sin 60°| = 8(\sqrt{3}/2) = 4\sqrt{3}.$$

**1.** $\alpha = 45°$, $\|\vec{v}\| = 5\sqrt{2}$.

**2.** $\alpha = 30°$, $\|\vec{v}\| = 6$.

**3.** $\alpha = 90°$, $\|\vec{v}\| = 15$.

**4.** $\alpha = 120°$, $\|\vec{v}\| = 2$.

In Problems 5 to 8, the initial points and magnitude of the horizontal and vertical components of a vector $\vec{v}$ are given. Find the magnitude $\|\vec{v}\|$, the direction angle $\alpha$, and sketch $\vec{v}$.

**Example.**

$\vec{v}_x$ :  initial point at $(1, 0)$, $\|\vec{v}_x\| = \sqrt{3}$.
$\vec{v}_y$ :  initial point at $(0, 2)$, $\|\vec{v}_y\| = 1$.

*Solution.*

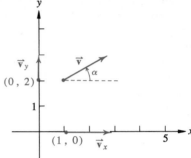

$$\|\vec{v}\| = \sqrt{(\sqrt{3})^2 + (1)^2} = \sqrt{4} = 2.$$

Because $\sin \alpha = \frac{1}{2}$, $\alpha = 30°$.

**°5.** $\vec{v}_x$ :   Initial point at $(2, 0)$, $\|\vec{v}_x\| = 2$.
    $\vec{v}_y$ :   Initial point at $(0, 4)$, $\|\vec{v}_y\| = 2$.

**6.** $\vec{v}_x$ :   Initial point at $(3, 0)$, magnitude $= 1$.
    $\vec{v}_y$ :   Initial point at $(0, \frac{1}{2})$, magnitude $= \sqrt{3}$.

**7.** $\vec{v}_x$ :   Initial point at $(\frac{3}{2}, 0)$, magnitude $= 15$.
    $\vec{v}_y$ :   Initial point at $(0, 4)$, magnitude $= 8$.

**8.** $\vec{v}_x$ :   Initial point at $(\sqrt{5}, 0)$, magnitude $= 24$.
    $\vec{v}_y$ :   Initial point at $(0, \sqrt{7})$, magnitude $= 7$.

In Problems 9 to 12, the two given forces act on a point in a plane at an angle with the specified measure. Find the magnitude of the resultant force and the angles the resultant force makes with the given forces.

**Example.**   Forces of 10 and 20 pounds act on a point at an angle of 45° with respect to each other.

*Solution.*   The figure shows the given forces $\vec{v}_1$ and $\vec{v}_2$ and the resultant $\vec{v}_3 = \vec{v}_1 + \vec{v}_2$. Since $\alpha + \beta = 45°$, the measure of the angle opposite $\vec{v}_3$ is 135°, and $\cos 135° = -\cos 45°$. Thus from the Law of Cosines,

$$\|\vec{v}_3\|^2 = 10^2 + 20^2 - 2(10)(20)(-\cos 45°)$$

$$\approx 100 + 400 + 283 \approx 783,$$

and by interpolation from Table I,

$$\|\vec{v}_3\| \approx 28.0.$$

From the Law of Sines

$$\frac{\|\vec{v}_1\|}{\sin \alpha} = \frac{\|\vec{v}_3\|}{\sin 135°}$$

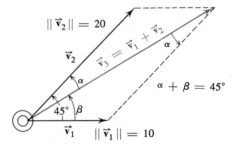

or equivalently,

$$\sin \alpha = \frac{\|\vec{v}_1\| \sin 135°}{\|\vec{v}_3\|} \approx \frac{10(0.7071)}{28.0} \approx 0.2525.$$

From Table IV, $\alpha \approx 14° \, 40'$, correct to the nearest ten minutes. If $\alpha + \beta = 45°$, then

$$\beta \approx 45° - 14° \, 40' \approx 30° \, 20'.$$

Thus, the resultant force is approximately 28.0 pounds. The angles this force makes with the force of 10 pounds and with the force of 20 pounds, have measures of approximately $30° \, 20'$ and $14° \, 40'$, respectively.

9. Forces of 3 and 4 pounds, acting at an angle of 90° with respect to each other.
10. Forces of 5 and 7 pounds, acting at an angle of 30° with respect to each other.
11. Forces of 15 and 20 pounds, acting at an angle of 75° with respect to each other.
12. Forces of 8 and 11 pounds, acting at an angle of 132° 50′ with respect to each other.
13. A crew can row a boat at a speed of 9.2 miles per hour in still water. If the crew rows directly across a river at right angles to the current, and finds its "drift angle" to be 6°, find the speed of the current.
14. A man can swim at a speed of 2 miles per hour in still water. If he swims directly across a river at right angles to a current of 5 miles per hour, find his speed in relation to the land and the direction in which he actually moves.
15. An airplane is headed southeast (bearing 135°) with an airspeed of 600 miles per hour, with the wind blowing from the northeast (bearing 225°) at a speed of 120 miles per hour. Find the drift angle, the ground speed, and the course of the airplane.
16. An airplane is headed on a bearing of 250° with an airspeed of 425 miles per hour. The course has a bearing of 262°. The ground speed is 475 miles per hour. Find the drift angle, the wind direction, and the wind speed.

**B.**

Some well-known theorems of plane geometry can be proved by using geometric vectors.

*Example.* Show that the diagonals of a parallelogram bisect each other.

*Solution.* The figure on page 200 shows a parallelogram with geometric vectors $\vec{v}_1$, $\vec{v}_2$, $\vec{v}_3$, and $\vec{v}_4$ as the four sides, so that $\vec{v}_1 = \vec{v}_3$ and $\vec{v}_2 = \vec{v}_4$. We assume the midpoint of $\overline{AC}$ is at $E$ and the midpoint of $\overline{DB}$ is at $F$. It must be shown that $E$ and $F$ are actually the same point; that is, $\overrightarrow{AE} = \overrightarrow{AF}$.

*(Solution continued on the next page.)*

Because $E$ is the midpoint of $\overrightarrow{AC}$, thus $\overrightarrow{AE} = \frac{1}{2}\overrightarrow{AC}$; because $F$ is the mid-point of $\overrightarrow{DB}$, thus $\overrightarrow{DF} = \frac{1}{2}\overrightarrow{DB}$. Because $\overrightarrow{AC} = \vec{v}_1 + \vec{v}_2$,

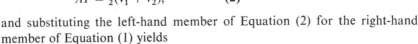

$$\overrightarrow{AE} = \frac{1}{2}(\vec{v}_1 + \vec{v}_2); \qquad (1)$$

on the other hand, $\overrightarrow{AF} = \vec{v}_4 + \frac{1}{2}\overrightarrow{DB}$. But $\overrightarrow{DB} = \vec{v}_1 - \vec{v}_4$, hence

$$\overrightarrow{AF} = \vec{v}_4 + \frac{1}{2}(\vec{v}_1 - \vec{v}_4) = \frac{1}{2}(\vec{v}_1 + \vec{v}_4).$$

Now, $\vec{v}_4 = \vec{v}_2$. Thus,

$$\overrightarrow{AF} = \frac{1}{2}(\vec{v}_1 + \vec{v}_2), \qquad (2)$$

and substituting the left-hand member of Equation (2) for the right-hand member of Equation (1) yields

$$\overrightarrow{AE} = \overrightarrow{AF}.$$

Since the vectors are equivalent, their magnitudes are equal and the proof is complete.

**17.** By using geometric vectors show that the line segment joining the mid-points of two sides of any triangle is equal in length to half the third side.

**18.** The point $P$ in the figure divides $\vec{v}_2$ in the ratio $a : b$. Express $\vec{v}_4$ in terms of $\vec{v}_1$, $\vec{v}_3$ and the numbers $a, b$.

**19.** Using the results and the figure of Problem 18, express $\vec{v}_4$ in terms of $\vec{v}_1$ and $\vec{v}_3$ if $P$ is the midpoint of $\vec{v}_2$.   *Hint: $a = b$.*

**20.** Using the result of Problem 19, show that two of the medians of an isosceles triangle are equal in length.

**21.** Using the result of Problem 19, show that the midpoint of the hypotenuse of any right tri-angle is equidistant from the three vertices.

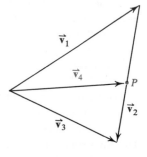

*Problem 18*

**C.**

**22.** Show that the medians of every triangle intersect at a point two-thirds of the distance from each vertex to the opposite side.

**23.** Show that a line drawn from the vertex of a parallelogram to a midpoint of an opposite side trisects a diagonal and is itself trisected by that diagonal.

## Chapter Summary

**1.** The following symbols have been introduced: $\vec{v}$, $\overrightarrow{V}$, $\|\vec{v}\|$, $\alpha$, $\vec{0}$, $\vec{v}_x$, and $\vec{v}_y$.

**2.** The six trigonometric function values of an acute angle can be expressed as ratios of the sides of a right triangle. In the right triangle in Figure 6.22:

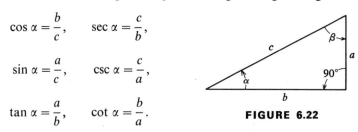

$$\cos \alpha = \frac{b}{c}, \qquad \sec \alpha = \frac{c}{b},$$

$$\sin \alpha = \frac{a}{c}, \qquad \csc \alpha = \frac{c}{a},$$

$$\tan \alpha = \frac{a}{b}, \qquad \cot \alpha = \frac{b}{a}.$$

**FIGURE 6.22**

3. The **Law of Sines**, which can be used to solve certain triangles, is stated thus:

   If $\alpha$, $\beta$, and $\gamma$ are the angles of any triangle, and if $a$, $b$, and $c$ are the lengths of the sides opposite $\alpha$, $\beta$, and $\gamma$, respectively, then

$$\frac{a}{\sin \alpha} = \frac{b}{\sin \beta} = \frac{c}{\sin \gamma}.$$

4. When solving a triangle, if the given information includes *the lengths of two sides and the measure of an angle opposite one of the two sides, certain ambiguities may arise.* Two, one, or no triangle(s) may exist.

5. The **Law of Cosines**, which can also be used to solve certain triangles, is stated thus:

   If $\alpha$, $\beta$, and $\gamma$ are the angles of any triangle, and $a$, $b$, and $c$ are the lengths of the sides opposite $\alpha$, $\beta$, and $\gamma$, respectively, then

$$c^2 = a^2 + b^2 - 2ab \cos \gamma, \tag{1}$$

$$b^2 = a^2 + c^2 - 2ac \cos \beta, \tag{2}$$

   and

$$a^2 = b^2 + c^2 - 2bc \cos \alpha. \tag{3}$$

6. If the lengths of three sides of a triangle are known, the measure of the angle $\alpha$ can be found by first rewriting Equation (3) in paragraph 5 above, equivalently, as

$$\cos \alpha = \frac{b^2 + c^2 - a^2}{2bc},$$

   solving for $\cos \alpha$, and then using the appropriate table.

7. The description of some physical quantities such as force, velocity, and acceleration, require both a reference to their *directions and magnitudes.* These can be represented graphically by a **geometric vector**, a line segment with a specified direction.

8. A geometric vector can be designated by different symbols. In general we have used the symbol $\vec{v}$. Its **norm** or **magnitude** has been designated by $\|\vec{v}\|$.

9. The **direction angle** of a geometric vector $\vec{v}$ is the angle $\alpha$, where $-180° < \alpha \le 180°$, such that:

   The initial side of $\alpha$ is the ray from the initial point of $\vec{v}$ parallel to the $x$ axis and directed in the positive $x$ direction; the terminal side of $\alpha$ is the ray from the initial point of $\vec{v}$ and containing $\vec{v}$.

10. The geometric vector $-\vec{v}$ has the same magnitude as $\vec{v}$ but is of the opposite direction.

11. Two geometric vectors are **equivalent** if they have the same magnitude and the same direction. They are **collinear** if they have the same or opposite directions regardless of their magnitudes.

12. The **sum** $\vec{v}_1 + \vec{v}_2$ is defined as the geometric vector having as its initial point the initial point of $\vec{v}_1$ and as its terminal point the terminal point of $\vec{v}_2$, and where the terminal point of $\vec{v}_1$ is the initial point of $\vec{v}_2$. When viewed in this way, geometric vectors are said to be added according to "the triangle law." Since $\vec{v}_1 + \vec{v}_2$ can be viewed as the diagonal of a parallelogram with adjacent sides $\vec{v}_1$ and $\vec{v}_2$, geometric vectors are also said to be added according to "the parallelogram law."

13. If $\vec{v}$ is any geometric vector and $c$ is any real number (scalar), then $c\vec{v}$ is a geometric vector collinear to $\vec{v}$ with magnitude multiplied by $c$.

14. The **zero geometric vector, $\vec{0}$,** has magnitude equal to zero and any direction may be assigned to it.

15. The following properties of geometric vectors follow from the definitions we have made. If $\vec{v}_1, \vec{v}_2, \vec{v}_3$ are geometric vectors and $c, d \in R$, then

   I. $\vec{v}_1 + \vec{v}_2$ is a geometric vector.        V. $\vec{v}_1 + (-\vec{v}_1) = \vec{0}$.
   II. $\vec{v}_1 + \vec{v}_2 = \vec{v}_2 + \vec{v}_1$.        VI. $(c + d)\vec{v}_1 = c\vec{v}_1 + d\vec{v}_1$.
   III. $(\vec{v}_1 + \vec{v}_2) + \vec{v}_3 = \vec{v}_1 + (\vec{v}_2 + \vec{v}_3)$.        VII. $c(\vec{v}_1 + \vec{v}_2) = c\vec{v}_1 + c\vec{v}_2$.
   IV. $\vec{v}_1 + \vec{0} = \vec{v}_1$.

16. The **difference** $\vec{v}_2 - \vec{v}_1$ of two geometric vectors is defined either by

$$\vec{v}_1 + (\vec{v}_2 - \vec{v}_1) = \vec{v}_2, \quad \text{or} \quad \vec{v}_2 - \vec{v}_1 = \vec{v}_2 + (-\vec{v}_1).$$

17. The **geometric vector projections,** $\vec{v}_x$ and $\vec{v}_y$, of $\vec{v}$ on the $x$ and $y$ axes, respectively, are called the **rectangular components** of $\vec{v}$; $\vec{v}_x$ is called the **horizontal component** of $\vec{v}$ and $\vec{v}_y$ is called the **vertical component.**

18. If $\vec{v}_x$ and $\vec{v}_y$ are the geometric vector projections of $\vec{v}$ on the $x$ and $y$ axes, respectively, then

$$\|\vec{v}_x\| = \|\vec{v}\| \cdot |\cos \alpha|, \quad \text{and} \quad \|\vec{v}_y\| = \|\vec{v}\| \cdot |\sin \alpha|.$$

19. Geometric vectors can help to visualize and thus facilitate the solution of problems involving such quantities as force, velocity, acceleration, and directed line segments, since these quantities have both magnitude and direction.

## Chapter Review

Solve each of the right triangles and find its area. In each case, $C = 90°$.

**1.** $c = 14$, $B = 51°$.           **3.** $c = 6.7$, $a = 5.4$.
**2.** $a = 2.5$, $B = 32° 40'$.       **4.** $a = 9.1$, $A = 27° 10'$.

Solve each of the triangles in Problems 5 to 12 and find its area.

**5.** $a = 1.8$, $\beta = 51° 10'$, $\gamma = 62° 40'$.   **9.** $b = 2.3$, $a = 1.9$, $\gamma = 58° 10'$.
**6.** $c = 6.2$, $\alpha = 75° 10'$, $\gamma = 91° 20'$.  **10.** $a = 4.7$, $c = 3.8$, $\beta = 101° 30'$.
**7.** $a = 7.2$, $b = 4.3$, $B = 27° 30'$.     **11.** $a = 7$, $b = 7.2$, $c = 4.3$.
**8.** $a = 4.9$, $c = 5.6$, $C = 53° 40'$.     **12.** $a = 3.6$, $b = 4.2$, $c = 6.1$.

**13.** A force of 15.2 pounds makes an angle of 30° with the horizontal. Find the horizontal and vertical components of the force.

**14.** A force of 4.6 pounds makes an angle of 50° with the vertical. Find the horizontal and vertical components of the force.

**15.** Forces of 5 and 6 pounds act on an object at an angle of 80° with respect to each other. Find the magnitude of the resultant force and the angles the resultant force makes with respect to the given forces.

**16.** An airplane has a heading of 280° with an airspeed of 450 miles per hour. The wind is blowing from a direction of 220° at a speed of 50 miles per hour. Find the drift angle, the ground speed, and the course of the airplane.

# 7

# Vectors

**I**N CHAPTER 6 you studied geometric vectors and then used them in some elementary applications. In this chapter you will be introduced to the more general notion of a vector and some of its properties. You will see the close relationship that exists between a two-dimensional vector and a geometric vector in a plane.

## 7.1
### Vectors as Ordered Pairs

Recall from Section 6.4 (page 186) that a geometric vector in a plane with its initial point at a fixed point is called a bound geometric vector. Thus, the set of all geometric vectors equivalent to a given geometric vector can be associated with a unique bound geometric vector with *initial point at the origin*. The coordinates of the terminal point of such a bound geometric vector are the components of a unique ordered pair $(a, b)$ of real numbers (see Figure 7.1) and every ordered pair $(a, b)$ corresponds to the terminal point of such a geometric vector and, in fact, can be associated with any two-dimensional geometric vector equivalent to it. This relationship suggests the following definition.

**Definition 7.1.** *A **two-dimensional vector** $\mathbf{v}$ is an ordered pair of real numbers. That is, for all $a, b \in R$,*

$$\mathbf{v} = (a, b).$$

*The set of all such vectors is designated by* **V**.

Notice that boldface type without an arrow above the symbol is being used to differentiate *a vector* **v**, *an ordered pair of real numbers, from a geometric vector* $\vec{v}$, *a directed line segment.* Since it is difficult to show boldface in script-writing, an alternative symbol should be devised to represent a nongeometric

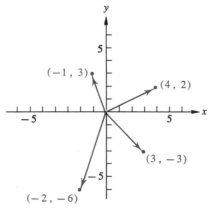

**FIGURE 7.1**

vector. One such device suggested is to use a symbol such as $\underset{\sim}{v}$ in which a wavy line is included to differentiate a vector $\underset{\sim}{v} \in V$ from the real number $v \in R$. It should be remembered that **v** itself may be regarded as a variable (vector). Although our discussion will be limited to two-dimensional vectors many of the notions concerning such vectors are also applicable to three-dimensional vectors, ordered triples $(a, b, c)$ of real numbers, four-dimensional vectors, and in general, $n$-dimensional vectors, ordered $n$-tuples $(a, b, c, d, \ldots, n)$. Of course, while three-dimensional vectors can be related to geometric vectors and visualized in a three-dimensional coordinate system, such a visualization for vectors of higher dimensions is difficult.

The following definitions for vectors correspond to similar definitions for bound geometric vectors.

**Definition 7.2.** *For all* $v_1 = (a_1, b_1)$ *and* $v_2 = (a_2, b_2)$, *where* $v_1, v_2 \in V$,

$$v_1 = v_2 \quad \text{if and only if} \quad a_1 = a_2 \text{ and } b_1 = b_2.$$

Notice that this definition refers to equal vectors and not to the notion of equivalence defined for geometric vectors.

**Definition 7.3.** *For all* $v = (a, b), v \in V$,

$$-v = -(a, b) = (-a, -b).$$

**Definition 7.4.** *The zero vector is given by*

$$0 = (0, 0).$$

**Definition 7.5.**   *For all* $\mathbf{v} = (a, b)$, $\mathbf{v} \in V$, *the **norm** or **magnitude** of* $\mathbf{v}$ *is given by*

$$\|\mathbf{v}\| = \|(a, b)\| = \sqrt{a^2 + b^2}.$$

**Definition 7.6.**   *For all* $\mathbf{v} = (a, b)$, $\mathbf{v} \in V$ *with* $\|\mathbf{v}\| \neq 0$, *the **direction angle** of* $\mathbf{v}$ *is the angle* $\alpha$ *such that*

$$\cos \alpha = \frac{a}{\|\mathbf{v}\|}, \qquad \sin \alpha = \frac{b}{\|\mathbf{v}\|},$$

*where* $-180° < \alpha \leq 180°$.

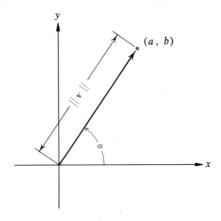

Thus the norm of a vector $\mathbf{v}$ is the length of the associated geometric vector $\vec{\mathbf{v}}$ and its direction angle is the angle $\alpha$, $-180° < \alpha \leq 180°$, measured from the positive $x$ axis to the geometric vector corresponding to $\mathbf{v}$ (see Figure 7.2).

**FIGURE 7.2**

***Example.***   Find the norm and the direction angle of $\mathbf{v} = (2, -5)$.

*Solution.*   The norm

$$\|\mathbf{v}\| = \sqrt{(2)^2 + (-5)^2} = \sqrt{29}.$$

Because $P(2, -5)$ is in the fourth quadrant, the direction angle $\alpha$ is given by

$$\sin \alpha = \frac{-5}{\sqrt{29}} \approx -0.928, \qquad -90° < \alpha < 0°.$$

To the nearest ten minutes, $\alpha = -68° \, 10'$.

The following operations on vectors are related to the corresponding operations on geometric vectors.

**Definition 7.7.**   *For all* $\mathbf{v}_1 = (a_1, b_1)$ *and* $\mathbf{v}_2 = (a_2, b_2)$, $\mathbf{v}_1, \mathbf{v}_2 \in V$,

$$\mathbf{v}_1 + \mathbf{v}_2 = (a_1, b_1) + (a_2, b_2) = (a_1 + a_2, b_1 + b_2).$$

*This operation is called **vector addition**;* $\mathbf{v}_1 + \mathbf{v}_2$ *is called the **sum** of* $\mathbf{v}_1$ *and* $\mathbf{v}_2$.

**Definition 7.8.**   *For all* $\mathbf{v}_1 = (a_1, b_1)$ *and* $\mathbf{v}_2 = (a_2, b_2)$, $\mathbf{v}_1, \mathbf{v}_2 \in V$,

$$\mathbf{v}_1 - \mathbf{v}_2 = \mathbf{v}_1 + (-\mathbf{v}_2).$$

*This operation is called **vector subtraction**;* $\mathbf{v}_1 - \mathbf{v}_2$ *is called the **difference** of* $\mathbf{v}_2$ *subtracted from* $\mathbf{v}_1$.

***Examples.***   If $\mathbf{v}_1 = (2, 3)$ and $\mathbf{v}_2 = (5, -6)$, then

(a)   $\mathbf{v}_1 + \mathbf{v}_2 = (2, 3) + (5, -6) = [2 + 5, 3 + (-6)] = (7, -3)$,

(b)   $\mathbf{v}_1 - \mathbf{v}_2 = (2, 3) + (-5, 6) = [2 + (-5), 3 + 6] = (-3, 9)$.

The corresponding geometric vectors are shown in the figure.

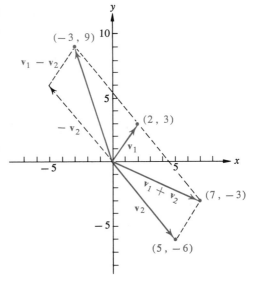

**Definition 7.9.** *For all*

$$\mathbf{v} = (a, b), \mathbf{v} \in V \text{ and } c \in R,$$

$$c\mathbf{v} = c(a, b) = (ca, cb).$$

*This operation is called **multiplication of a vector by a scalar.***

**Definition 7.10.** *For all*

$$\mathbf{v} \in V \text{ and } c \in R, (c \neq 0),$$

$$\frac{\mathbf{v}}{c} = \frac{1}{c} \mathbf{v}.$$

**Example.** If $\mathbf{v} = (2, 4)$, $c_1 = 2$, and $c_2 = -2$, then

$$c_1 \mathbf{v} = 2(2, 4) = (4, 8),$$

$$\frac{\mathbf{v}}{c_1} = \frac{(2, 4)}{2} = \frac{1}{2}(2, 4) = (1, 2),$$

$$c_2 \mathbf{v} = -2(2, 4) = (-4, -8),$$

and

$$\frac{\mathbf{v}}{c_2} = \frac{(2, 4)}{-2} = -\frac{1}{2}(2, 4)$$

$$= (-1, -2).$$

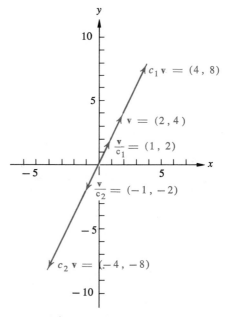

The corresponding geometric vectors are shown in the figure.

Similar to the meaning given to collinear geometric vectors, two vectors are said to be **collinear** if they have the same direction angle or their direction angles differ by 180°. Notice that the vectors in the above example are all collinear.

The difference of two vectors has now been defined in terms of a sum, and the quotient of a vector divided by a scalar has been defined in terms of a product in much the same way that the difference and quotient of two real numbers were expressed in Section 1.2, pages 7 and 8.

The following properties (similar to those for geometric vectors, page 188) of the set of vectors under the operations of *vector addition* and *multiplication of a vector by a scalar* follow from the properties of real numbers (see Appendix B, page 288) and the definitions in this section.

**Theorem 7.1.**  *If* $\mathbf{v}_1, \mathbf{v}_2, \mathbf{v}_3 \in V$, *and* $c, d \in R$, *then*

I. $\mathbf{v}_1 + \mathbf{v}_2 \in V$,      $V$ *is closed for vector addition.*

II. $\mathbf{v}_1 + \mathbf{v}_2 = \mathbf{v}_2 + \mathbf{v}_1$,      *the operation is commutative.*

III. $(\mathbf{v}_1 + \mathbf{v}_2) + \mathbf{v}_3 = \mathbf{v}_1 + (\mathbf{v}_2 + \mathbf{v}_3)$,      *the operation is associative.*

IV. $\mathbf{v}_1 + \mathbf{0} = \mathbf{v}_1$,      *there is an additive identity element.*

V. $\mathbf{v}_1 + (-\mathbf{v}_1) = \mathbf{0}$,      *each element has an additive inverse.*

VI. $(c + d)\,\mathbf{v}_1 = c\mathbf{v}_1 + d\mathbf{v}_1$,      *a scalar sum is distributive over a vector product.*

VII. $c(\mathbf{v}_1 + \mathbf{v}_2) = c\mathbf{v}_1 + c\mathbf{v}_2$,      *a scalar product is distributive over a vector sum.*

We shall only prove part IV here. The other parts of the theorem are left for you to complete in a similar way in the exercises.

*Proof of IV.*   Let $\mathbf{v}_1 = (a_1, b_1)$. Then,

$$\mathbf{v}_1 + \mathbf{0} = (a_1, b_1) + (0, 0),$$

and from Definition 7.7,

$$\mathbf{v}_1 + \mathbf{0} = (a_1 + 0, b_1 + 0) = (a_1, b_1) = \mathbf{v}_1.$$

**EXERCISE 7.1**

A.

In Problems 1 to 8,
  (a) show each corresponding geometric vector graphically.
  (b) find the norm and the direction angle (to the nearest ten minutes) of each vector.

*Example.*   $(-2, 3)$.

*Solution.*                                    (a)

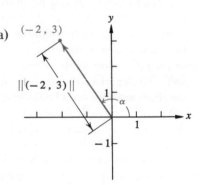

(b) The norm, $\|(-2, 3)\|$

$$= \sqrt{(-2)^2 + (3)^2} = \sqrt{13}.$$

Because $(-2, 3)$ is in the second quadrant, the direction angle $\alpha$ is given by

$$\sin \alpha = \frac{3}{\sqrt{13}} \approx 0.8319, \quad 90° < \alpha < 180°.$$

To the nearest ten minutes, $\alpha = 123° \, 40'$.

**˙1.** $(3, 4)$.            **3.** $(1, 1)$.            **˙5.** $(-3, -3)$.            **7.** $(4, 0)$.

**2.** $(-3, 4)$.            **4.** $(-1, 1)$.            **6.** $(0, -5)$.            **8.** $(\sqrt{3}, 1)$.

Given the following norms and direction angles, find the vector **v**, and show the corresponding geometric vector with initial point at the origin.

*Example.*   $\|\mathbf{v}\| = 5; \alpha = 30°$.

*Solution.*   See figure. $l(\overline{OA}) = 5 \cos 30°$;
$l(\overline{OB}) = 5 \sin 30°$. Thus,

$$\mathbf{v} = (5 \cos 30°, 5 \sin 30°) = \left(\frac{5\sqrt{3}}{2}, \frac{5}{2}\right).$$

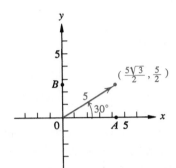

**˙9.** $\|\mathbf{v}\| = 2\sqrt{2}; \alpha = 45°$.
**10.** $\|\mathbf{v}\| = 2; \alpha = 60°$.
**11.** $\|\mathbf{v}\| = 1; \alpha = 90°$.
**12.** $\|\mathbf{v}\| = 4; \alpha = 150°$.
**˙13.** $\|\mathbf{v}\| = 5; \alpha = -150°$.
**14.** $\|\mathbf{v}\| = 3\sqrt{2}; \alpha = -30°$.

Let $\mathbf{v}_1 = (3, 6)$, $\mathbf{v}_2 = (2, -5)$, $\mathbf{v}_3 = (-4, 2)$, $c_1 = 2$, and $c_2 = -3$. Write each expression in Problems 15 to 26 as an ordered pair and show the corresponding geometric vector.

*Example.*   $\mathbf{v}_2 + \mathbf{v}_3$.

*Solution.*

$$\begin{aligned}\mathbf{v}_2 + \mathbf{v}_3 &= (2, -5) + (-4, 2) \\ &= [2 + (-4), -5 + 2] \\ &= (-2, -3).\end{aligned}$$

**15.** $\mathbf{v}_1 + \mathbf{v}_2$.          **22.** $\mathbf{v}_2 - c_1\mathbf{v}_3$.
**16.** $\mathbf{v}_1 + \mathbf{v}_3$.          **23.** $c_1\mathbf{v}_2 + c_2\mathbf{v}_3$.
**˙17.** $\mathbf{v}_1 - \mathbf{v}_3$.          **24.** $c_2\mathbf{v}_3 - c_1\mathbf{v}_1$.

**18.** $\mathbf{v}_3 - \mathbf{v}_2$.       **˙25.** $\dfrac{\mathbf{v}_1}{c_1}$.

**19.** $c_1(\mathbf{v}_1 + \mathbf{v}_2)$.
**20.** $c_2(\mathbf{v}_3 - \mathbf{v}_1)$.       **26.** $\dfrac{\mathbf{v}_2}{\|\mathbf{v}_2\|}$.
**˙21.** $c_1\mathbf{v}_1 + \mathbf{v}_2$.

**B.**

**27.** Prove Theorem 7.1-I.          **30.** Prove Theorem 7.1-V.
**28.** Prove Theorem 7.1-II.          **31.** Prove Theorem 7.1-VI.
**29.** Prove Theorem 7.1-III.          **32.** Prove Theorem 7.1-VII.

**C.**

**33.** What is the norm of the vector in Problem 26 ? Can you make a conjecture as to the method of finding a vector one unit in length collinear with any given vector?

## 7.2

### Vectors in the Form  *ai* + *bj*

It can be shown (see Exercise 7.2, Problem 35; also see page 191 for a geometric interpretation) that if $v_1$ and $v_2$ are noncollinear vectors, then for each $v_3$ there exist scalars $c_1$ and $c_2$ such that

$$v_3 = c_1 v_1 + c_2 v_2.$$

Vectors $v_1$ and $v_2$ are said to form a **basis** for the set of all two-dimensional vectors. Although $v_1$ and $v_2$ can be any two vectors, you will study a special basis in this section which is particularly useful.

**Definition 7.11.** *A vector* **v** *with norm equal to* 1 *(*$\|v\| = 1$*) is called a* **unit vector**.

Unit vectors $v_1 = (1, 0)$ and $v_2 = (0, 1)$ whose corresponding geometric vectors are in the direction of the positive $x$ and $y$ axes, respectively, form an **orthogonal** (perpendicular) **basis**. If unit vectors $v_1 = (1, 0)$ and $v_2 = (0, 1)$ are designated **i** and **j**, respectively, then each vector

$$v = (a, b) = (a, 0) + (0, b) = a(1, 0) + b(0, 1),$$

from which

$$v = ai + bj.$$

The real numbers $a$ and $b$ are called the **scalar components** of the corresponding geometric vector along the $x$ and $y$ axes, respectively. Several examples of such vectors are shown in Figure 7.3.

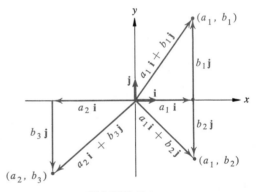

**Examples.** Write each vector as an ordered pair and sketch the corresponding geometric vector.

(a) 5i.        (b) 5i − 3j.

**FIGURE 7.3**

*Solutions.*  (a)  $5i = 5(1, 0) = (5, 0)$.

(b)  $5i - 3j = 5(1, 0) - 3(0, 1)$
$$= (5, 0) - (0, 3) = (5 + 0, 0 - 3)$$
$$= (5, -3).$$

**Example.** Find the norm and the direction angle to the nearest ten minutes of $5i - 3j$.

*Solution.*  From the preceding example, we have that $5i - 3j = (5, -3)$. Thus,

$$\|5i - 3j\| = \|(5, -3)\|,$$

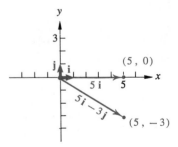

from which by Definition 7.5 and substituting appropriately,

$$\|5\mathbf{i} - 3\mathbf{j}\| = \sqrt{5^2 + (-3)^2} = \sqrt{34}.$$

From Definition 7.6,

$$\sin \alpha = \frac{-3}{\sqrt{34}} \approx \frac{-3}{5.83} \approx -0.515.$$

Because $-90° < \alpha < 0°$,

$$\alpha \approx -31° \, 00'.$$

**Examples.** Express each vector as the sum of scalar multiples of $\mathbf{i}$ and $\mathbf{j}$ and sketch the corresponding geometric vectors.
(a) $(-4, 0)$.     (b) $(-4, 3)$.

*Solutions.* (a) $(-4, 0) = -4(1, 0) = -4\mathbf{i}$.
    (b) $(-4, 3) = (-4, 0) + (0, 3)$
               $= -4(1, 0) + 3(0, 1)$
               $= -4\mathbf{i} + 3\mathbf{j}$.

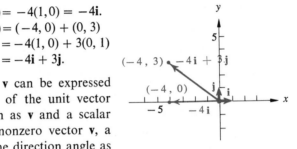

Any nonzero vector $\mathbf{v}$ can be expressed as the scalar product of the unit vector with the same direction as $\mathbf{v}$ and a scalar equal to $\|\mathbf{v}\|$. Given a nonzero vector $\mathbf{v}$, a unit vector with the same direction angle as $\mathbf{v}$ can always be found. First, consider the special case for the vector $(a, 0)$ with direction angle $0°$. We shall argue that $(a, 0)/\|(a, 0)\|$ is the unit vector of $(a, 0)$. From Definitions 7.10, 7.5, and 7.9, in turn, the quotient

$$\frac{(a, 0)}{\|(a, 0)\|} = \frac{1}{\|(a, 0)\|}(a,0) = \frac{1}{\sqrt{a^2 + 0^2}}(a, 0) = \frac{1}{a}(a, 0), \qquad (a \neq 0),$$

$$= \left(\frac{1}{a} \cdot a, \frac{1}{a} \cdot 0\right) = (1, 0).$$

Thus, the unit vector with the same direction angle as $(a, 0)$, $a \neq 0$, is given by $(a, 0)/\|(a, 0)\|$. It can be shown in a similar way that $(0, b)/\|(0, b)\|$ is the unit vector with the same direction angle as $(0, b)$, $b \neq 0$. For the general case we have the following theorem.

**Theorem 7.2.** *For any two-dimensional vector* $\mathbf{v} = (a, b)$, *the unit vector with the same direction angle as* $\mathbf{v}$ *is given by*

$$\frac{\mathbf{v}}{\|\mathbf{v}\|} = \frac{(a, b)}{\|(a, b)\|}, \qquad [(a, b) \neq (0, 0)].$$

The proof of Theorem 7.2 is left as an exercise.

*Examples.*  Find the unit vector with the same direction angle as that of each vector.

(a) (2, 3).          (b) 4i − 5j.

*Solutions.*  From Theorem 7.2, the unit vectors in each case are given by

(a)  $\dfrac{(2, 3)}{\|(2, 3)\|} = \dfrac{(2, 3)}{\sqrt{2^2 + 3^2}} = \dfrac{(2, 3)}{\sqrt{13}} = \dfrac{1}{\sqrt{13}}(2, 3) = \left(\dfrac{2}{\sqrt{13}}, \dfrac{3}{\sqrt{13}}\right).$

(b)  $\dfrac{4i - 5j}{\|4i - 5j\|} = \dfrac{4i - 5j}{\sqrt{4^2 + 5^2}} = \dfrac{4i - 5j}{\sqrt{41}} = \dfrac{1}{\sqrt{41}}(4i - 5j) = \dfrac{4}{\sqrt{41}}i - \dfrac{5}{\sqrt{41}}j.$

The components of $\mathbf{v} = (a, b)$ can be expressed in terms of its norm and direction angle:

$a = \|\mathbf{v}\| \cos \alpha$ and $b = \|\mathbf{v}\| \sin \alpha$

(see Figure 7.4). Because

$$\mathbf{v} = (a, b) = a\mathbf{i} + b\mathbf{j},$$

we have that

$$\mathbf{v} = \|\mathbf{v}\| \cos \alpha\mathbf{i} + \|\mathbf{v}\| \sin \alpha\mathbf{j}.$$

This is called the **trigonometric form** of a vector.

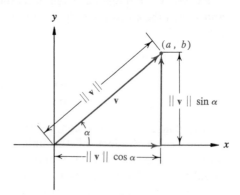

**FIGURE 7.4**

### EXERCISE 7.2

**A.**

Write each vector as an ordered pair and sketch the corresponding geometric vector.

*Examples.*  (a) 2j.          (b) 3i + 2j.

*Solutions.*  (a) $2j = 2(0, 1) = (0, 2)$.
   (b) $3i + 2j = 3(1, 0) + 2(0, 1)$
      $= (3, 0) + (0, 2)$
      $= (3, 2)$.

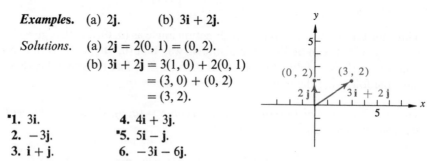

*1. 3i.
2. −3j.
3. i + j.

4. 4i + 3j.
*5. 5i − j.
6. −3i − 6j.

Express each vector as the sum of scalar multiples of **i** and **j** and sketch the corresponding geometric vectors.

*Example.* (2, 9).

*Solution.* $(2, 9) = (2, 0) + (0, 9)$
$$= 2(1, 0) + 9(0, 1)$$
$$= 2\mathbf{i} + 9\mathbf{j}.$$

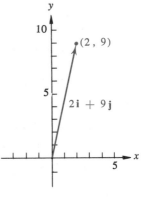

7. $(-4, 7)$.                    10. $(2, -2)$.
8. $(3, -5)$.                    11. $(3, -3\sqrt{3})$.
•9. $(\sqrt{3}, 1)$.             12. $(-2, -8)$.

Find the norm and the direction angle, to the nearest ten minutes, of each vector. Sketch the corresponding geometric vector.

*Example.* $2\mathbf{i} - 3\mathbf{j}$.

*Solution.* Because $2\mathbf{i} - 3\mathbf{j} = 2(1, 0) - 3(0, 1) = (2, 0) - (0, 3) = (2, -3)$, then, by Definition 7.5,

$$\|2\mathbf{i} - 3\mathbf{j}\| = \|(2, -3)\| = \sqrt{2^2 + (-3)^2} = \sqrt{13}.$$

From Definition 7.6,

$$\sin \alpha = \frac{-3}{\sqrt{13}} \approx \frac{-3}{3.61} \approx -0.8310.$$

Because $-90° < \alpha < 0°$,

$$\alpha \approx -56° \ 10'.$$

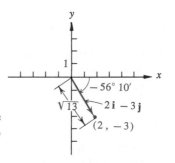

Thus, the norm is equal to $\sqrt{13}$ units and the direction angle is approximately equal to $-56° \ 10'$.

•13. $4\mathbf{i} - 4\mathbf{j}$.          16. $\mathbf{i} + 5\mathbf{j}$.          19. $3\mathbf{i} - 2\mathbf{j}$.
14. $\sqrt{3}\mathbf{i} + \mathbf{j}$.     •17. $-\mathbf{i} - 3\mathbf{j}$.         20. $-\mathbf{i} - \sqrt{3}\mathbf{j}$.
15. $2\mathbf{i} + \mathbf{j}$.            18. $-6\mathbf{i} + \mathbf{j}$.

Find the unit vector with the same direction angle as that of each vector.

*Examples.* (a) $(-6, -2)$.          (b) $9\mathbf{i} + 4\mathbf{j}$.

*Solutions.* From Theorem 7.2, the unit vector is given by:

(a) $\dfrac{(-6, -2)}{\|(-6, -2)\|} = \dfrac{(-6, -2)}{\sqrt{(-6)^2 + (-2)^2}} = \dfrac{(-6, -2)}{\sqrt{40}} = \dfrac{1}{\sqrt{40}}(-6, -2)$

$$= \left(\frac{-6}{\sqrt{40}}, \frac{-2}{\sqrt{40}}\right) = \left(\frac{-3}{\sqrt{10}}, \frac{-1}{\sqrt{10}}\right).$$

(b) $\dfrac{9\mathbf{i} + 4\mathbf{j}}{\|9\mathbf{i} + 4\mathbf{j}\|} = \dfrac{9\mathbf{i} + 4\mathbf{j}}{\sqrt{9^2 + 4^2}} = \dfrac{9\mathbf{i} + 4\mathbf{j}}{\sqrt{97}} = \dfrac{1}{\sqrt{97}}(9\mathbf{i} + 4\mathbf{j}) = \dfrac{9}{\sqrt{97}}\mathbf{i} + \dfrac{4}{\sqrt{97}}\mathbf{j}.$

**21.** $(4, 3)$.                **23.** $(7, -6)$.              **25.** $5\mathbf{i} + \sqrt{2}\mathbf{j}$.

**22.** $(-8, -15)$.            **24.** $(-9, 3)$.              **26.** $-4\mathbf{i} + \sqrt{3}\mathbf{j}$.

Resolve each vector **v** with given norm and given direction angle $\alpha$ into a linear combination of vectors **i** and **j**. Sketch the geometric vector corresponding to the given vector and its scalar components along the $x$ and $y$ axes.

*Example.* $\|\mathbf{v}\| = 2; \alpha = 30°$.

*Solution.* The scalar component of the given vector along the $x$ axis is $2 \cos 30°$, and along the $y$ axis is $2 \sin 30°$. Thus,

$$\mathbf{v} = (2 \cos 30°)\mathbf{i} + (2 \sin 30°)\mathbf{j}$$

$$= \sqrt{3}\mathbf{i} + \mathbf{j}.$$

**•27.** $\|\mathbf{v}\| = 4; \alpha = 90°$.

**28.** $\|\mathbf{v}\| = 8\sqrt{2}; \alpha = 45°$.

**29.** $\|\mathbf{v}\| = 5; \alpha = 60°$.

**30.** $\|\mathbf{v}\| = 3; \alpha = -150°$.

**•31.** $\|\mathbf{v}\| = \frac{7}{8}; \alpha = 79°$.

**32.** $\|\mathbf{v}\| = \frac{1}{2}; \alpha = 332°$.

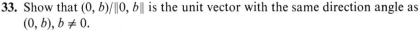

**B.**

**33.** Show that $(0, b)/\|0, b\|$ is the unit vector with the same direction angle as $(0, b), b \neq 0$.

**34.** Prove Theorem 7.2. *Hint:* Let $\mathbf{v} = (a, b)$. Find $\mathbf{v}/\|\mathbf{v}\|$. What are its norm and its direction angle?

**C.**

**35.** Show that if $\mathbf{v}_1$ and $\mathbf{v}_2$ are noncollinear vectors, then for each $\mathbf{v}_3$, there exists scalars $c_1$ and $c_2$ such that $\mathbf{v}_3 = c_1\mathbf{v}_1 + c_2\mathbf{v}_2$. *Hint:* Let $\mathbf{v}_1 = (a_1, b_1)$, $\mathbf{v}_2 = (a_2, b_2)$, and $\mathbf{v}_3 = (a_3, b_3)$. If $c_1$ and $c_2$ exist, then $(a_3, b_3) = (c_1a_1 + c_2a_2, c_1b_1 + c_2b_2)$. Then show that solutions for $c_1$ and $c_2$ in terms of $a_1, b_1, a_2$, and $b_2$ exist.

## 7.3
### Inner Products

In Section 7.1 you considered two basic operations, *vector addition* and *scalar multiplication* (Definitions 7.7 and 7.9). The result of either operation is a vector and the set of vectors **V** is said to be closed for each of these operations. Another operation on two vectors is defined in such a way as to be useful in a variety of problems related to physical and geometric concepts.

**Definition 7.12.** *For all* $\mathbf{v}_1 = (a_1, b_1)$ *and* $\mathbf{v}_2 = (a_2, b_2)$, $\mathbf{v}_1, \mathbf{v}_2 \in \mathbf{V}$,

$$\mathbf{v}_1 \cdot \mathbf{v}_2 = (a_1, b_1) \cdot (a_2, b_2) = a_1a_2 + b_1b_2. \tag{1}$$

$\mathbf{v}_1 \cdot \mathbf{v}_2$ *is called the* **inner product** *of* $\mathbf{v}_1$ *and* $\mathbf{v}_2$.

Since the symbol for the operation is a dot, the inner product is also called a **dot product**. Notice that the dot is used with a different meaning than the way it is used for the operation of multiplication in the set of real numbers.

**Examples.** Find $v_1 \cdot v_2$ for the given vectors.
(a) $v_1 = (2, 3)$, $v_2 = (-5, 4)$.          (b) $v_1 = 3i - 4j$, $v_2 = i + 2j$.

*Solutions.* (a) $v_1 \cdot v_2 = (2, 3) \cdot (-5, 4) = (2)(-5) + (3)(4) = 2$.
(b) $v_1 \cdot v_2 = (3i - 4j) \cdot (i + 2j) = (3)(1) + (-4)(2) = -5$.

Notice that the set of vectors $V$ *is not closed* for this operation. The inner product of two vectors is a *scalar*, not a vector. For this reason the inner product is also called a **scalar product**.

Another useful form of the inner product can be obtained from the Law of Cosines.

**Theorem 7.3.** *If* $v_1 = (a_1, b_1)$, $v_2 = (a_2, b_2)$, $v_1$ *and* $v_2$ *are nonzero vectors, and* $\gamma$ *is the angle between the geometric vectors corresponding to* $v_1$ *and* $v_2$, *then*

$$v_1 \cdot v_2 = \|v_1\| \, \|v_2\| \cos \gamma.$$

*Proof.* Consider the vectors $v_1 = (a_1, b_1)$ and $v_2 = (a_2, b_2)$, their corresponding geometric vectors and the line segment with length $c$ drawn between the points corresponding to $(a_1, b_1)$ and $(a_2, b_2)$, as shown in Figure 7.5. From the Law of Cosines,

$$c^2 = \|v_1\|^2 + \|v_2\|^2$$
$$- 2\|v_1\| \, \|v_2\| \cos \gamma,$$

from which, for nonzero vectors $v_1$ and $v_2$,

$$\cos \gamma = \frac{\|v_1\|^2 + \|v_2\|^2 - c^2}{2\|v_1\| \, \|v_2\|}.$$

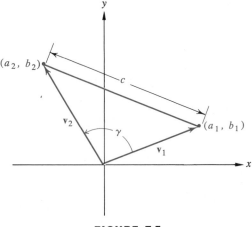

**FIGURE 7.5**

Substituting $\sqrt{a_1^2 + b_1^2}$ for $\|v_1\|$, $\sqrt{a_2^2 + b_2^2}$ for $\|v_2\|$, and $\sqrt{(a_2 - a_1)^2 + (b_2 - b_1)^2}$ for $c$ in the numerator of the right-hand member yields

$$\cos \gamma = \frac{a_1^2 + b_1^2 + a_2^2 + b_2^2 - [(a_2 - a_1)^2 + (b_2 - b_1)^2]}{2\|v_1\| \, \|v_2\|}$$

$$= \frac{a_1^2 + b_1^2 + a_2^2 + b_2^2 - a_2^2 + 2a_1a_2 - a_1^2 - b_2^2 + 2b_1b_2 - b_1^2}{2\|v_1\| \, \|v_2\|}$$

$$= \frac{2(a_1a_2 + b_1b_2)}{2\|v_1\| \, \|v_2\|}.$$

*(Proof continued on the next page.)*

Now, since $a_1a_2 + b_1b_2$ is precisely the inner product $\mathbf{v}_1 \cdot \mathbf{v}_2$, we have

$$\cos \gamma = \frac{\mathbf{v}_1 \cdot \mathbf{v}_2}{\|\mathbf{v}_1\| \, \|\mathbf{v}_2\|},$$

from which

$$\mathbf{v}_1 \cdot \mathbf{v}_2 = \|\mathbf{v}_1\| \, \|\mathbf{v}_2\| \cos \gamma. \tag{2}$$

Thus, you can now express the inner product of two vectors $\mathbf{v}_1 = (a_1, b_1)$ and $\mathbf{v}_2 = (a_2, b_2)$ either in terms of their scalar components in Equation (1), or in terms of the norms of the two vectors and the cosine of the angle $\gamma$ formed by their corresponding geometric vectors in Equation (2).

***Example.*** Find $\mathbf{v}_1 \cdot \mathbf{v}_2$ if $\|\mathbf{v}_1\| = 3$, and $\mathbf{v}_1$ has a direction angle, $\alpha_1 = 20°$; $\|\mathbf{v}_2\| = 8$, and $\mathbf{v}_2$ has a direction angle, $\alpha_2 = 170°$.

*Solution.* From Theorem 7.3,

$$\mathbf{v}_1 \cdot \mathbf{v}_2 = (3)(8) \cos (20° - 170°) = (3)(8) \cos (-150°),$$

and, since $\cos (-\alpha) = \cos \alpha$,

$$\mathbf{v}_1 \cdot \mathbf{v}_2 = 24 \cos 150° = 24 \left( -\frac{\sqrt{3}}{2} \right) = -12\sqrt{3}.$$

The property of inner products exhibited for two nonzero vectors in Equation (2) is particularly useful in several equivalent forms:

$$7.3\text{-I.} \qquad \cos \gamma = \frac{\mathbf{v}_1 \cdot \mathbf{v}_2}{\|\mathbf{v}_1\| \, \|\mathbf{v}_2\|}, \tag{2a}$$

$$7.3\text{-II.} \qquad \|\mathbf{v}_1\| \cos \gamma = \frac{\mathbf{v}_1 \cdot \mathbf{v}_2}{\|\mathbf{v}_2\|}, \tag{2b}$$

$$7.3\text{-III.} \qquad \|\mathbf{v}_2\| \cos \gamma = \frac{\mathbf{v}_1 \cdot \mathbf{v}_2}{\|\mathbf{v}_1\|}. \tag{2c}$$

Theorem 7.3-I enables you to find the angle $\gamma$ between any two nonzero vectors.

***Example.*** Find the angle $\gamma$, where $0° \le \gamma \le 180°$, between the geometric vectors corresponding to $(2, 3)$ and $(0, 2)$.

*Solution.* From Theorem 7.3-I,

$$\cos \gamma = \frac{(2, 3) \cdot (0, 2)}{\|(2, 3)\| \, \|(0, 2)\|} = \frac{2 \cdot 0 + 3 \cdot 2}{\sqrt{4 + 9} \, \sqrt{0 + 4}} = \frac{3}{\sqrt{13}} \approx 0.832.$$

Thus, from Table IV, $\gamma \approx 33° \, 40'$.

The product $\|\mathbf{v}_1\| \cos \gamma$ is called the **scalar projection of $\mathbf{v}_1$ on $\mathbf{v}_2$** and the product $\|\mathbf{v}_2\| \cos \gamma$ is called the **scalar projection of $\mathbf{v}_2$ on $\mathbf{v}_1$**. In a geometric sense, this can be interpreted as the directed length of the line segments shown in Figures 7.6a and 7.6b.

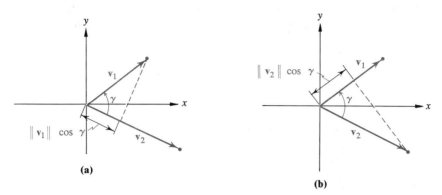

(a)

(b)

**FIGURE 7.6**

***Example.*** If $\mathbf{v}_1 = (-3, -2)$ and $\mathbf{v}_2 = (-5, 5)$, find the scalar projection of $\mathbf{v}_1$ on $\mathbf{v}_2$.

*Solution.* From Theorem 7.3-II,

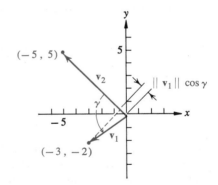

$$\|\mathbf{v}_1\| \cos \gamma = \frac{\mathbf{v}_1 \cdot \mathbf{v}_2}{\|\mathbf{v}_2\|}$$

$$= \frac{(-3, -2) \cdot (-5, 5)}{\|(-5, 5)\|}$$

$$= \frac{15 - 10}{\sqrt{5^2 + 5^2}}$$

$$= \frac{5}{\sqrt{50}} = \frac{5}{5\sqrt{2}} = \frac{1}{\sqrt{2}}.$$

Notice that if $\mathbf{v}_1$ and $\mathbf{v}_2$ are nonzero vectors, then $\|\mathbf{v}_1\| \neq 0$, $\|\mathbf{v}_2\| \neq 0$, and

$$\mathbf{v}_1 \cdot \mathbf{v}_2 = \|\mathbf{v}_1\| \, \|\mathbf{v}_2\| \cos \gamma = 0,$$

if and only if $\cos \gamma$ equals zero, or if $\gamma = \pi/2$. Since, from Equation (1),

$$\mathbf{v}_1 \cdot \mathbf{v}_2 = a_1 a_2 + b_1 b_2,$$

you can test for the orthogonality (perpendicularity) of two vectors directly from their scalar components.

**Theorem 7.4.** *The vectors* $\mathbf{v}_1 = (a_1, b_1)$ *and* $\mathbf{v}_2 = (a_2, b_2)$, $[\mathbf{v}_1, \mathbf{v}_2 \neq (0, 0)]$, *are orthogonal if and only if their inner product*

$$\mathbf{v}_1 \cdot \mathbf{v}_2 = a_1 a_2 + b_1 b_2 = 0. \tag{3}$$

The proof is established by the previous argument.

**Examples.**   Verify that the following pairs of vectors are orthogonal.
(a) $(2, -4)$ and $(6, 3)$.          (b) $-i - 2j$ and $4i - 2j$.

**Solutions.**   (a) $(2, -4) \cdot (6, 3) = (2)(6) + (-4)(3) = 12 - 12 = 0$.
          (b) $(-i - 2j) \cdot (4i - 2j) = (-1)(4) + (-2)(-2) = 0$.

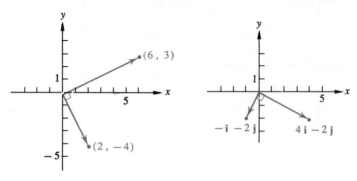

From Theorem 7.4, the vectors are orthogonal. See the figures for the corresponding geometric vectors.

The test for the orthogonality of two vectors can be used to establish a test for collinearity.

**Theorem 7.5.**   *The vectors* $v_1 = (a_1, b_1)$ *and* $v_2 = (a_2, b_2)$, $[v_1, v_2 \neq (0, 0]$, *are collinear if and only if*

$$(-b_1, a_1) \cdot (a_2, b_2) = 0.$$

**Proof.**   Because

$$(-b_1, a_1) \cdot (a_1, b_1) = -a_1 b_1 + a_1 b_1 = 0,$$

$(-b_1, a_1)$ and $(a_1, b_1)$ are orthogonal. Now, if $(-b_1, a_1) \cdot (a_2, b_2) = 0$ (the given information), then $(-b_1, a_1)$ and $(a_2, b_2)$ are orthogonal. Since $(a_1, b_1)$ and $(a_2, b_2)$ are both orthogonal to $(-b_1, a_1)$, it follows that $(a_1, b_1)$ and $(a_2, b_2)$ are collinear (they have the same direction angle). A similar argument proceeds in the reverse direction and the theorem is proved.

**Examples.**   Verify that each pair of vectors is collinear.
(a) $(3, 2)$ and $(6, 4)$.          (b) $2i - 6j$ and $3i - 9j$.

**Solutions.**   (a) $(-2, 3) \cdot (6, 4) = -12 + 12 = 0$.
          (b) $(6, 2) \cdot (3, -9) = 18 - 18 = 0$.

Therefore, by Theorem 7.5, the vectors in each pair are collinear.

### EXERCISE 7.3

A.

Find the inner product, $v_1 \cdot v_2$, for the given vectors.

***Examples.***   (a) $v_1 = (7, -9)$; $v_2 = (-6, -5)$.
           (b) $v_1 = -9i + 8j$; $v_2 = 3i - 2j$.

*Solutions.*   (a) $v_1 \cdot v_2 = (7, -9) \cdot (-6, -5) = (7)(-6) + (-9)(-5) = 3$.
           (b) $v_1 \cdot v_2 = (-9i + 8j) \cdot (3i - 2j) = (-9)(3) + (8)(-2) = -43$.

1. $v_1 = (-6, 0)$; $v_2 = (-8, 10)$.          5. $v_1 = 9i - 4j$; $v_2 = 5i$.
2. $v_1 = (-9, 3)$; $v_2 = (8, -5)$.          6. $v_1 = 8j$; $v_2 = 7i + 7j$.
3. $v_1 = (7, -8)$; $v_2 = (2, 2)$.          7. $v_1 = i - 2j$; $v_2 = 3i - 5j$.
4. $v_1 = (3, 0)$; $v_2 = (0, -5)$.          8. $v_1 = -4i + 3j$; $v_2 = 8i + 10j$.

Find $v_1 \cdot v_2$, given the norms and direction angles $\alpha_1$ and $\alpha_2$ for $v_1$ and $v_2$.

***Example.***   $\|v_1\| = 10$, $\alpha_1 = 75°$; $\|v_2\| = 3$, $\alpha_2 = 45°$.

*Solution.*   From Theorem 7.3,

$$v_1 \cdot v_2 = 10 \cdot 3 \cdot \cos (75° - 45°)$$

$$= 30 \cos 30° = 30\left(\frac{\sqrt{3}}{2}\right) = 15\sqrt{3}.$$

9. $\|v_1\| = 8$, $\alpha_1 = 60°$; $\|v_2\| = 9$, $\alpha_2 = 15°$.
10. $\|v_1\| = 7$, $\alpha_1 = 160°$; $\|v_2\| = 6$, $\alpha_2 = 190°$.
11. $\|v_1\| = 5$, $\alpha_1 = 205°$; $\|v_2\| = \frac{3}{10}$, $\alpha_2 = 145°$.
12. $\|v_1\| = \frac{1}{2}$, $\alpha_1 = 300°$; $\|v_2\| = 10$, $\alpha_2 = 210°$.

Determine the angle $\gamma$, where $0° \le \gamma \le 180°$, to the nearest ten minutes, between each pair of vectors.

***Example.***   $v_1 = (1, -7)$; $v_2 = (9, -1)$.

*Solution.*   From Theorem 7.3-I,

$$\cos \gamma = \frac{(1, -7) \cdot (9, -1)}{\sqrt{1^2 + (-7)^2} \sqrt{9^2 + (-1)^2}}$$

$$= \frac{16}{\sqrt{50} \sqrt{82}} = \frac{16}{5\sqrt{2} \cdot \sqrt{2} \cdot \sqrt{41}} = \frac{8}{5\sqrt{41}} \approx 0.2499,$$

from which, we find that, $\gamma \approx 75° 30'$.

13. $v_1 = (3, 0)$; $v_2 = (-7, 3)$.          15. $v_1 = 9i + 6j$; $v_2 = 2i$.
14. $v_1 = (1, -3)$; $v_2 = (8, -2)$.          16. $v_1 = -4i - 8j$; $v_2 = i - 5j$.

Find the scalar projection of $v_1$ on $v_2$ and of $v_2$ on $v_1$ for the given vectors. Sketch the corresponding geometric vectors and show the scalar projections.

***Example.***   $v_1 = (1, 2)$; $v_2 = (-3, 4)$.

*Solution.*   From Theorem 7.3-II, the scalar projection of $v_1$ on $v_2$ is

$$\frac{v_1 \cdot v_2}{\|v_2\|} = \frac{(1)(-3) + (2)(4)}{\sqrt{(-3)^2 + (4)^2}} = \frac{-3 + 8}{\sqrt{25}} = \frac{5}{5} = 1.$$

*(Solution continued on the next page.)*

From Theorem 7.3-III, the scalar projection of $\mathbf{v}_2$ on $\mathbf{v}_1$ is

$$\frac{\mathbf{v}_1 \cdot \mathbf{v}_2}{\|\mathbf{v}_1\|} = \frac{5}{\sqrt{1^2 + 2^2}} = \frac{5}{\sqrt{5}} = \sqrt{5}.$$

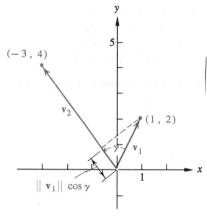

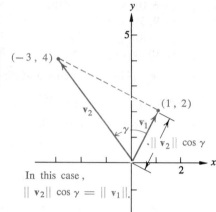

In this case,

$\|\mathbf{v}_2\| \cos \gamma = \|\mathbf{v}_1\|$.

**17.** $\mathbf{v}_1 = -4\mathbf{i}$; $\mathbf{v}_2 = -5\mathbf{i} - 2\mathbf{j}$.   **20.** $\mathbf{v}_1 = (-9, -10)$; $\mathbf{v}_2 = (2, -4)$.
**18.** $\mathbf{v}_1 = \mathbf{i} - 6\mathbf{j}$; $\mathbf{v}_2 = -5\mathbf{j}$.   **21.** $\mathbf{v}_1 = (-3, -5)$; $\mathbf{v}_2 = (3, -1)$.
**19.** $\mathbf{v}_1 = (2, 3)$; $\mathbf{v}_2 = (-4, 5)$.   **22.** $\mathbf{v}_1 = (-7, 6)$; $\mathbf{v}_2 = (3, -6)$.

Verify that the following pairs of vectors are orthogonal.

***Example.***   $\mathbf{i} + 4\mathbf{j}$ and $8\mathbf{i} - 2\mathbf{j}$.

*Solution.*   Because $(\mathbf{i} + 4\mathbf{j}) \cdot (8\mathbf{i} - 2\mathbf{j}) = (1)(8) + (4)(-2) = 8 - 8 = 0$, then by Theorem 7.4, the vectors are orthogonal.

**23.** $(0, 5)$ and $(-3, 0)$.   **26.** $(9, 1)$ and $(-1, 9)$.
**24.** $(-2, 2)$ and $(5, 5)$.   **27.** $(a, b)$ and $(-b, a)$.
**25.** $6\mathbf{i} - 9\mathbf{j}$ and $3\mathbf{i} + 2\mathbf{j}$.   **28.** $-3\mathbf{i} + \sqrt{3}\mathbf{j}$ and $\mathbf{i} + \sqrt{3}\mathbf{j}$.

Find all values for $k$ such that the given pairs of vectors are orthogonal.

***Example.***   $(2, k)$ and $(-3, 1)$.

*Solution.*   By Theorem 7.4, the vectors are orthogonal if

$$(2, k) \cdot (-3, 1) = (2)(-3) + (k)(1) = 0,$$

$$k = 6.$$

**29.** $(k, 4)$ and $(-2, 3)$.   **32.** $(k, -6)$ and $(1, k)$.
**30.** $(5, -k)$ and $(3, 3)$.   **33.** $(15, k)$ and $(2, -k)$.
**31.** $(9, k)$ and $(-1, k)$.   **34.** $(k, k)$ and $(k, -4)$.

Verify that the following pairs of vectors are collinear.

***Example.***   (1, 5) and (2, 10).

*Solution.*   Because $(-5, 1) \cdot (2, 10) = -10 + 10 = 0$, then by Theorem 7.5, the given vectors are collinear.

35. $(6, -3)$ and $(4, -2)$.                    37. $(9, -8)$ and $(-2, \frac{16}{9})$.
36. $(-10, -4)$ and $(6, \frac{12}{5})$.        38. $(1, \frac{10}{3})$ and $(2, \frac{20}{3})$.

In Problems 39 to 42, find all values of $k$ such that the given pairs of vectors are collinear.

39. $(k, 1)$ and $(6, -9)$.                     41. $(-k, 2)$ and $(k, 8)$.
40. $(2, 3)$ and $(5, -k)$.                     42. $(-k, 4)$ and $(16, -k)$.

**B.**

43. Show that $\mathbf{v}_1 \cdot \mathbf{v}_1 = \|\mathbf{v}_1\|^2$.                    44. Show that $\mathbf{v}_1 \cdot \mathbf{v}_2 = \mathbf{v}_2 \cdot \mathbf{v}_1$.
45. Show that $c(\mathbf{v}_1 \cdot \mathbf{v}_2) = (c\mathbf{v}_1) \cdot \mathbf{v}_2$.
46. Show that $\mathbf{v}_1 \cdot (\mathbf{v}_2 + \mathbf{v}_3) = \mathbf{v}_1 \cdot \mathbf{v}_2 + \mathbf{v}_1 \cdot \mathbf{v}_3$.
•47. In physics, the work done by a force is defined by

$$W = \mathbf{F} \cdot \mathbf{d},$$

where $\mathbf{F}$ is the force vector and $\mathbf{d}$ is the displacement vector. Find the work done if $\mathbf{F}$ has a magnitude of 10 pounds and has a direction angle of 60°, and $\mathbf{d} = 8\mathbf{i}$ (8 feet along the $x$ axis).

48. Find the work done (see Problem 47) if $\mathbf{F} = \sqrt{3}\mathbf{i} + \mathbf{j}$ and $\mathbf{d} = \mathbf{i} + \mathbf{j}$.

# 7.4
## Polar Coordinates

The basis you used to graph relations and functions in Chapters 1 and 3 was the fact that you associated a point in the plane with an ordered pair of real numbers called Cartesian coordinates, the first component being the directed distance of the point from the $y$ axis and the second component being the directed distance from the $x$ axis. In this chapter you have also described the location of a point in a plane by specifying the terminal point of a bound geometric vector $\bar{\mathbf{v}}$ in terms of its magnitude $\|\bar{\mathbf{v}}\|$ and its direction angle $\alpha$, $-180° < \alpha \leq 180°$.

The location of a point $A$ in the plane can also be specified by giving the distance $\rho$ (sometimes designated by $r$) from the origin $O$ to $A$ and the measure of the angle that the ray through point $A$ makes with the positive $x$ axis (see Figure 7.7 on page 222). Here *the angle is not restricted* and we shall use the symbol $\theta$ instead of $\alpha$. The symbol $\rho$ is read "rho"; the symbol $\theta$ is read "theta." Thus the ordered pair $[\rho, m°(\theta)]$ or the ordered pair $[\rho, m^R(\theta)]$ describes, or locates, the point $A$. The components of either of these ordered pairs are called **polar coordinates** of $A$. We shall, as is customary, simply use the notation $(\rho, \theta)$ for either ordered pair.

Now, while there is a one-to-one correspondence between the set of ordered pairs in a Cartesian coordinate system and the points in the plane, this is not the case in a polar coordinate system. For example, although a specified

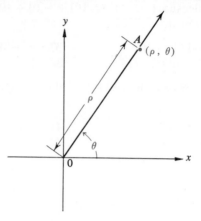

**FIGURE 7.7**

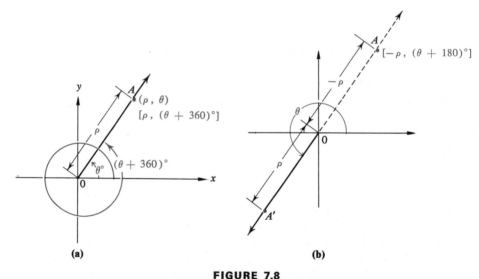

**(a)**                                              **(b)**

**FIGURE 7.8**

ordered pair of polar coordinates determines a unique point, each point in the plane has infinitely many pairs of polar coordinates. Thus, as illustrated in Figure 7.8a, if $(\rho, \theta)$ are polar coordinates of $A$, then so are $[\rho, (\theta + k \cdot 360)°]$, $k \in J$. Also, if $-\rho$ $(\rho > 0)$ denotes the directed distance from $O$ to $A$ along the negative extension of the ray $OA'$, as shown in Figure 7.8b, then $[-\rho, (\theta + 180 + k \cdot 360)°]$, are also infinitely many polar coordinates of $A$. For example, as shown in Figure 7.9, the point having polar coordinates $(4, 30°)$ also has polar coordinates $(4, 390°)$ and $(4, -330°)$ for positive values of $\rho$, and $(-4, 210°)$ for a negative value of $\rho$. Of course, there are infinitely many other possible polar coordinates for this point.

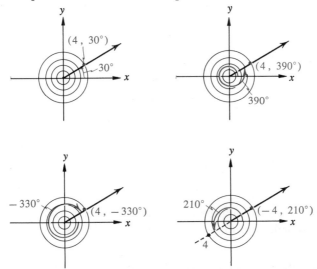

**FIGURE 7.9**

Because $\cos \theta = x/\rho$ and $\sin \theta = y/\rho$, Cartesian and polar coordinates can be related as follows:

$$\rho = \pm \sqrt{x^2 + y^2},$$

$$\left.\begin{array}{l} x = \rho \cos \theta, \\ y = \rho \sin \theta. \end{array}\right\} (1) \qquad \begin{array}{l} \cos \theta = \dfrac{x}{\rho} = \dfrac{x}{\pm\sqrt{x^2 + y^2}}, \\[2mm] \sin \theta = \dfrac{y}{\rho} = \dfrac{y}{\pm\sqrt{x^2 + y^2}}, \ (x, y) \neq (0, 0). \end{array}\left.\right\} (2)$$

Equations (1) can be used to find the rectangular coordinates for a point if the polar coordinates are known, and Equations (2) can be used to find the polar coordinates for a point if the Cartesian coordinates are known. The positive or negative radical expression is chosen depending upon the quadrant in which the point A represented by $(x, y)$ lies, and whether $\rho$ is specified as greater than zero or less than zero.

**Example.** Find the Cartesian coordinates of the point with polar coordinates $(3, 60°)$. Graph the point.

*Solution.* From Equations (1), it follows that

$$x = 3 \cos 60° = 3 \cdot \frac{1}{2} = \frac{3}{2},$$

$$y = 3 \sin 60° = 3 \cdot \frac{\sqrt{3}}{2} = \frac{3\sqrt{3}}{2}.$$

(*Figure on the next page.*)

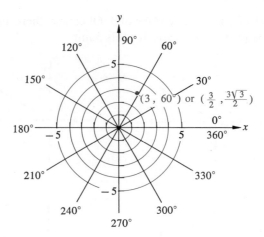

**Example.** Find an approximation for the polar coordinates for the point corresponding to $(5, -3)$ where $0° < \theta < 360°$ and $\rho > 0$. Graph the point.

*Solution.* From Equations (2),

$$\rho = \pm\sqrt{x^2 + y^2} = \pm\sqrt{25 + 9} = \pm\sqrt{34}.$$

Because $(5, -3)$ is in the fourth quadrant, $\cos \theta > 0$. Arbitrarily taking the positive value for $\rho$ as stated above,

$$\cos \theta = \frac{5}{\sqrt{34}} \approx \frac{5}{5.83} \approx 0.8576.$$

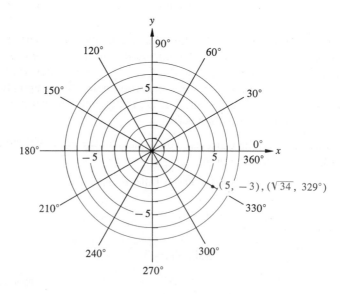

From Table IV and the appropriate reduction formula, $\theta \approx 329° \, 00'$ (to the nearest ten minutes). Therefore, the pair of polar coordinates is $(\sqrt{34}, \approx 329°)$.

In Chapters 1 and 3 you have seen that an equation in two variables, say $x$ and $y$, serves to pair elements from the replacement sets of the variables and has a graph on the geometric plane in a Cartesian coordinate system. An equation in $\rho$ and $\theta$ also has a graph on the geometric plane in a polar coordinate system. Equations (1) and (2) on page 223 can be used to transform an equation in Cartesian form to polar form, and vice versa.

***Example.*** Rewrite $x^2 + y^2 - 4x = 0$ in polar form.

*Solution.* Substituting $\rho \cos \theta$ for $x$ and $\rho^2$ for $x^2 + y^2$ in

$$x^2 + y^2 - 4x = 0$$

yields

$$\rho^2 - 4\rho \cos \theta = 0,$$
$$\rho(\rho - 4 \cos \theta) = 0.$$

Multiplying each member by $1/\rho$ $(\rho \neq 0)$ yields

$$\rho - 4 \cos \theta = 0,$$

or equivalently

$$\rho = 4 \cos \theta,$$

the polar form of the equation $x^2 + y^2 - 4x = 0$.

Frequently, the polar form of an equation is more useful than the rectangular form. Their graphs may also be simpler to sketch. An example is shown in the exercise.

### EXERCISE 7.4

A.

Find four sets of polar coordinates $(-360° < \theta \leq 360°)$ for each point whose coordinates are given.

1. $(3, 380°)$.     3. $(5, -450°)$.     5. $(-4, 420°)$.
2. $(6, 840°)$.     4. $(4, -540°)$.     6. $(-5, -600°)$.

Find the Cartesian coordinates of each point with polar coordinates as given.

7. $\left(4, \dfrac{\pi}{6}^{R}\right)$.     9. $\left(3, -\dfrac{3\pi}{4}^{R}\right)$.     11. $\left(-3, \dfrac{5\pi}{6}^{R}\right)$.

8. $\left(6, \dfrac{2\pi}{3}^{R}\right)$.     10. $\left(0, \dfrac{13\pi}{3}^{R}\right)$.     12. $\left(-4, \dfrac{\pi}{4}^{R}\right)$.

Find the pair of polar coordinates ($\theta$ in radians) using the angle of least positive measure for each point with Cartesian coordinates as given.

**13.** $(4, 0)$. **15.** $(2, 2)$. **17.** $\left(-\dfrac{\sqrt{3}}{2}, \dfrac{1}{2}\right)$.

**14.** $(0, -3)$. **16.** $(-1, -\sqrt{3})$. **18.** $\left(\dfrac{1}{2}, -\dfrac{\sqrt{3}}{2}\right)$.

Transform each equation given in Cartesian form to an equation in polar form.

*Example.* $y = 3x$.

*Solution.* Substituting $\rho \sin \theta$ for $y$ and $\rho \cos \theta$ for $x$ yields

$$\rho \sin \theta = 3\rho \cos \theta.$$

Multiplying each member by $1/\rho$ ($\rho \neq 0$) yields

$$\sin \theta = 3 \cos \theta.$$

Since $\cos \theta$ does not equal zero for values of $\theta$ for which $\sin \theta$ equals zero, each member can be multiplied by $1/\cos \theta$ to obtain

$$\frac{\sin \theta}{\cos \theta} = 3, \qquad (\cos \theta \neq 0),$$

$$\tan \theta = 3.$$

Notice that for values of $\theta$ for which the equation is satisfied, $\tan \theta$ is a constant, namely, 3. Thus the graph of the equation is a straight line with slope 3.

**19.** $x = 4$. **21.** $x^2 + y^2 = 16$. **23.** $x^2 + y^2 - 3y = 0$.
**20.** $y = -3$. **22.** $4x^2 + y^2 = 4$. **24.** $x^2 + y^2 + 4x = 0$.

Transform each equation given in polar form to an equation in Cartesian form.

*Example.* $\rho = 2 \sec \theta$.

*Solution.* Substituting $1/\cos \theta$ for $\sec \theta$ yields

$$\rho = \frac{2}{\cos \theta}$$

or, equivalently,

$$\rho \cos \theta = 2.$$

Since $\rho \cos \theta = x$, the Cartesian form of the equation is

$$x = 2.$$

**25.** $\rho = -\csc \theta$.          **27.** $\rho = 4 \sin \theta$.          **29.** $\rho = \dfrac{4}{1 - \sin \theta}$.

**26.** $\rho = 9$.                **28.** $\rho = 3 \cos \theta$.          **30.** $\rho = \dfrac{4}{1 + 2 \cos \theta}$.

**B.**

Graph each function for $0° \leq \theta < 360°$.

   ***Example.***   $\{(\rho, \theta) \,|\, \rho = 4 \sin \theta\}$.

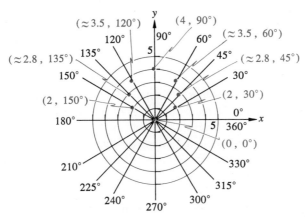

   *Solution.*   Some ordered pairs with commonly used angles $30°, 45°$, etc. are first obtained. These ordered pairs are shown in the figure. Assuming that the curve is continuous and smooth between these points, the graph is completed. Of course, additional ordered pairs could have been obtained from

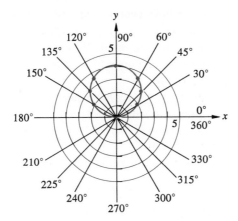

Table IV, to locate points between those obtained. Notice that the complete curve is obtained for $0° \leq \theta < 180°$. For values of $\theta$ where $180° \leq \theta < 360°$,

*(Solution continued on the next page.)*

the components of ordered pairs $(\rho, \theta)$ are also coordinates of points on the curve shown. Verify this statement by graphing several ordered pairs $(\rho, \theta)$ for values of $\theta$ in the interval, $180° \le \theta < 360°$.

**·31.** $\{(\rho, \theta) \,|\, \rho = 4\}$.   *Hint*: Consider the equation $\rho = 0 \cdot \theta + 4$.
**32.** $\{(\rho, \theta) \,|\, \theta = 30°\}$.   *Hint*: Consider the equation $\theta = 0 \cdot \rho + 30°$.
**33.** $\{(\rho, \theta) \,|\, \rho = 4 \cos \theta\}$.        **·35.** $\{(\rho, \theta) \,|\, \rho = 2 \sin 2\theta\}$.
**34.** $\{(\rho, \theta) \,|\, \rho = 4(1 + \cos \theta)\}$.        **36.** $\{(\rho, \theta) \,|\, \rho^2 = 4 \cos 2\theta\}$.

Graph each of the given pairs of equations and find coordinates for the points of intersection.

**37.** $\rho = \sin \theta$,                    **·39.** $\rho = 1 + \sin \theta$,
$\qquad \rho = \cos \theta$.                     $\qquad\quad \rho = 1 + \cos \theta$.
**38.** $\rho = 1 + \sin \theta$,                **40.** $\rho = 2 \sin \theta$,
$\qquad \rho = 1$.                               $\qquad\;\; \rho = 2 \sin 2\theta$.

**41.** Solve the system in Problem 37 analytically. Explain why the coordinates of the origin are not in the solution set of the system.
**42.** Solve the system in Problem 38 analytically.
**43.** Solve the system in Problem 39 analytically.
**44.** Solve the system in Problem 40 analytically.

**C.**

**45.** Show that the distance between the points with coordinates $(\rho_1, \theta_1)$ and $(\rho_2, \theta_2)$ is given by

$$d = \sqrt{\rho_1^2 + \rho_2^2 - 2\rho_1\rho_2 \cos (\theta_1 - \theta_2)}.$$

## Chapter Summary

1. The following symbols have been introduced: $\mathbf{v}$, $\mathbf{V}$, $\mathbf{0}$, $\|\mathbf{v}\|$, $\mathbf{i}$, $\mathbf{j}$, $\rho$, and $\theta$.
2. A **two-dimensional vector** $\mathbf{v}$ is an ordered pair of real numbers. That is, if $a, b \in R$, then $\mathbf{v} = (a, b)$. The set of all such vectors is designated by $\mathbf{V}$. Each vector in the set corresponds to a unique bound geometric vector (initial point at the origin).
3. For all $\mathbf{v}_1 = (a_1, b_1)$ and $\mathbf{v}_2 = (a_2, b_2)$, where $\mathbf{v}_1, \mathbf{v}_2 \in V$,
   (a) $\mathbf{v}_1 = \mathbf{v}_2$ if and only if $a_1 = a_2$ and $b_1 = b_2$;
   (b) $-\mathbf{v}_1 = -(a_1, b_1) = (-a_1, -b_1)$;
   (c) the zero vector $\mathbf{0} = (0, 0)$;
   (d) the norm, or magnitude, of $\mathbf{v}_1$ is the real number (scalar)

$$\|\mathbf{v}_1\| = \sqrt{a_1^2 + b_1^2};$$

   (e) and for $\|\mathbf{v}_1\| \ne 0$, the direction angle of $\mathbf{v}_1$ is the angle $\alpha$ such that

$$\cos \alpha = \frac{a_1}{\|\mathbf{v}_1\|}, \qquad \sin \alpha = \frac{b_1}{\|\mathbf{v}_1\|}, \qquad -180° < \alpha \le 180°.$$

4. From the preceding definitions, it follows that the **norm** (or **magnitude**) of a vector $\mathbf{v}$ is the length of the associated geometric vector $\mathbf{v}$ and the **direction angle** of the vector is the angle $\alpha$, $-180° < \alpha \le 180°$, from the positive $x$ axis to the geometric vector corresponding to $\mathbf{v}$.

5. Two vectors are said to be **collinear** if their corresponding geometric vectors are collinear; that is, if they have the same direction angle $\alpha$ or their direction angles differ by $180°$.

6. For all $\mathbf{v}_1 = (a_1, b_1)$ and $\mathbf{v}_2 = (a_2, b_2)$; $\mathbf{v}_1, \mathbf{v}_2 \in V$ and $c \in R$,
   (a) $\mathbf{v}_1 + \mathbf{v}_2 = (a_1, b_1) + (a_2, b_2) = (a_1 + a_2, b_1 + b_2)$, the **sum** of $\mathbf{v}_1$ and $\mathbf{v}_2$;
   (b) $\mathbf{v}_1 - \mathbf{v}_2 = \mathbf{v}_1 + (-\mathbf{v}_2)$, the **difference** of $\mathbf{v}_2$ subtracted from $\mathbf{v}_1$;
   (c) $c\mathbf{v} = c(a_1, b_1) = (ca_1, cb_1)$, the **product** of a scalar $c$ and $\mathbf{v}$;
   (d) $\dfrac{\mathbf{v}}{c} = \dfrac{1}{c} v$, $(c \ne 0)$, the **quotient** of a vector divided by a scalar.

7. Vector addition and multiplication of a vector by a scalar have the following properties. If $\mathbf{v}_1, \mathbf{v}_2, \mathbf{v}_3 \in V$ and $c, d \in R$, then

   I. $\mathbf{v}_1 + \mathbf{v}_2 \in V$.  
   II. $\mathbf{v}_1 + \mathbf{v}_2 = \mathbf{v}_2 + \mathbf{v}_1$.  
   III. $(\mathbf{v}_1 + \mathbf{v}_2) + \mathbf{v}_3 = \mathbf{v}_1 + (\mathbf{v}_2 + \mathbf{v}_3)$.  
   IV. $\mathbf{v}_1 + \mathbf{0} = \mathbf{v}_1$.  
   V. $\mathbf{v}_1 + (-\mathbf{v}_2) = \mathbf{0}$.  
   VI. $(c + d)\mathbf{v}_1 = c\mathbf{v}_1 + d\mathbf{v}_1$.  
   VII. $c(\mathbf{v}_1 + \mathbf{v}_2) = c\mathbf{v}_1 + c\mathbf{v}_2$.

8. If $\mathbf{v}_1$ and $\mathbf{v}_2$ are noncollinear vectors, then for each $\mathbf{v}_3$ there exist scalars $c_1$ and $c_2$ such that

$$\mathbf{v}_3 = c_1\mathbf{v}_1 + c_2\mathbf{v}_2.$$

   Vectors $\mathbf{v}_1$ and $\mathbf{v}_2$ are said to form a **basis** for the set of all two-dimensional vectors.

9. A vector $\mathbf{v}$ with norm equal to 1 ($\|\mathbf{v}\| = 1$) is called a **unit vector**.

10. The unit vectors $\mathbf{v}_1 = (1, 0)$ and $\mathbf{v}_2 = (0, 1)$, designated as $\mathbf{i}$ and $\mathbf{j}$, respectively, whose corresponding geometric vectors are in the direction of the $x$ and $y$ axes, form an **orthogonal basis**.

11. Each vector

$$\mathbf{v} = (a, b) = (a, 0) + (0, b) = a(1, 0) + b(0, 1) = a\mathbf{i} + b\mathbf{j}.$$

12. For each nonzero vector $\mathbf{v} = (a, b)$,

$$\frac{\mathbf{v}}{\|\mathbf{v}\|} = \frac{(a, b)}{\|(a, b)\|}$$

   is the *unit vector having the same direction angle as* $\mathbf{v}$.

13. The **inner (dot, scalar) product** of vectors $\mathbf{v}_1$ and $\mathbf{v}_2$ is defined by

$$\mathbf{v}_1 \cdot \mathbf{v}_2 = (a_1, b_1) \cdot (a_2, b_2) = a_1a_2 + b_1b_2.$$

14. If vector $\mathbf{v}_1 = (a_1, b_1)$, vector $\mathbf{v}_2 = (a_2, b_2)$ and $\gamma$ is the angle between the geometric vectors corresponding to $\mathbf{v}_1$ and $\mathbf{v}_2$, then

$$\mathbf{v}_1 \cdot \mathbf{v}_2 = \|\mathbf{v}_1\| \, \|\mathbf{v}_2\| \cos \gamma.$$

Equivalently,

$$\cos \gamma = \frac{\mathbf{v}_1 \cdot \mathbf{v}_2}{\|\mathbf{v}_1\| \, \|\mathbf{v}_2\|};$$

$$\|\mathbf{v}_1\| \cos \gamma = \frac{\mathbf{v}_1 \cdot \mathbf{v}_2}{\|\mathbf{v}_2\|}, \qquad \text{the scalar projection of } \mathbf{v}_1 \text{ on } \mathbf{v}_2 \, ;$$

$$\|\mathbf{v}_2\| \cos \gamma = \frac{\mathbf{v}_1 \cdot \mathbf{v}_2}{\|\mathbf{v}_1\|}, \qquad \text{the scalar projection of } \mathbf{v}_2 \text{ on } \mathbf{v}_1.$$

**15.** The vectors $\mathbf{v}_1 = (a_1, b_1)$ and $\mathbf{v}_2 = (a_2, b_2)$ are *orthogonal* if and only if the inner product

$$\mathbf{v}_1 \cdot \mathbf{v}_2 = a_1 a_2 + b_1 b_2 = 0.$$

**16.** The vectors $\mathbf{v}_1 = (a_1, b_1)$ and $\mathbf{v}_2 = (a_2, b_2)$ are collinear if and only if

$$\mathbf{v}_1 \cdot \mathbf{v}_2 = (-b_1, a_1) \cdot (a_2, b_2) = 0.$$

**17.** The components of the ordered pairs $[\rho, m^\circ(\theta)]$ or $[\rho, m^R(\theta)]$ locate a point $A$ in the plane and are called the **polar coordinates** of $A$. Each such ordered pair specifies a unique point; however, each point in the plane corresponds to infinitely many ordered pairs of polar coordinates.

**18.** Cartesian and polar coordinates are related as follows.

$$x = \rho \cos \theta \qquad \rho = \pm \sqrt{x^2 + y^2},$$

$$y = \rho \sin \theta \qquad \cos \theta = \frac{x}{\rho} = \frac{x}{\pm \sqrt{x^2 + y^2}},$$

$$\sin \theta = \frac{x}{\rho} = \frac{y}{\pm \sqrt{x^2 + y^2}}, \qquad (x, y) \neq (0, 0).$$

## Chapter Review

In Problems 1 and 2, (a) show each corresponding geometric vector graphically; (b) find the norm and the direction angle (to the nearest ten minutes) of each vector.

**·1.** $(-4, 3)$.                          **2.** $(5, -5)$.

Given the following norms and direction angles, find the vector $\mathbf{v}$.

**3.** $\|\mathbf{v}\| = 10; \ \alpha = 225^\circ$.             **4.** $\|\mathbf{v}\| = 2; \ \alpha = -210^\circ$.

Let $\mathbf{v}_1 = (3, -2)$, $\mathbf{v}_2 = (5, 4)$, $\mathbf{v}_3 = (1, -3)$, $c_1 = 2$, and $c_2 = -1$. Express each of the following as an ordered pair and sketch the corresponding geometric vectors graphically.

**·5.** $\mathbf{v}_2 - c_2 \mathbf{v}_1$.                      **6.** $c_1(\mathbf{v}_3 + \mathbf{v}_2)$.

7. $\dfrac{\mathbf{v_3}}{c_2}$.                                    8. $\dfrac{\mathbf{v_2}}{c_1 \|\mathbf{v_1}\|}$.

Write each vector as an ordered pair and sketch the corresponding geometric vector.

**•9.** $-8\mathbf{i}$.                                     **10.** $2\mathbf{i} + 5\mathbf{j}$.

Express each vector as the sum of scalar multiples of $\mathbf{i}$ and $\mathbf{j}$ and sketch the corresponding geometric vectors.

**11.** $(7, -9)$.                                     **12.** $(-8, -1)$.

Find the norm and the direction angle of each vector.

**•13.** $-10\mathbf{i} + 4\mathbf{j}$.                              **14.** $7\mathbf{i} - 9\mathbf{j}$.

Find the unit vector with the same direction angle corresponding to each vector.

**15.** $(4, -1)$.                                     **16.** $-\mathbf{i} + 8\mathbf{j}$.

Resolve each vector $\mathbf{v}$ with given norm and given direction angle $\alpha$ into a linear combination of vectors $\mathbf{i}$ and $\mathbf{j}$. Sketch the given vector and its scalar components along the $x$ and $y$ axes.

**•17.** $\|\mathbf{v}\| = 3;\ \alpha = 120°$.                        **18.** $\|\mathbf{v}\| = 2;\ \alpha = 229°$.

Find the inner product $\mathbf{v_1} \cdot \mathbf{v_2}$ for the given vectors.

**19.** $\mathbf{v_1} = (6, 6);\ \mathbf{v_2} = (-5, -4)$.        **20.** $\mathbf{v_1} = 8\mathbf{i} - \mathbf{j};\ \mathbf{v_2} = 9\mathbf{i} - 3\mathbf{j}$.
**21.** $\|\mathbf{v_1}\| = 8,\ \alpha_1 = 70°;\quad \|\mathbf{v_2}\| = 4,\ \alpha_2 = 25°$.
**22.** $\|\mathbf{v_1}\| = 10,\ \alpha_1 = 315°;\quad \|\mathbf{v_2}\| = 5,\ \alpha_2 = 345°$.

Determine the angle $\gamma$, where $0° < \gamma < 180°$, to the nearest ten minutes, formed by each pair of vectors.

**23.** $\mathbf{v_1} = (-7, 5);\ \mathbf{v_2} = (1, -5)$.        **24.** $\mathbf{v_1} = -7\mathbf{i} - \mathbf{j};\ \mathbf{v_2} = -8\mathbf{i} + 9\mathbf{j}$.

Find the scalar projection of $\mathbf{v_1}$ on $\mathbf{v_2}$ and of $\mathbf{v_2}$ on $\mathbf{v_1}$ for the given vectors

**25.** $\mathbf{v_1} = (1, 7);\ \mathbf{v_2} = (10, -1)$.        **26.** $\mathbf{v_1} = 9\mathbf{i} + 2\mathbf{j};\ \mathbf{v_2} = -8\mathbf{i} - 5\mathbf{j}$.

In Problems 27 and 28, verify that the following pairs of vectors are orthogonal.

**27.** $(-4, -5)$ and $(4, -\frac{16}{5})$.            **28.** $7\mathbf{i} + 6\mathbf{j}$ and $5\mathbf{i} - \frac{35}{6}\mathbf{j}$.
**29.** Find the inner product $(\mathbf{i} + 0\mathbf{j}) \cdot (\mathbf{i} + 0\mathbf{j})$ or $\mathbf{i} \cdot \mathbf{i}$.
**30.** Find the inner product $(0\mathbf{i} + \mathbf{j}) \cdot (0\mathbf{i} + \mathbf{j})$ or $\mathbf{j} \cdot \mathbf{j}$.

Find three additional sets of polar coordinates $(\rho, \theta)$,    $-360° < \theta < 360°$, for each point whose coordinates are given.

**31.** $(2, 250°)$.                                    **32.** $(-3, 390°)$.

Find the Cartesian coordinates of each point whose polar coordinates are given.

**33.** $\left(2, \dfrac{\pi}{3}^{R}\right)$.

**34.** $\left(-3, \dfrac{3\pi}{4}^{R}\right)$.

In Problems 35 and 36, find the set of polar coordinates $(\rho, \theta^{R})$ involving the angle of least positive measure for each point with Cartesian coordinates as given.

**35.** $(-\sqrt{3}, -1)$.

**36.** $(2, -2)$.

**37.** Write $x^2 + 3y^2 = 5$ as an equation in polar form.

**38.** Write $\rho = \dfrac{2}{1 - \cos \theta}$ as an equation in Cartesian form.

**B.**

**•39.** Graph $\{(\rho, \theta) \mid \rho = 3 \sin \theta\}$ for $0 \leq \theta < 2\pi$.

**40.** Solve the system

$$\rho = 3 - \cos \theta,$$
$$\rho = 2,$$

using analytic methods.

# 8

# Complex Numbers

**R**ECALL FROM CHAPTER 1 that the set of real numbers $R$ does not contain an element which when multiplied by itself equals a negative real number $-a$, where $a > 0$. Thus, any symbol of the form $\sqrt{-a}$, $a > 0$, is undefined in the system of real numbers. We now consider a set of numbers, called the **set of complex numbers $C$**, which contains all such numbers and in addition contains the set of real numbers.*

## 8.1

### Complex Numbers as Ordered Pairs

In this section we view elements in the set $C$ as ordered pairs of real numbers and when convenient designate such a pair of numbers by $z$.

**Definition 8.1.** *A **complex number** $z$ is an ordered pair of real numbers. That is, for all $a, b, \in R$,*

$$z = (a, b).$$

*The set of all such complex numbers is designated by $C$.*

As you will see, the elements of the set of vectors $\mathbf{V}$, which have also been defined in a similar way in Chapter 7, have many properties similar to the properties of the elements of the set $C$.

---

* More precisely, the set of complex numbers $C$ does not contain the set of real numbers $R$, but contains a subset whose elements can be identified with the elements of $R$ for the operations defined. Such an identification between the elements of two sets is called an **isomorphism.**

Since we shall be working with the equals relation also designated by " $=$ " in $C$, we first establish criteria for equality and then note some properties that follow from similar properties for equality in $R$.

**Definition 8.2.**   *For all complex numbers* $z_1 = (a_1, b_1)$ *and* $z_2 = (a_2, b_2)$,

$$z_1 = z_2 \quad \textit{if and only if} \quad a_1 = a_2 \textbf{ and } b_1 = b_2.$$

Since the equality axioms (see Appendix B, page 288) hold in the set of real numbers $R$, similar properties hold in $C$.

**Theorem 8.1.**   *If* $z_1, z_2$, *and* $z_3 \in C$,

| | | |
|---|---|---|
| E-1. | *then* $z_1 = z_1$; | *Reflexive Law* |
| E-2. | *and if* $z_1 = z_2$, *then* $z_2 = z_1$; | *Symmetric Law* |
| E-3. | *and if* $z_1 = z_2$ *and* $z_2 = z_3$, *then* $z_1 = z_3$. | *Transitive Law* |

*Proof of E-1.*   Since, from Definition 8.2, $(a_1, b_1) = (a_1, b_1)$, it follows directly that $z_1 = z_1$.

The proof of parts E-2 and E-3 are equally as direct and are left as exercises.

The meanings given below to sums and products of the elements in $C$ are motivated by the fact that we want $C$ to contain the set of real numbers and also to contain elements which when multiplied by each other give every element in the set.

**Definition 8.3.**   *For all complex numbers* $z_1 = (a_1, b_1)$ *and* $z_2 = (a_2, b_2)$,

I.  $z_1 + z_2 = (a_1, b_1) + (a_2, b_2) = (a_1 + a_2, b_1 + b_2)$,

II.  $z_1 z_2 = (a_1, b_1) \cdot (a_2, b_2) = (a_1 a_2 - b_1 b_2, a_1 b_2 + a_2 b_1)$.

This establishes two operations, addition and multiplication, for the elements in the set $C$. Notice that I is similar to vector addition, but that no operation analogous to II was defined for vectors. Furthermore, do not confuse the operation for multiplication of two complex numbers with the inner (dot) product of two vectors.

**Example.**   Write each expression as an ordered pair.
(a) $(4, 7) + (2, -3)$.        (b) $(4, 7) \cdot (2, -3)$.

*Solution.*   (a) By Definition 8.3-I,
$$(4, 7) + (2, -3) = (4 + 2, 7 - 3) = (6, 4).$$
(b) By Definition 8.3-II,
$$(4, 7) \cdot (2, -3) = [4 \cdot 2 - 7(-3), (4)(-3) + 2 \cdot 7]$$
$$= (8 + 21, -12 + 14) = (29, 2).$$

Although we shall not formalize the statement of a substitution law for the set of complex numbers, it does follow from Theorem 8.1 that a complex number can be substituted for an equal complex number in either the operation of addition or of multiplication.

Now, consider the sum and product of two complex numbers of the form $(a, 0)$. From Definition 8.3,

$$(a_1, 0) + (a_2, 0) = (a_1 + a_2, 0 + 0)$$

$$= (a_1 + a_2, 0), \tag{1}$$

and

$$(a_1, 0) \cdot (a_2, 0) = (a_1 \cdot a_2 - 0 \cdot 0, \quad a_1 \cdot 0 + a_2 \cdot 0)$$

$$= (a_1 a_2, 0). \tag{2}$$

Recall that the operations of addition and multiplication on two real numbers $a_1$ and $a_2$ yield

$$a_1 + a_2, \tag{1'}$$

and

$$a_1 a_2. \tag{2'}$$

By comparing the results of (1') with (1) and (2') with (2) you can observe a *correspondence* between the elements in $\{(a, 0) \mid a \in R\}$ and the elements in $\{a \mid a \in R\}$ for the operations of addition and multiplication. Thus, you can identify the complex number $(a, 0)$ with the real number $a$ and write

$$(a, 0) = a,$$

at least for these basic operations.* In Section 8.2, you will see that you can also take this view for the operations of subtraction and division. One consequence of viewing $(a, 0) = a$ is the following.

**Theorem 8.2.**  *If $a_1 \in R$ and $(a_2, b_2) \in C$, then*

$$a_1(a_2, b_2) = (a_1 a_2, a_1 b_2).$$

*Proof.*  From Definition 8.3-II,

$$(a_1, 0) \cdot (a_2, b_2) = (a_1 a_2 - 0 \cdot b_2, a_1 b_2 + a_2 \cdot 0)$$

$$= (a_1 a_2, a_1 b_2).$$

Since $(a_1, 0) = a_1$, we have that

$$a_1(a_2, b_2) = (a_1 a_2, a_1 b_2).$$

**Examples.**  Write each product (complex number) as an ordered pair.
(a) $3(4, -5)$.        (b) $-2(-4, 1)$.

*Solutions.*  (a) $3(4, -5) = [3 \cdot 4, (3)(-5)] = (12, -15)$.
            (b) $-2(-4, 1) = [(-2)(-4), (-2)(1)] = (8, -2)$.

* We have noted earlier in the footnote on page 233 that such an identification determines an isomorphism between the two sets.

Now, consider the product

$$(0, b) \cdot (0, b) = (0 \cdot 0 - b \cdot b, 0 \cdot b + 0 \cdot b)$$

$$= (-b^2, 0),$$

where $(-b^2, 0)$ corresponds to the negative real number $-b^2$, for $b > 0$ or $b < 0$. Thus, you see that unlike the set $R$, *the set $C$ provides a square root for each negative real number* $-b^2$. In fact, there are two such square roots $(0, b)$ and $(0, -b)$. In particular $(0, 1)$ and $(0, -1)$ are elements of $C$ such that

$$(0, 1) \cdot (0, 1) = (-1^2, 0) = (-1, 0) = -1,$$

and

$$(0, -1) \cdot (0, -1) = (-1^2, 0) = (-1, 0) = -1.$$

*Thus, in $C$, the square roots of $-1$ are $(0, 1)$ and $(0, -1)$.* The complex number $(0, 1)$ is the principal square root.

The set $C$ of complex numbers can be partitioned into two subsets:

**1.** $\{(a, 0) \mid a \in R\}$ or $\{a \mid a \in R\}$ called the **set of real numbers** and designated by $R$.

**2.** $\{(a, b) \mid a \in R \text{ and } b \in R, b \neq 0\}$ called the **set of imaginary numbers** and designated by $I$. If $a = 0$, the element $(0, b)$ is called a **pure imaginary number**.

The set $C$ and its subsets are related as shown in Figure 8.1.

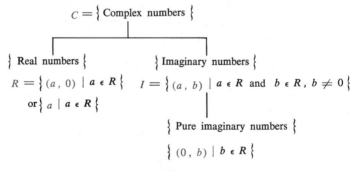

**FIGURE 8.1**

**EXERCISE 8.1**

**A.**

Write each sum as an ordered pair.

**Example.**   $(6, 2) + (0, -2)$.

*Solution.*   From Definition 8.3-I,

$$(6, 2) + (0, -2) = [6 + 0, 2 + (-2)] = (6, 0).$$

**1.** $(1, 8) + (-3, 0)$.

**2.** $(-9, -7) + (-9, -6)$.

**3.** $(2, -5) + (-4, 0)$.

**4.** $(-9, 6) + (4, 5)$.

**5.** $(-4, 6) + (0, 0)$.

**6.** $(-1, 5) + (5, -3)$.

Write each product as an ordered pair.

***Example.*** $(-2, 0) \cdot (1, 6)$.

*Solution.* From Definition 8.3-II,

$$(-2, 0) \cdot (1, 6) = [(-2)(1) - (0)(6), (-2) \cdot (6) + (1)(0)] = (-2, -12).$$

**7.** $(-3, 8) \cdot (3, 1)$.

**8.** $(-6, 4) \cdot (-9, 8)$.

**9.** $(-6, -6) \cdot (1, 1)$.

**10.** $(-9, -1) \cdot (-8, -2)$.

**11.** $(-3, -9) \cdot (2, -4)$.

**12.** $(7, 1) \cdot (0, 0)$.

Write each expression as an ordered pair.

***Examples.*** (a) $3(4, -5)$.  (b) $2(-1, 3) + 5(1, -2)$.

*Solutions.* (a) From Theorem 8.2,

$$3(4, -5) = (12, -15).$$

(b) From Theorem 8.2 and Definition 8.3-I,

$$2(-1, 3) + 5(1, -2) = (-2, 6) + (5, -10) = (3, -4).$$

**13.** $-9(8, -5)$.

**14.** $6(-2, 7)$.

**15.** $2(-2, 8)$.

**16.** $\sqrt{2}(-5, -\sqrt{2})$.

**17.** $6(-3, 1) + 2(5, 0)$.

**18.** $3(0, 7) + (-4)(-7, 8)$.

**19.** $-3(-2, 0) \cdot 9(5, 8)$.

**20.** $-9(4, 2) \cdot (-3)(-5, -7)$.

Solve for $x$ and $y$.

***Example.*** $(7, 1) = (x + y, x - y)$.

*Solution.* From Definition 8.2, if $(7, 1) = (x + y, x - y)$, then

$$7 = x + y \quad \text{and} \quad 1 = x - y,$$

from which

$$x = 4 \quad \text{and} \quad y = 3.$$

**21.** $(6, 3) = (2x - 3y, x + 3y)$.

**22.** $(7, -1) = (3x + y, 2x - 5y)$.

**23.** $(3, 0) = (5x + 2y, x)$.

**24.** $(0, -5) = (2x - y, y)$.

**B.**

In Problems 25 to 28, find $z = (a, b)$ for which each statement is true.

**25.** $(a, b)^2 = -1$.

**26.** $(a, b)^2 = -3$.

**27.** $(a, b)^2 = 1$.

**28.** $(a, b)^2 = 3$.

**29.** If $(a, b) \neq (0, 0)$, solve the equation $(a, b) \cdot (x, y) = (1, 0)$ for $(x, y)$, where both $x$ and $y$ are expressed in terms of $a$ and $b$.

**30.** Find the conditions on $a$, $b$, $c$, and $d$ such that

$$(a, b) \cdot (c, d) = (k, 0),$$

where $a, b, c, d, k \in R$.   *Hint:* Consider three cases, $k > 0$, $k = 0$, and $k < 0$.

**31.** Prove Theorem 8.1, E-2.

**32.** Prove Theorem 8.1, E-3.

## 8.2
### Properties of *C*

Since complex numbers have been defined in terms of ordered pairs of real numbers, complex numbers have many properties similar to the properties of real numbers. These properties in the set $C$ are given the same names that are used in Appendix B, page 288, for the corresponding properties in $R$.

**Theorem 8.3.**   *If $z = (a, b)$, $z_1 = (a_1, b_1)$, $z_2 = (a_2, b_2)$, and $z_3 = (a_3, b_3)$ are elements in $C$, then*

F-1   $z_1 + z_2 \in C$.                                      *Closure Law for Addition*

F-2   $z_1 + z_2 = z_2 + z_1$.                               *Commutative Law of Addition*

F-3   $(z_1 + z_2) + z_3 = z_1 + (z_2 + z_3)$.               *Associative Law of Addition*

F-4   *There exists an element $(0, 0)$ such that*          *Identity Law of Addition*

$$z + (0, 0) = z.$$

F-5   *There exists an element $-z = (-a, -b)$*             *Additive Inverse Law*
      *such that*

$$z + (-z) = (0, 0).$$

F-6   $z_1 z_2 \in C$.                                       *Closure Law for Multiplication*

F-7   $z_1 z_2 = z_2 z_1$.                                   *Commutative Law of Multiplication*

F-8   $(z_1 z_2) z_3 = z_1 (z_2 z_3)$.                       *Associative Law of Multiplication*

F-9   *There exists an element $(1, 0)$*                    *Identity Law of Multiplication*
      *such that*

$$z(1, 0) = z.$$

F-10  *There exists an element*                             *Multiplicative Inverse Law*

$$\frac{1}{z} = \left( \frac{a}{a^2 + b^2}, \frac{-b}{a^2 + b^2} \right) \quad [z \neq (0, 0)]$$

*such that*

$$z \frac{1}{z} = (1, 0).$$

F-11  $z_1(z_2 + z_3) = z_1 z_2 + z_1 z_3$.                  *Distributive Law*

Because the sum of two complex numbers is defined (Definition 8.3-I) in the same way as the sum of two vectors (Definition 7.7), the properties of vector addition stated in Theorem 7.1 are valid for the operation of addition in $C$ in F-1 to F-5. The proofs of F-6 to F-8 are similar to the proofs of F-1 to F-3, with multiplication used in place of addition. These are left as exercises.

*Proof of F-9.*   Let $z = (a, b)$, then

$$z \cdot (1, 0) = (a, b) \cdot (1, 0) = (a \cdot 1 - b \cdot 0, a \cdot 0 + 1 \cdot b) = (a, b).$$

Because $(a, b) = z$, then $z \cdot (1, 0) = z$ and the theorem is proved.

*Proof of F-10.*   Let $z = (a, b)$, and $1/z = 1/(a, b) = (x, y)$. Now, we wish to find an ordered pair $(x, y)$ in terms of $(a, b)$ such that

$$(a, b) \cdot (x, y) = (1, 0).$$

Because

$$(a, b) \cdot (x, y) = (ax - by, ay + xb),$$

then by Definition 8.1, $ax - by = 1$ and $ay + xb = 0$. Solving these equations for $x$ and $y$ in terms of $a$ and $b$ (see Problem 29, Exercise 8.1) yields

$$x = \frac{a}{a^2 + b^2} \quad \text{and} \quad y = \frac{-b}{a^2 + b^2}.$$

Therefore,

$$\frac{1}{z} = \frac{1}{(a, b)} = (x, y) = \left( \frac{a}{a^2 + b^2}, \frac{-b}{a^2 + b^2} \right) \quad [(a, b) \neq (0, 0)],$$

and the theorem is proved.

The proof of F-11 is left as an exercise.

Recall from page 7 that any mathematical system which consists of a set of elements and two operations for which F-1 to F-11 are valid is called a field. Thus, the complex number system is a field; although not an ordered field. Unlike the members of the set of real numbers $R$, the members of the set of complex numbers cannot be ordered. That is, we cannot assert that one complex number, say $(1, 5)$, is less than or greater than another, say $(4, 2)$.

As in the system of real numbers, two more operations defined in terms of addition and multiplication are very useful.

**Definition 8.4.**   *For all $z_1$ and $z_2$, elements in $C$,*

$$z_1 - z_2 = z_1 + (-z_2).$$

*The number $z_1 - z_2$ is called the **difference** of $z_1$ and $z_2$, and is viewed as the result of subtracting $z_2$ from $z_1$.*

**Definition 8.5.**   *For all $z_1$ and $z_2$, elements in C, where $z_2 \neq (0, 0)$,*

$$\frac{z_1}{z_2} = z_1 \cdot \frac{1}{z_2} \, .$$

*The number $z_1/z_2$ is called the **quotient** of $z_1$ and $z_2$ and is viewed as the result of dividing $z_1$ by $z_2$.*

Notice that these definitions are similar to the definitions for a difference and a quotient in the set of real numbers (see pages 7 and 8).

**Examples.**   Write each difference and each quotient as an ordered pair.
(a) $(3, 4) - (5, 2)$.          (b) $(3, 4)/(5, 2)$.

*Solutions.*   (a) From Definition 8.4,

$$(3, 4) - (5, 2) = (3, 4) + (-5, -2),$$

$$= (3 - 5, 4 - 2) = (-2, 2).$$

(b) From Definition 8.5,

$$\frac{(3, 4)}{(5, 2)} = (3, 4) \frac{1}{(5, 2)},$$

$$\frac{(3, 4)}{(5, 2)} = (3, 4) \left( \frac{5}{5^2 + 2^2}, \frac{-2}{5^2 + 2^2} \right) = (3, 4) \left( \frac{5}{29}, \frac{-2}{29} \right)$$

$$= \left[ 3 \cdot \frac{5}{29} - 4 \left( \frac{-2}{29} \right), 3 \left( \frac{-2}{29} \right) + 4 \left( \frac{5}{29} \right) \right]$$

$$= \left( \frac{15}{29} + \frac{8}{29}, \frac{-6}{29} + \frac{20}{29} \right) = \left( \frac{23}{29}, \frac{14}{29} \right).$$

One special case of a quotient of two complex numbers is especially useful.

**Theorem 8.4.**   *If $(a_1, b_1)$ and $(a_2, 0)$ are elements in C, where $a_2 \neq 0$, then*

$$\frac{(a_1, b_1)}{(a_2, 0)} = \left( \frac{a_1}{a_2}, \frac{b_1}{a_2} \right).$$

*Proof.*   From Definition 8.5 and Theorem 8.3, F-7,

$$\frac{(a_1, b_1)}{(a_2, 0)} = (a_1, b_1) \frac{1}{(a_2, 0)} = \frac{1}{(a_2, 0)} (a_1, b_1).$$

Since $(a_2, 0) = a_2$ and from Theorem 8.2, we have

$$\frac{1}{(a_2, 0)} (a_1, b_1) = \frac{1}{a_2} (a_1, b_1)$$

$$= \left( \frac{a_1}{a_2}, \frac{b_1}{a_2} \right).$$

Thus,

$$\frac{(a_1, b_1)}{(a_2, 0)} = \left(\frac{a_1}{a_2}, \frac{b_1}{a_2}\right).$$

**Examples.** Write each quotient as an ordered pair.
(a) $(2, 5)/(3, 0)$.        (b) $(8, -4)/(-2, 0)$.

*Solutions.* From Theorem 8.4,

$$\text{(a)} \quad \frac{(2, 5)}{(3, 0)} = \left(\frac{2}{3}, \frac{5}{3}\right).$$

$$\text{(b)} \quad \frac{(8, -4)}{(-2, 0)} = \left(\frac{8}{-2}, \frac{-4}{-2}\right) = (-4, 2).$$

Since Theorem 8.3 was derived exclusively from the corresponding axioms for the real numbers along with equality axioms, for each theorem that follows from these properties in the set $R$ (Theorems 1 to 16 in Appendix B, page 289) there is a corresponding theorem in the set $C$. We will not rewrite these theorems for elements in the set $C$. You can, however, use the theorems stated for elements in the set $R$, reworded appropriately for elements in the set $C$, as necessary.

One such theorem stated as:    For all $z_1, z_2, z_3 \in C$, where $z_2, z_3 \neq 0$,

$$\frac{z_1}{z_2} = \frac{z_1 z_3}{z_2 z_3}, \tag{1}$$

has an interesting application in the set $C$. To show such an application, we first make the following definition.

**Definition 8.6.**    *For all $z = (a, b)$ where $z \in C$, the **conjugate** of $z$ is given by*

$$\bar{z} = (a, -b).$$

For example, the conjugate of $(3, 4)$ is $(3, -4)$ and the conjugate of $(-2, -1)$ is $(-2, 1)$.

**Example.**    Write the quotient $(4, -1)/(2, 3)$ as an ordered pair, by first multiplying both the numerator and the denominator by the conjugate of the denominator.

*Solution.* From Equation (1) above and Definition 8.3-II,

$$\frac{(4, -1)}{(2, 3)} = \frac{(4, -1)(2, -3)}{(2, 3)(2, -3)} = \frac{(8 - 3, -12 - 2)}{(4 + 9, -6 + 6)} = \frac{(5, -14)}{(13, 0)}.$$

Now, from Theorem 8.4,

$$\frac{(5, -14)}{(13, 0)} = \left(\frac{5}{13}, \frac{-14}{13}\right).$$

In the preceding example, note that the product of the complex number $(2, 3)$ and its conjugate is the real number 13. (See Problem 29, Exercise 8.2.)

## EXERCISE 8.2

**A.**

Let $z_1 = (-8, -5)$, $z_2 = (4, 2)$, and $z_3 = (-4, -6)$. Show that the following statements are true.

**1.** $z_1 + z_2 \in C.$

**2.** $z_2 + z_3 = z_3 + z_2.$

**3.** $z_1 + (z_2 + z_3) = (z_1 + z_2) + z_3.$

**4.** $z_2 + (0, 0) = z_2.$

**5.** $(0, 0) + z_3 = z_3.$

**6.** $z_1 + (-z_1) = (0, 0).$

**7.** $z_1 + \bar{z}_1 \in R.$

**8.** $z_2 \cdot \bar{z}_2 \in R.$

Write each difference as an ordered pair.

**Example.** $(0, -4) - (-1, 10).$

*Solution.* From Definition 8.4 and Definition 8.3-I,

$$(0, -4) - (-1, 10) = [(0, -4) + (1, -10)]$$
$$= [0 + 1, -4 + (-10)]$$
$$= (1, -14).$$

**9.** $(5, 0) - (-7, 7).$

**10.** $(3, 1) - (1, -3).$

**11.** $(5, 7) - (0, 6).$

**12.** $(-7, -2) - (-3, 5).$

**13.** $(3, 1) - (-6, 5).$

**14.** $(-6, 4) - (-9, 0).$

Write each quotient as an ordered pair by first using Definition 8.5.

**Example.** $(-8, -3)/(9, 5).$

*Solution.*  $\dfrac{(-8, -3)}{(9, 5)} = (-8, -3)\dfrac{1}{(9, 5)}$

$$= (-8, -3)\left(\frac{9}{9^2 + 5^2}, \frac{-5}{9^2 + 5^2}\right)$$

$$= \left[-8\left(\frac{9}{106}\right) - (-3)\left(\frac{-5}{106}\right), -8\left(\frac{-5}{106}\right) - 3\left(\frac{9}{106}\right)\right]$$

$$= \left(\frac{-87}{106}, \frac{13}{106}\right).$$

**15.** $\dfrac{(9, 8)}{(4, 6)}.$

**16.** $\dfrac{(2, -7)}{(8, 3)}.$

**17.** $\dfrac{(-1, 6)}{(-3, 0)}.$

**18.** $\dfrac{(8, 0)}{(-3, 5)}.$

**19.** $\dfrac{(-1, 2)}{(-5, -4)}.$

**20.** $\dfrac{(0, -5)}{(0, -1)}.$

Write each quotient as an ordered pair by first using Equation (1), page 241.

**Example.** $(7, 0)/(-6, -4)$.

*Solution.* Multiplying numerator and denominator by the conjugate of the denominator yields

$$\frac{(7, 0)}{(-6, -4)} = \frac{(7, 0)(-6, 4)}{(-6, -4)(-6, 4)}$$

$$= \frac{[(7)(-6) - (0)(4), (7)(4) + (-6)(0)]}{[(-6)(-6) - (-4)(4), (-6)(4) + (-6)(-4)]} = \frac{(-42, 28)}{(52, 0)}.$$

By Theorem 8.4,

$$\frac{(-42, 28)}{(52, 0)} = \left(\frac{-42}{52}, \frac{28}{52}\right) = \left(\frac{-21}{26}, \frac{7}{13}\right).$$

21. $\dfrac{(-6, 0)}{(7, 2)}$.     22. $\dfrac{(-2, -7)}{(-9, 7)}$.     23. $\dfrac{(-8, 2)}{(-8, -1)}$.     24. $\dfrac{(-5, -4)}{(3, -8)}$.

**B.**

25. Prove Theorem 8.3, F-6.      27. Prove Theorem 8.3, F-8.
26. Prove Theorem 8.3, F-7.      28. Prove Theorem 8.3, F-11.

29. Show that if $z = (a, b)$, then

$$z + \bar{z} = (2a, 0) = 2a \quad \text{and} \quad z \cdot \bar{z} = (a^2 + b^2, 0) = a^2 + b^2.$$

30. Prove that for all $z \in C$, $\quad z \cdot (0, 0) = (0, 0)$.
31. Prove that if $z_1, z_2$, and $z_3 \in C$, and if $z_2$ and $z_3$ are not $(0, 0)$, then

$$\frac{z_1}{z_2} = \frac{z_1 z_3}{z_2 z_3}.$$

32. Show that Theorem 4 in $R$ (Appendix B, page 289) is also valid in $C$.

## 8.3
## Complex Numbers in the Form a + bi

There are various representations for complex numbers. As you will see in this and the following sections, sometimes one form can be more useful than another. We first designate a special symbol for the pure imaginary number $(0, b)$ which, as you observed in Section 8.1, has the property that

$$(0, b) \cdot (0, b) = (-b^2, 0) = -b^2, \tag{1}$$

a negative real number for $b > 0$ or $b < 0$.

Definition 8.7. *In the set C, for all b,*

$$(0, b) = bi.$$

*The number i is called the **imaginary unit**.*

For $b \geq 0$, $bi$ can be named by $\sqrt{-b^2}$ and for $b < 0$, $bi$ can be named by $-\sqrt{-b^2}$.

In particular, if $b = 1$,

$$(0, 1) = \sqrt{-1} = 1 \cdot i = i,$$

and if $b = -1$,

$$(0, -1) = -\sqrt{-1} = -1 \cdot i = -i.$$

From this definition and Equation (1) above

$$(0, b) \cdot (0, b) = (\pm\sqrt{-b^2})^2 = (bi)^2 = -b^2$$

and as a special case in the set of complex numbers, when $b = \pm 1$,

$$(\pm\sqrt{-1})^2 = (\pm 1 \cdot i)^2 = i^2 = -1.$$

Since any ordered pair $(a, b)$ can be represented by the sum $(a, 0) + (0, b)$, we have that for all $a, b \in R$,

$$(a, b) = (a, 0) + (0, b) = a + bi = a + b\sqrt{-1}.$$

Definitions which were adopted in Sections 8.1 and 8.2 for elements of $C$ in the form of ordered pairs can now be expressed in the form $a + bi$.

**Definition 8.8.**   *For all $a_1, b_1, a_2, b_2 \in R$ and $i = (0, 1) = \sqrt{-1}$:*

I.   $a_1 + b_1 i = a_2 + b_2 i$ *if and only if* $a_1 = a_2$ *and* $b_1 = b_2$.

II.   $(a_1 + b_1 i) + (a_2 + b_2 i) = (a_1 + a_2) + (b_1 + b_2)i$.

III.   $(a_1 + b_1 i) \cdot (a_2 + b_2 i) = (a_1 a_2 - b_1 b_2) + (a_1 b_2 + a_2 b_1)i$.

IV.   $(a_1 + b_1 i) - (a_2 + b_2 i) = (a_1 - a_2) + (b_1 - b_2)i$.

V.   $\dfrac{a_1 + b_1 i}{a_2 + b_2 i} = \dfrac{a_1 a_2 + b_1 b_2}{a_2{}^2 + b_2{}^2} + \dfrac{b_1 a_2 - a_1 b_2}{a_2{}^2 + b_2{}^2} i$,   $(a_2, b_2$ *both not* $\mathbf{0})$.

VI.   *The **conjugate** of $a_1 + b_1 i$ is given by* $a_1 - b_1 i$.

Now, expressions involving complex numbers in the form $a + bi$ can be treated as polynomials and rewritten in the same way that expressions involving real numbers are rewritten; in addition, $i^2$ *can be replaced with* $-1$.

**Examples.**   Write each of the following in the form $a + bi$.
(a) $(3 - i) + (2 + 3i)$.        (b) $(4 + 2i) - (2 - 3i)$.        (c) $(3 + i) \cdot (2 + i)$.

*Solutions.*

(a)   $(3 - i) + (2 + 3i) = (3 + 2) + (-1 + 3)i = 5 + 2i$.

(b)   $(4 + 2i) - (2 - 3i) = (4 - 2) + [(2 - (-3)]i = 2 + 5i$.

(c)   $(3 + i) \cdot (2 + i) = 6 + 5i + i^2 = 6 + 5i - 1 = 5 + 5i$.

In expressing the quotient of two complex numbers in the form $a + bi$, it is generally simpler to use Equation (1) page 241, and multiply the numerator and the denominator by the unit $i$ or the conjugate of the denominator as appropriate rather than to use Definition 8.8-V directly.

**Examples.** Write each quotient in the form $a + bi$.

(a) $\dfrac{2 - i}{3i}$.      (b) $\dfrac{3}{1 - 2i}$.      (c) $\dfrac{4 - i}{2 + 3i}$.

*Solutions.*

(a) $\dfrac{2 - i}{3i} = \dfrac{(2 - i)i}{3i \cdot i} = \dfrac{2i - i^2}{3i^2} = \dfrac{2i + 1}{-3} = -\dfrac{1}{3} - \dfrac{2}{3}i.$

(b) $\dfrac{3}{1 - 2i} = \dfrac{3(1 + 2i)}{(1 - 2i)(1 + 2i)} = \dfrac{3 + 6i}{1 - 4i^2} = \dfrac{3}{5} + \dfrac{6}{5}i.$

(c) $\dfrac{4 - i}{2 + 3i} = \dfrac{(4 - i)(2 - 3i)}{(2 + 3i)(2 - 3i)} = \dfrac{8 - 14i + 3i^2}{4 - 9i^2} = \dfrac{5 - 14i}{13} = \dfrac{5}{13} - \dfrac{14}{13}i.$

Relationships involving the square root symbol which are valid for real numbers are not necessarily valid when the symbol does not represent a real number. For example, for $a, b > 0$, the product of two real numbers

$$\sqrt{a} \cdot \sqrt{b} = \sqrt{a \cdot b},$$

however, the product of two imaginary numbers

$$\sqrt{-a}\,\sqrt{-b} \neq \sqrt{(-a)(-b)},$$

a positive real number. By Definition 8.7,

$$\sqrt{-a}\,\sqrt{-b} = i\sqrt{a} \cdot i\sqrt{b} = i^2\sqrt{ab} = -\sqrt{ab},$$

a negative real number. To avoid difficulty in rewriting products or quotients, *expressions of the form* $\sqrt{-b}$ $(b > 0)$ *should first be expressed in the form* $i\sqrt{b}$.

**Examples.** Write each expression in the form $a + bi$.

(a) $(3 - \sqrt{-5})(3 + \sqrt{-5}).$      (b) $\dfrac{1}{2 - \sqrt{-9}}$.

*Solutions.*

(a) $(3 - \sqrt{-5})(3 + \sqrt{-5}) = (3 - i\sqrt{5})(3 + i\sqrt{5})$

$$= 9 - i^2 \cdot 5 = 9 - (-1)(5) = 14.$$

(b) $\dfrac{1}{2 - \sqrt{-9}} = \dfrac{1}{2 - 3i} = \dfrac{1(2 + 3i)}{(2 - 3i)(2 + 3i)} = \dfrac{2 + 3i}{4 - 9i^2} = \dfrac{2}{13} + \dfrac{3}{13}i.$

The powers of $i$ exhibit an interesting periodic property. For example,

$$i = i,$$
$$i^2 = -1,$$
$$i^3 = i^2 \cdot i = -1 \cdot i = -i,$$
$$i^4 = i^2 \cdot i^2 = (-1)(-1) = 1,$$
$$i^5 = i^4 \cdot i = 1 \cdot i = i, \text{ etc.}$$

Thus, any power of $i$ can be expressed as one of the complex numbers $i$, $-1$, $-i$, or 1.

You probably used complex numbers in the form $a + bi$ in previous algebra courses when you wrote solutions for certain quadratic equations in one variable which were complex numbers. For example, using the quadratic formula (see page 27) to solve the equation

$$x^2 - x + 1 = 0,$$

the solution set is found to be

$$\left\{ \frac{1 + \sqrt{1-4}}{2}, \frac{1 - \sqrt{1-4}}{2} \right\} = \left\{ \frac{1}{2} + \frac{\sqrt{3}}{2}i, \frac{1}{2} - \frac{\sqrt{3}}{2}i \right\},$$

whose members are elements in $C$.

## EXERCISE 8.3

**A.**

Write each complex number in the form $a + bi$.

1. $(-8, 7)$.           3. $(-4, 0)$.           5. $(7, -\sqrt{2})$.
2. $(0, 8)$.            4. $(6, -5)$.           6. $(-\sqrt{3}, 1)$.

Write each complex number as an ordered pair.

7. $6i$.                9. $5 - 9i$.            11. $8 + i$.
8. $-2 + 7i$.           10. $-1 + 6i$.          12. $-5$.

Write the conjugate $\bar{z}$ of each complex number $z$.

**Example.**  (a) $z = -9 - 4i$.       (b) $z = 5 + 8i$.

Solutions.   (a) $\bar{z} = -9 + 4i$.      (b) $\bar{z} = 5 - 8i$.

13. $6i$.               15. $3 - 5i$.           17. $-3 - 2i$.
14. $-1 - i$.           16. $8 - 3i$.           18. $7$.

Find real numbers for $x$ and $y$ for which each statement is true.

**Example.**   $x + 2yi = -3 + 6i$.

Solution.   By Definition 8.8-I,   $x + 2yi = -3 + 6i$   if and only if

$$x = -3 \quad \text{and} \quad 2y = 6,$$

or equivalently, if

$$x = -3 \quad \text{and} \quad y = 3.$$

**19.** $-5x - 6yi = -5 - i.$                    **21.** $(x - yi)^2 = 2yi.$

**20.** $y + 4xi = 9 - 6i.$                     **22.** $x - 4i = 2y - x^2 i.$

Write each expression in the form $a + bi$.

***Examples.***   (a) $(6 - 8i) + (-2 - 5i).$        (b) $(3 - 9i) \cdot (-4 + 6i).$

*Solutions.*

(a)   $(6 - 8i) + (-2 - 5i) = [6 + (-2)] + [-8 + (-5)]i = 4 - 13i.$

(b)   $(3 - 9i) \cdot (-4 + 6i) = -12 + 18i + 36i - 54i^2$
$$= -12 + 54i + 54 = 42 + 54i.$$

**23.** $(5 - 9i) + (-4 + 5i).$                  **28.** $(-1 + \sqrt{-8}) \cdot (-7 + \sqrt{-16}).$

**24.** $(4 + 7i) - (6 + 9i).$                   **29.** $(8i)^2.$

**25.** $(2 + 8i) \cdot (-8 - 7i).$               **30.** $(-4 - 6i)^2.$

**26.** $(6 - \sqrt{-4}) + (-2 + \sqrt{-9}).$      **31.** $(-2 - \sqrt{-9})^2.$

**27.** $(4 - \sqrt{-5}) - (-1 + \sqrt{-1}).$      **32.** $(8 + i)^2 \cdot (-7 + \sqrt{-4}).$

Write each quotient in the form $a + bi$.

***Examples.***   (a) $\dfrac{8 - 2i}{8i}.$        (b) $\dfrac{-4}{-3 + i}.$        (c) $\dfrac{-6 + \sqrt{-4}}{7 - \sqrt{-9}}.$

*Solutions.*

(a)   $\dfrac{8 - 2i}{8i} = \dfrac{(8 - 2i)i}{8i \cdot i} = \dfrac{8i - 2i^2}{8i^2} = \dfrac{8i + 2}{-8} = \dfrac{-1}{4} - i.$

(b)   $\dfrac{-4}{-3 + i} = \dfrac{(-4)(-3 - i)}{(-3 + i)(-3 - i)} = \dfrac{12 + 4i}{9 + 3i - 3i - i^2} = \dfrac{12 + 4i}{9 + 1} = \dfrac{6}{5} + \dfrac{2}{5}i.$

(c)   $\dfrac{-6 + \sqrt{-4}}{7 - \sqrt{-9}} = \dfrac{-6 + 2i}{7 - 3i} = \dfrac{(-6 + 2i)(7 + 3i)}{(7 - 3i)(7 + 3i)} = \dfrac{-42 - 18i + 14i + 6i^2}{49 + 21i - 21i - 9i^2}$
$$= \dfrac{-42 - 4i - 6}{49 + 9} = \dfrac{-48 - 4i}{58} = \dfrac{-24}{29} - \dfrac{2}{29}i.$$

**33.** $\dfrac{-9 - 6i}{-4i}.$       **35.** $\dfrac{4}{2 - 9i}.$       **37.** $\dfrac{3 + i}{-8 + 5i}.$       **39.** $\dfrac{6 + \sqrt{-9}}{\sqrt{-25}}.$

**34.** $\dfrac{-2 + 3i}{3i}.$        **36.** $\dfrac{-5i}{-1 + 6i}.$      **38.** $\dfrac{-5 + 9i}{6 - 2i}.$       **40.** $\dfrac{5 - \sqrt{-2}}{-4 - \sqrt{-6}}.$

Write each expression as one of the complex numbers $i$, $-1$, $-i$, or 1.

**41.** $i^7$.          **42.** $i^{29}$.          **43.** $\dfrac{1}{i^6}$.          **44.** $\dfrac{1}{i^{25}}$.

Write the elements in the solution set of each equation in the form $a + bi$.

**Example.**   $x^2 + 2x + 2 = 0$.

*Solution.*   From the quadratic formula (see page 27),

$$x = \frac{-2 \pm \sqrt{4 - 8}}{2},$$

from which the solution set is

$$\{-1 + i, \ -1 - i\}.$$

**45.** $x^2 + 6x + 13 = 0$.                    **47.** $x^2 + 9 = -2x$.
**46.** $x^2 - 8x + 25 = 0$.                    **48.** $x^2 = x - \frac{17}{4}$.

**B.**

**49.** Show that $z - \bar{z}$ is a pure imaginary number.

**50.** Show that $\bar{z}_1 + \bar{z}_2 = \overline{z_1 + z_2}$.

**51.** Show that $\bar{z}_1 \cdot \bar{z}_2 = \overline{z_1 \cdot z_2}$.

**52.** Show that $\dfrac{\bar{z}_1}{\bar{z}_2} = \overline{\left(\dfrac{z_1}{z_2}\right)}$.

**53.** Under what conditions are the elements in the solution set of the quadratic equation

$$ax^2 + bx + c = 0, \qquad a, b, c, x \in R$$

not elements of $R$?

## 8.4
### Graphical Representation

Since the elements in $\{(a, b) \mid a, b \in R\}$ are in a one-to-one correspondence with the set of points in the geometric plane, each point in the plane can be viewed as the graph of a complex number. A plane on which complex numbers are thus represented is called a complex plane or an Argand plane, after the French mathematician Jean Argand (1768–1822). Since the elements in $\{(a, 0) \mid a \in R\}$, or simply $\{a \mid a \in R\}$, are associated with the points on the $x$ axis, it is called the **real axis** in the complex plane. Similarly, the elements in $\{(0, b) \mid b \in R\}$ are associated with points on the $y$ axis and this axis is called the **imaginary axis**.

**Examples.**   Graph each complex number, its negative, and its conjugate.
(a) $(2, -6)$.          (b) $-3 - 4i$.

*Solutions.* (a) The negative is $(-2, 6)$; the conjugate is $(2, 6)$.
 (b) The negative is $3 + 4i$; the conjugate is $-3 + 4i$.

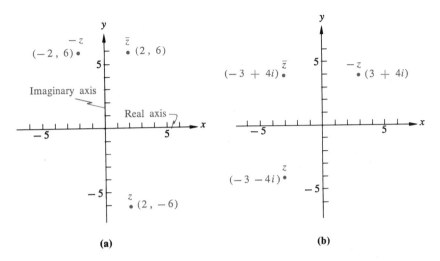

(a)   (b)

Since the operation for addition of complex numbers (Definition 8.3-I) is analogous to vector addition (Definition 7.7), complex numbers can be added graphically by the parallelogram law applicable to geometric vectors.

**Examples.** Represent each sum graphically and check the results by analytical methods.
(a) $(2, 3) + (4, -5)$.   (b) $(3 + 5i) + (-6 + 4i)$.

*Solutions.*

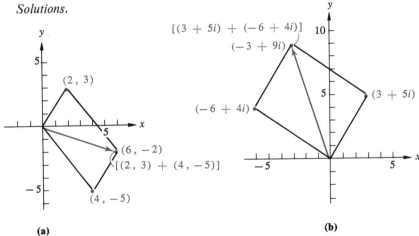

(a)   (b)

(a) Check: $(2, 3) + (4, -5) = (6, -2)$.
(b) Check: $(3 + 5i) + (-6 + 4i) = -3 + 9i$.

Corresponding to the norm $\|\mathbf{v}\|$ (a real number) and direction angle $\alpha$ for each vector $\mathbf{v}$, there are a corresponding real number and a set of angles associated with each complex number.

**Definition 8.9.**   *For all $z \in C$, the **absolute value** or **modulus** of $z = (a, b) = a + bi$ is given by*

$$\rho = |z| = |(a, b)| = |a + bi| = \sqrt{a^2 + b^2}.$$

Thus the modulus is the distance from the origin to the graph of $(a, b)$ (see Figure 8.2).

**Definition 8.10.**   *For all $z \in C$, an **argument**, or **amplitude** of $z = (a, b) = a + bi$, denoted by **arg**$(a, b)$ or **arg**$(a + bi)$ is an angle $\theta$ given by*

$$\cos \theta = \frac{a}{\sqrt{a^2 + b^2}} = \frac{a}{\rho} \quad \text{or} \quad \sin \theta = \frac{b}{\sqrt{a^2 + b^2}} = \frac{b}{\rho}, \quad (a^2 + b^2 \neq 0).$$

Thus, the arg$(a + bi)$ is the angle $\theta$ with initial side the positive $x$ axis and ter- minal side the ray from the origin through the graph of $(a, b)$, (see Figure 8.2). If $\theta$ is an argument of $a + bi$, then for $k \in J$, so is $(\theta + k \cdot 360)°$. For $a + bi = (0, 0)$; that is, for $a^2 + b^2 = 0$, any angle $\theta$ can be used as an argument. The angle $\theta$ with the least positive measure is sometimes called the **principal argument**.

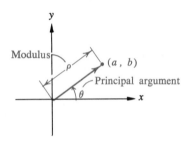

**FIGURE 8.2**

*Example.*   Find the absolute value and principal argument of $(\sqrt{3}, 1)$.

*Solution.*   $\rho = |(\sqrt{3}, 1)| = \sqrt{(\sqrt{3})^2 + 1^2} = \sqrt{3 + 1} = 2.$

$$\cos \theta = \frac{a}{\rho} = \frac{\sqrt{3}}{2};$$

because $\theta$ is in Quadrant I, $\theta = 30°$.

### EXERCISE 8.4

A.

Graph each complex number $z$, its negative $-z$, and its conjugate $\bar{z}$.

*Examples.*   (a) $(3, 2)$.        (b) $-5 - 4i$.

*Solutions.*   (a) The negative is $(-3, -2)$; the conjugate is $(3, -2)$.
(b) The negative is $5 + 4i$; the conjugate is $-5 + 4i$.

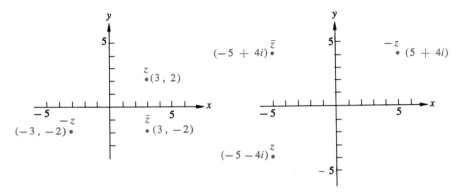

**¹1.** $(4, 6)$.

**2.** $(-5, -3)$.

**3.** $(-1, 0)$.

**4.** $2 - 7i$.

**⁵5.** $-6 + 6i$.

**6.** $4 + 3i$.

**7.** $5$.

**8.** $2i$.

Represent each sum graphically. Check the results by analytical methods.

**Example.** $(8 + 3i) + (1 - 5i)$.

*Solution.*

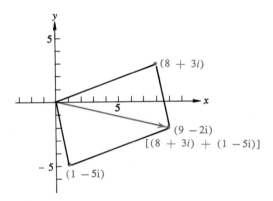

Check: $(8 + 3i) + (1 - 5i) = (8 + 1) + (3 - 5)i = 9 - 2i$.

**⁹9.** $2 + (-3 + 6i)$.

**10.** $(2 + 4i) + (1 - 5i)$.

**11.** $(7 - i) + 2i$.

**12.** $(-6 + 7i) + (5 - 3i)$.

Find the absolute value and principal argument of each complex number.

**Examples.** (a) $(3, 9)$.      (b) $-7 + 5i$.

*Solutions.* From Definitions 8.9 and 8.10,

$$(a) \quad \rho = |(3, 9)| = \sqrt{3^2 + 9^2} = \sqrt{90} = 3\sqrt{10};$$

*(Solution continued on the next page.)*

$$\cos \theta = \frac{3}{3\sqrt{10}} \approx \frac{1}{3.162} \approx 0.316.$$

Because $\theta$ is in Quadrant I, $\theta \approx 71° \ 30'$.

(b)   $\rho = |-7 + 5i| = \sqrt{(-7)^2 + 5^2} = \sqrt{74};$

$$\cos \theta = -\frac{7}{\sqrt{74}} \approx -\frac{7}{8.602} \approx -0.814.$$

Because $\theta$ is in Quadrant II, $\theta \approx 144° \ 30'$.

**13.** $(3, -9)$.          **17.** $-5 + 2i$.          **21.** $7$.

**14.** $(-7, -4)$.          **18.** $6 + 8i$.          **22.** $-4$.

**15.** $(4, 0)$.          **19.** $-2 - 2i$.          **23.** $-5i$.

**16.** $(0, -7)$.          **20.** $8 - 3i$.          **24.** $6i$.

**25.** Determine the conditions that $a$ and $b$ must satisfy if the graph of the complex number $a + bi$ is:
(a) on the real axis,                  (b) on the imaginary axis,
(c) above the real axis,               (d) below the real axis,
(e) to the right of the imaginary axis,
(f) to the left of the imaginary axis.

**26.** What condition does $a^2 + b^2$ satisfy if the graph of $a + bi$ is:
(a) on a circle with radius of length 5,
(b) inside a circle with radius of length 5,
(c) outside a circle with radius of length 5.

**B.**

Graphically, the multiplication or division of any complex number $a + bi$ by the complex number $i$ produces a 90° rotation of the geometric vector corresponding to $a + bi$, counterclockwise for multiplication and clockwise for division. (The proof of this statement follows more readily after we have discussed the trigonometric form of a complex number in Section 8.5. See Exercise 8.5, Problems 35 and 36.)

In Problems 27 to 32, graph:
(a) the given complex number $a + bi$.
(b) $i \cdot (a + bi)$.          (c) $(a + bi)/i$.

**Example.**   $2 + 3i$.

Solution.   $i \cdot (2 + 3i) = 2i + 3i^2$
$\qquad\qquad\qquad = -3 + 2i;$

$$\frac{2 + 3i}{i} = \frac{(2 + 3i) \cdot i}{i \cdot i}$$

$$= \frac{2i + 3i^2}{i^2} = \frac{2i - 3}{-1} = 3 - 2i.$$

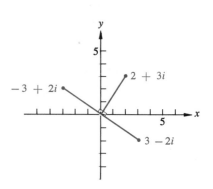

**◦27.** $5 + 4i$.      **30.** $-3 - 4i$.

**28.** $1 - 2i$.      **◦31.** $2$.

**29.** $-6 + i$.      **32.** $-5i$.

**33.** Prove that $|z_1 \cdot z_2| = |z_1| \cdot |z_2|$.

**34.** Prove that $\left|\dfrac{z_1}{z_2}\right| = \dfrac{|z_1|}{|z_2|}$,      $[z_2 \neq (0, 0)]$.

## 8.5

### Trigonometric Form

Because $\cos \theta = a/\rho$ and $\sin \theta = b/\rho$ for $\rho \neq 0$ (see Figure 8.3), then

$$a = \rho \cos \theta \quad \text{and} \quad b = \rho \sin \theta.$$

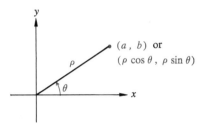

Thus, the components of a nonzero complex number $(a, b)$ can be expressed in terms of its absolute value $\rho = \sqrt{a^2 + b^2}$ and its argument $\theta$. Also, since $(a, b) = a + bi$, then

**FIGURE 8.3**

$$(a, b) = \rho \cos \theta + \rho \sin \theta \cdot i,$$

from which

$$(a, b) = \rho(\cos \theta + i \sin \theta).*$$

The right-hand member of either equation is called the **trigonometric form** for a complex number.

In general, for angles expressed in degree measure

$$a + bi = \rho[\cos (\theta + k \cdot 360°) + i \sin (\theta + k \cdot 360°)], \; k \in J,$$

where $\theta$ is the angle of smallest nonnegative measure.

***Example.*** Represent $\sqrt{3} - i$ in trigonometric form.

*Solution.* $\rho = |\sqrt{3} - i| = \sqrt{3 + 1} = 2$; $\cos \theta = \sqrt{3}/2$, and $\theta = 330°$.

Therefore, in general,

$$\sqrt{3} - i = 2[\cos (330° + k \cdot 360°) + i \sin (330° + k \cdot 360°)], \; k \in J.$$

For $k = 0$,

$$\sqrt{3} - i = 2 (\cos 330° + i \sin 330°).$$

* The expression $\cos \theta + i \sin \theta$ is sometimes abbreviated as cis $\theta$.

*Example.*  Represent $6 (\cos 315° + i \sin 315°)$ graphically and express the number in the form $a + bi$.

*Solution.*  Since

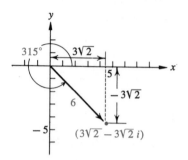

$$a = \rho \cos \theta = 6\left(\frac{\sqrt{2}}{2}\right) = 3\sqrt{2}, \text{ and}$$

$$b = \rho \sin \theta = 6\left(\frac{-\sqrt{2}}{2}\right) = -3\sqrt{2},$$

$$a + bi = 3\sqrt{2} - 3\sqrt{2}i.$$

Products and quotients of complex numbers written in trigonometric form can be found quite easily. Consider the following result which will be very useful in Section 8.6.

**Theorem 8.5.**  *If $z_1, z_2 \in C$,*

$$z_1 = \rho_1(\cos \theta_1 + i \sin \theta_1) \text{ and } z_2 = \rho(\cos \theta_2 + i \sin \theta_2), \text{ then}$$

$$\text{I. } z_1 \cdot z_2 = \rho_1\rho_2[\cos (\theta_1 + \theta_2) + i \sin (\theta_1 + \theta_2)],$$

$$\text{II. } \frac{z_1}{z_2} = \frac{\rho_1}{\rho_2}[\cos (\theta_1 - \theta_2) + i \sin (\theta_1 - \theta_2)], \qquad z_2 \neq (0, 0).$$

*Proof of I.*  Because

$$z_1 = \rho_1(\cos \theta_1 + i \sin \theta_1) \qquad \text{and} \qquad z_2 = \rho_2(\cos \theta_2 + i \sin \theta_2),$$

$$z_1 \cdot z_2 = \rho_1(\cos \theta_1 + i \sin \theta_1) \cdot \rho_2(\cos \theta_2 + i \sin \theta_2)$$

$$= \rho_1\rho_2[(\cos \theta_1 \cos \theta_2 - \sin \theta_1 \sin \theta_2) + i(\cos \theta_1 \sin \theta_2 + \sin \theta_1 \cos \theta_2)].$$

From Theorems 2.3 and 2.7, the equation can be written

$$z_1 \cdot z_2 = \rho_1\rho_2[\cos (\theta_1 + \theta_2) + i \sin (\theta_1 + \theta_2)].$$

The proof of part II is similar and is left as an exercise.

*Example.*  Rewrite the product $6(\cos 75° + i \sin 75°) \cdot 3(\cos 15° + i \sin 15°)$ in the form $a + bi$.

*Solution.*  From Theorem 8.5-I,

$$6(\cos 75° + i \sin 75°) \cdot 3(\cos 15° + i \sin 15°)$$
$$= 6 \cdot 3[\cos (75° + 15°) + i \sin (75° + 15°)]$$
$$= 18(\cos 90° + i \sin 90°)$$
$$= 18(0 + i) = 18i.$$

*Example.*  Rewrite the quotient $\dfrac{6(\cos 75° + i \sin 75°)}{3(\cos 15° + i \sin 15°)}$ in the form $a + bi$.

*Solution.* From Theorem 8.5-II,

$$\frac{6(\cos 75° + i \sin 75°)}{3(\cos 15° + i \sin 15°)} = \frac{6}{3}[\cos (75° - 15°) + i \sin (75° - 15°)]$$

$$= 2(\cos 60° + i \sin 60°)$$

$$= 2\left(\frac{1}{2} + i\frac{\sqrt{3}}{2}\right) = 1 + \sqrt{3}i.$$

### EXERCISE 8.5

**A.**

Represent each complex number in trigonometric form, using the angle with least positive measure.

*Example.* $-2 + 2i$.

*Solution.* From Definitions 8.9 and 8.10,

$$\rho = |-2 + 2i| = \sqrt{(-2)^2 + 2^2} = \sqrt{8} = 2\sqrt{2}; \quad \cos \theta = \frac{-2}{2\sqrt{2}} = \frac{-1}{\sqrt{2}}.$$

Because $\theta$ is in Quadrant II, $\theta = 135°$. Thus

$$-2 + 2i = 2\sqrt{2}\,[\cos (135° + k \cdot 360°) + i \sin (135° + k \cdot 360°], \ k \in J.$$

For $k = 0$,

$$-2 + 2i = 2\sqrt{2}(\cos 135° + i \sin 135°).$$

**1.** $-5\sqrt{3} - 5i$.     **4.** $6 + 6i$.     **7.** $-8$.
**2.** $2 - 2i$.     **5.** $-4 - 4i$.     **8.** $-9i$.
**3.** $-2 + 2\sqrt{3}i$.     **6.** $3\sqrt{3} + 3i$.

Represent each complex number graphically and write it in the form $a + bi$.

*Example.* $3(\cos 60° + i \sin 60°)$.

*Solution.* Because

$$a = \rho \cos \theta = 3 \cos 60° = 3\left(\frac{1}{2}\right) = \frac{3}{2}, \text{ and}$$

$$b = \rho \sin \theta = 3 \sin 60° = 3\left(\frac{\sqrt{3}}{2}\right) = \frac{3\sqrt{3}}{2},$$

$$3(\cos 60° + i \sin 60°) = \frac{3}{2} + \frac{3\sqrt{3}}{2}i.$$

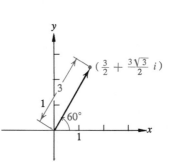

**°9.** $6(\cos 135° + i \sin 135°)$.

**10.** $4(\cos 30° + i \sin 30°)$.

**11.** $5(\cos 240° + i \sin 240°)$.

**12.** $\cos 90° + i \sin 90°$.

**°13.** $7[\cos(-45°) + i \sin(-45°)]$.

**14.** $\frac{1}{4}(\cos 150° + i \sin 150°)$.

**15.** $\frac{8}{5}(\cos 330° + i \sin 330°)$.

**16.** $\frac{2}{3}[\cos(-210°) + i \sin(-210°)]$.

Rewrite each product in the form $a + bi$.

**Example.**   $9(\cos 160° + i \sin 160°) \cdot 3(\cos 50° + i \sin 50°)$.

**Solution.**   From Theorem 8.5-I,

$$9(\cos 160° + i \sin 160°) \cdot 3(\cos 50° + i \sin 50°)$$

$$= 9 \cdot 3[\cos(160° + 50°) + i \sin(160° + 50°)]$$

$$= 27(\cos 210° + i \sin 210°)$$

$$= 27\left[\frac{-\sqrt{3}}{2} + i\left(-\frac{1}{2}\right)\right] = \frac{-27\sqrt{3}}{2} - \frac{27}{2}i.$$

**17.** $3(\cos 170° + i \sin 170°) \cdot (\cos 145° + i \sin 145°)$.

**18.** $6(\cos 110° + i \sin 110°) \cdot \frac{2}{3}(\cos 115° + i \sin 115°)$.

**19.** $2(\cos 75° + i \sin 75°) \cdot \frac{9}{4}(\cos 105° + i \sin 105°)$.

**20.** $\sqrt{5}(\cos 80° + i \sin 80°) \cdot \sqrt{5}(\cos 40° + i \sin 40°)$.

**21.** $7(\cos 10° + i \sin 10°) \cdot 10(\cos 35° + i \sin 35°)$.

**22.** $6(\cos 5° + i \sin 5°) \cdot 4(\cos 25° + i \sin 25°)$.

In Problems 23 to 28, rewrite each quotient in the form $a + bi$.

**Example.**   $\dfrac{10(\cos 65° + i \sin 65°)}{2(\cos 20° + i \sin 20°)}$.

**Solution.**   From Theorem 8.5-II,

$$\frac{10(\cos 65° + i \sin 65°)}{2(\cos 20° + i \sin 20°)} = \frac{10}{2}[\cos(65° - 20°) + i \sin(65° - 20°)]$$

$$= 5(\cos 45° + i \sin 45°)$$

$$= \frac{5}{\sqrt{2}} + \frac{5}{\sqrt{2}}i.$$

**23.** $\dfrac{8(\cos 410° + i \sin 410°)}{4(\cos 80° + i \sin 80°)}$.

**24.** $\dfrac{3(\cos 215° + i \sin 215°)}{4(\cos 35° + i \sin 35°)}$.

**25.** $\dfrac{2(\cos 135° + i \sin 135°)}{3(\cos 15° + i \sin 15°)}$.

**26.** $\dfrac{\cos 280° + i \sin 280°}{6(\cos 55° + i \sin 55°)}$.

**27.** $\dfrac{\sqrt{15}(\cos 70° + i \sin 70°)}{\sqrt{5}(\cos 100° + i \sin 100°)}$.

**28.** $\dfrac{4(\cos 19° + i \sin 19°)}{5(\cos 259° + i \sin 259°)}$.

**B.**

**29.** Write $[2(\cos 15° + i \sin 15°)]^3$ in the form $a + bi$.

**30.** Write $\left(\dfrac{1}{2} + \dfrac{\sqrt{3}}{2} i\right)^3$ in the form $a + bi$.

**31.** Show that if $a + bi = \rho(\cos \theta + i \sin \theta)$, then

$$(a + bi)^2 = \rho^2(\cos 2\theta + i \sin 2\theta).$$

**32.** Use the result of Problem 31 to show that if $a + bi = \rho(\cos \theta + i \sin \theta)$, then $(a + bi)^3 = \rho^3(\cos 3\theta + i \sin 3\theta)$.

**33.** Using the results of Problems 31 and 32, what can you conjecture about $(a + bi)^n$, $n \in N$, if $a + bi = \rho(\cos \theta + i \sin \theta)$?

**34.** Prove Theorem 8.5-II.

**35.** Show that multiplication of a complex number $\rho(\cos \theta + i \sin \theta)$ by $i$ leaves the modulus unchanged, but changes the direction of the corresponding geometric vector by 90° in a counterclockwise direction. *Hint:* $i = 1(\cos 90° + i \sin 90°)$.

**36.** Show that division of a complex number $\rho(\cos \theta + i \sin \theta)$ by $i$ leaves the modulus unchanged, but changes the direction of the corresponding geometric vector by 90° in a clockwise direction.

**C.**

In the study of a certain function called the exponential function (see page 33), there is an irrational number $e$, approximately equal to 2.71818, and it is shown in the calculus that if $z \in C$ and $n \in N$, then

$$e^z = 1 + z + \frac{z^2}{2!} + \frac{z^3}{3!} + \cdots + \frac{z^{n-1}}{(n-1)!} + \cdots,$$

and in particular for $z = ix$, $x \in R$,

$$e^{ix} = 1 + ix - \frac{x^2}{2!} - i\frac{x^3}{3!} + \cdots + \frac{(ix)^{n-1}}{(n-1)!} + \cdots.$$

On page 79, you observed that for $x \in R$ and $n \in N$,

$$\cos x = 1 - \frac{x^2}{2!} + \frac{x^4}{4!} + \cdots + (-1)^{n-1}\frac{x^{2n-2}}{(2n-2)!} + \cdots,$$

and

$$\sin x = x - \frac{x^3}{3!} + \frac{x^5}{5!} + \cdots + (-1)^{n-1}\frac{x^{2n-1}}{(2n-1)!} + \cdots.$$

Thus, the terms in the series for $e^{ix}$ are alternately the terms in the series for $\cos x$ and $i$ times the terms in the series for $\sin x$ and hence $e^{ix}$ can be written as

$$e^{ix} = \cos x + i \sin x.$$

This formula is known as **Euler's formula** in honor of its discoverer, the Swiss mathematician Leonhard Euler (1701–1783).

In Problems 37 and 38, write each power in the form $a + bi$.

**37.** $e^{\pi i}$.                                     **38.** $e^{2\pi i}$.

**39.** Show that

$$\sin x = \frac{e^{ix} - e^{-ix}}{2i} \quad \text{and} \quad \cos x = \frac{e^{ix} + e^{-ix}}{2}.$$

*Hint*: First use Euler's formula to show that $e^{-ix} = \cos x - i \sin x$.

**40.** Show that

$$\cos ix = \frac{e^x + e^{-x}}{2} \quad \text{and} \quad \sin ix = \frac{e^{-x} - e^x}{2i}.$$

**41.** Show that $\cos^2 ix + \sin^2 ix = 1$.   *Hint*: Use the results of Problem 40.

## 8.6

### De Moivre's Theorem; Powers and Roots

Consider the complex number $z = a + bi = \rho(\cos \theta + i \sin \theta)$. From Theorem 8.5-I,

$$[\rho(\cos \theta + i \sin \theta)]^2 = \rho(\cos \theta + i \sin \theta) \cdot \rho(\cos \theta + i \sin \theta)$$

$$= \rho^2(\cos 2\theta + i \sin 2\theta),$$

$$[\rho(\cos \theta + i \sin \theta)]^3 = \rho^2(\cos 2\theta + i \sin 2\theta) \cdot \rho(\cos \theta + i \sin \theta)$$

$$= \rho^3(\cos 3\theta + i \sin 3\theta), \quad \text{etc.}$$

It can be shown that similar results are valid for each such power for $n \in N$. That is, if $z \in C$, $z = \rho(\cos \theta + i \sin \theta)$, and $n \in N$, then

$$z^n = \rho^n(\cos n\theta + i \sin n\theta). \tag{1}$$

The proof of this statement, known as **De Moivre's theorem**, after the French mathematician Abraham De Moivre (1667–1754), and extensions of this statement which are presented in this section, involve the process of mathematical induction and will not be shown. However, the informal argument above should make the validity of Equation (1) seem plausible.

***Example.***   Express $(\sqrt{2} + \sqrt{2}i)^5$ in the form $a + bi$.

*Solution.*   From Definitions 8.9 and 8.10,

$$\rho = |\sqrt{2} + \sqrt{2}i| = \sqrt{2 + 2} = 2; \qquad \cos \theta = \sqrt{2}/2 = 1/\sqrt{2}.$$

Because $\theta$ is in Quadrant I, $\theta = 45°$.

Thus,

$$(\sqrt{2} + \sqrt{2}i)^5 = [2(\cos 45° + i \sin 45°)]^5$$

$$= 2^5(\cos 5 \cdot 45° + i \sin 5 \cdot 45°)$$

$$= 32(\cos 225° + i \sin 225°)$$

$$= 32\left(-\frac{\sqrt{2}}{2} - \frac{\sqrt{2}}{2}i\right) = -16\sqrt{2} - 16\sqrt{2}i.$$

Equation (1) was stated for powers with natural number exponents. If this statement is to be valid for integral exponents, what meanings must be assigned to $z^0$ and $z^{-n}$ for $n \in N$? Consider the following.

Substituting 0 for $n$ in Equation (1), we have

$$z^0 = \rho^0(\cos 0 \cdot \theta + i \sin 0 \cdot \theta) = 1 \cdot (\cos 0 + i \sin 0)$$

$$= 1 \cdot (1 + 0) = 1.$$

Substituting $-n$ for $n$ in Equation (1), we have

$$z^{-n} = \rho^{-n}[\cos(-n\theta) + i \sin(-n\theta)].$$

For $z \neq 0 + 0i$, $\rho \in R$ and $\rho^{-n} = \dfrac{1}{\rho^n}$,

$$z^{-n} = \frac{1}{\rho^n}[\cos(-n\theta) + i \sin(-n\theta)].$$

From Theorem 2.2, $\cos(-n\theta) = \cos n\theta$ and $\sin(-n\theta) = -\sin n\theta$. Therefore,

$$z^{-n} = \frac{1}{\rho^n} \cdot (\cos n\theta - i \sin n\theta).$$

Multiplying the numerator and the denominator of the right-hand member by the conjugate of the numerator yields

$$z^{-n} = \frac{1}{\rho^n} \cdot \frac{(\cos n\theta - i \sin n\theta) \cdot (\cos n\theta + i \sin n\theta)}{\cos n\theta + i \sin n\theta}$$

$$= \frac{1}{\rho^n} \cdot \frac{\cos^2 n\theta + \sin^2 n\theta}{\cos n\theta + i \sin n\theta}$$

$$= \frac{1}{\rho^n} \cdot \frac{1}{\cos n\theta + i \sin n\theta} = \frac{1}{\rho^n(\cos n\theta + i \sin n\theta)}$$

Substituting $z^n$ for $\rho^n(\cos n\theta + i \sin n\theta)$ yields

$$z^{-n} = \frac{1}{z^n}.$$

Thus, if we want Equation (1) to be valid for every exponent $n \in J$, $z^0$ and $z^{-n}$ must be given the following interpretations.

**Definition 8.11.** *For all $z \neq 0 + 0i$,*

$$\text{I. } z^0 = 1 + 0i = 1.$$

$$\text{II. } z^{-n} = \frac{1}{z^n}, \quad \text{for } n \in N.$$

De Moivre's theorem can now be expressed for powers with integral exponents.

**Theorem 8.6.** *If $z \in C$, $z \neq 0 + 0i$, $n \in J$, and $z = \rho(\cos \theta + i \sin \theta)$, then*

$$z^n = \rho^n(\cos n\theta + i \sin n\theta).$$

*Example.* Express $(-\sqrt{3} + i)^{-3}$ in the form $a + bi$.

*Solution.* From Definitions 8.9 and 8.10,

$$\rho = |-\sqrt{3} + i| = \sqrt{3 + 1} = 2; \quad \cos \theta = -\sqrt{3}/2.$$

Because $\theta$ is in Quadrant II, $\theta = 150°$. Thus,

$$(-\sqrt{3} + i)^{-3} = [2(\cos 150° + i \sin 150°)]^{-3},$$

from which by Theorem 8.6 we have

$$(-\sqrt{3} + i)^{-3} = \frac{1}{2^3} [\cos(-3 \cdot 150°) + i \sin(-3 \cdot 150°)]$$

$$= \tfrac{1}{8}[\cos(-450°) + i \sin(-450°)]$$

$$= \tfrac{1}{8}[0 - 1 \cdot i] = -\tfrac{1}{8}i.$$

Another extension of De Moivre's theorem is possible for rational number exponents $1/n$, $n \in N$. Consider the consequences of substituting $z^{1/n}$ for $z$ in the equation $z^n = [\rho(\cos \theta + i \sin \theta)]^n$. We have that

$$(z^{1/n})^n = ([\rho(\cos \theta + i \sin \theta)]^{1/n})^n$$

$$= \left[ \rho^{1/n}\left(\cos \frac{\theta}{n} + i \sin \frac{\theta}{n}\right) \right]^n$$

$$= (\rho^{1/n})^n \left(\cos n \cdot \frac{\theta}{n} + i \sin n \cdot \frac{\theta}{n}\right)$$

$$= \rho(\cos \theta + i \sin \theta)$$

$$= z.$$

Therefore, for De Moivre's theorem to be valid for rational number exponents $1/n$, $n \in N$, the complex number $z^{1/n}$ is given an appropriate meaning.

**Definition 8.12.** *For all $z \in C$, $n \in N$, $z^{1/n}$ is an* **nth root of z**. *That is,*

$$(z^{1/n})^n = \quad .$$

We now have the following extension of De Moivre's theorem.

**Theorem 8.7.**   *If $z \in C$, $n \in N$, and $z = \rho(\cos \theta + i \sin \theta)$, then*

$$z^{1/n} = \rho^{1/n}\left(\cos \frac{\theta}{n} + i \sin \frac{\theta}{n}\right).$$

Since

$$\cos \frac{\theta}{n} + i \sin \frac{\theta}{n} = \cos\left(\frac{\theta + k \cdot 360°}{n}\right) + i \sin\left(\frac{\theta + k \cdot 360°}{n}\right), \qquad k \in J,$$

*it is possible to find n distinct complex nth roots for each $z \in C$, $z \neq 0 + 0i$.*

***Example.***  Express each of the five fifth roots of $z = (\sqrt{2} + \sqrt{2}i)$ in trigonometric form.

*Solution.*   From Definitions 8.9 and 8.10,

$$\rho = |\sqrt{2} + \sqrt{2}i| = \sqrt{2 + 2} = 2; \qquad \cos \theta = \frac{\sqrt{2}}{2} = \frac{1}{\sqrt{2}}.$$

Since $\theta$ is in Quadrant I, $\theta = 45°$. Thus, in trigonometric form, for $k \in J$,

$$z = 2(\cos 45° + i \sin 45°) = 2[\cos (45° + k \cdot 360°) + i \sin (45° + k \cdot 360°)].$$

From Theorem 8.7, and the periodic property of cosine and sine functions, each number

$$2^{1/5}\left[\cos\left(\frac{45° + k \cdot 360°}{5}\right) + i \sin\left(\frac{45° + k \cdot 360°}{5}\right)\right], \qquad k \in J$$

is a fifth root of $z$. Taking $k = 0, 1, 2, 3,$ and $4$ in turn yields the roots

$$2^{1/5}(\cos 9° + i \sin 9°),$$
$$2^{1/5}(\cos 81° + i \sin 81°),$$
$$2^{1/5}(\cos 153° + i \sin 153°),$$
$$2^{1/5}(\cos 225° + i \sin 225°),$$
$$2^{1/5}(\cos 297° + i \sin 297°).$$

The substitution of any other integer for $k$ will simply produce one of these five complex numbers.  (Why?)

It can be shown from Theorem 8.7 (although we will not do so here) that the graphs of the $n$ nth roots of a complex number lie on a circle with center at the origin and radius $\rho^{1/n} > 0$. The graphs are equally spaced on the circle, starting with the graph of a complex number whose principal argument is $\theta/n$ (see, Figure 8.4, page 262). In particular, the graphs of the roots of $z = 1 + 0i$ corresponding to the real number 1, lie on a unit circle as shown in Figure 8.5.

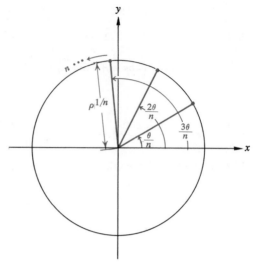

**FIGURE 8.4**

The graphs of the four fourth roots, the six sixth roots, and the eight eighth
roots of 1 are shown in Figure 8.5 (a), (b), and (c), respectively.

### EXERCISE 8.6

A.

Express each power in the form $a + bi$. Use Table IV as necessary.

**Example.**  $(\sqrt{3} + i)^{-2}$.

*Solution.*   From Definitions 8.9 and 8.10,

$$\rho = |\sqrt{3} + i| = \sqrt{(\sqrt{3})^2 + (1)^2} = \sqrt{4} = 2; \cos \theta = \frac{\sqrt{3}}{2}.$$

Because $\theta$ is in Quadrant I, $\theta = 30°$. Thus,

$$(\sqrt{3} + i)^{-2} = [2(\cos 30° + i \sin 30°)]^{-2}.$$

From Theorem 8.6,

$$(\sqrt{3} + i)^{-2} = \frac{1}{2^2} [\cos(-2 \cdot 30°) + i \sin(-2 \cdot 30°)]$$

$$= \left[\frac{1}{4} \cos(-60°) + i \sin(-60°)\right]$$

$$= \frac{1}{4}\left(\frac{1}{2} - \frac{\sqrt{3}}{2} i\right) = \frac{1}{8} - \frac{\sqrt{3}}{8} i.$$

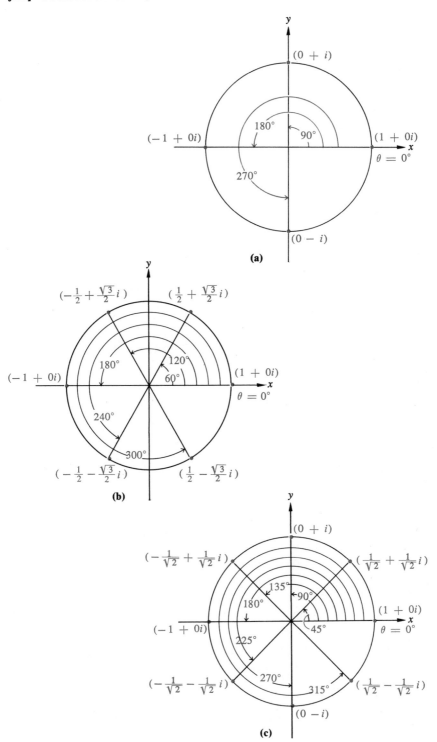

**FIGURE 8.5**

1. $[5(\cos 210° + i \sin 210°)]^3$.

2. $[6(\cos 180° + i \sin 180°)]^4$.

3. $\left(\frac{1}{4} + \frac{\sqrt{3}}{4}i\right)^3$.

4. $(2\sqrt{3} - 2i)^5$.

5. $\left(-\frac{\sqrt{3}}{2} + \frac{1}{2}i\right)^6$.

6. $(-1 + i)^7$.

7. $(1 + i)^{10}$.

8. $(-4i)^4$.

9. $[8(\cos 315° + i \sin 315°)]^{-2}$.

10. $\frac{1}{4}[\cos(-120°) + i \sin(-120°)]^{-3}$.

11. $(-2 - 2i)^{-4}$.

12. $\left(-\frac{1}{5} - \frac{\sqrt{3}}{5}i\right)^{-5}$.

13. $(-5 - 12i)^{-3}$

14. $(3 - 4i)^{-3}$.

Write each expression as a complex number in the form $a + bi$.

15. $\dfrac{(1 - i)^3}{(1 + i)^{-4}}$.

16. $\dfrac{(2 - 2\sqrt{3})^{-3}}{(-1 - i)^3}$.

17. $\dfrac{(-1 + i)^{-5}}{(1 + \sqrt{3}i)^{-2}}$.

18. $\dfrac{(-3i)^{-5}}{(3 + 4i)^{-2}}$.

Find all the roots of each complex number and sketch a graph for each problem.

•19. $[64(\cos 120° + i \sin 120°)]^{1/3}$.

20. $[81(\cos 240° + i \sin 240°)]^{1/4}$.

21. $(-3i)^{1/5}$.

22. $\left(\frac{1}{2} + \frac{\sqrt{3}}{2}i\right)^{1/3}$.

•23. $(4 - 4i)^{-1/3}$.

24. $(5 + 12i)^{-1/4}$.

Solve each equation, $x \in C$.

25. $x^3 + 1 = 0$.   *Hint*: $x^3 = -1 = -1 + 0i$. Thus, $x = (-1 + 0i)^{1/3}$.

26. $x^5 + 1 = 0$.

27. $x^3 - 1 = 0$.

28. $x^5 - 1 = 0$.

29. $x^3 = \dfrac{\sqrt{3}}{4} - \dfrac{1}{4}i$.

30. $x^5 = 12 - 5i$.

**B.**

31. Show that the sum of the four fourth roots of 1 is $(0, 0)$.

32. Show that the sum of the six sixth roots of 1 is $(0, 0)$.

## Chapter Summary

1. The following symbols have been introduced: $C, z, -z, \bar{z}, |z|, i, \rho, \theta,$ $\arg(a,b), z^0, z^{-n},$ and $z^{1/n}$.

2. The set of complex numbers $C$ is defined by

$$C = \{z = (a, b) \mid a \in R \text{ and } b \in R\}.$$

3. If $z_1 = (a_1, b_1)$ and $z_2 = (a_2, b_2)$ are elements of $C$, then

$$z_1 = z_2 \qquad \text{if and only if} \qquad a_1 = a_2 \text{ and } b_1 = b_2.$$

**4.** Properties of equality for elements in $C$ follow from corresponding properties for equality for elements in $R$. Thus, for $z_1, z_2, z_3 \in C$,

$$z_1 = z_1,$$
$$\text{if } z_1 = z_2, \quad \text{then } z_2 = z_1,$$
$$\text{if } z_1 = z_2 \text{ and } z_2 = z_3, \quad \text{then } z_1 = z_3.$$

Unlike the set $R$, the elements in $C$ cannot be ordered and the properties of order in the real number system are not applicable in the complex number system.

**5.** If $z_1 = (a_1, b_1)$ and $z_2 = (a_2, b_2)$, where $z_1, z_2 \in C$, then

$$z_1 + z_2 = (a_1, b_1) + (a_2, b_2) = (a_1 + a_2, b_1 + b_2),$$
$$z_1 \cdot z_2 = (a_1, b_1) \cdot (a_2, b_2) = (a_1 a_2 - b_1 b_2, a_1 b_2 + a_2 b_1).$$

These statements define the operations of addition and multiplication for elements in the set $C$.

**6.** One consequence of viewing the real number $a$ as the complex number $(a, 0)$ is that if $a_1 \in R$ and $(a_2, b_2) \in C$, then

$$a_1(a_2, b_2) = (a_1 a_2, a_1 b_2).$$

**7.** Unlike the set $R$, the set $C$ provides two square roots $(0, b)$ and $(0, -b)$ for each negative real number $(-b^2, 0)$ or $-b^2$.

**8.** The set $C$ can be partitioned into two subsets:
  (a) $\{(a, 0) \mid a \in R\}$ or $\{a \mid a \in R\}$ called **the set of real numbers** and designated by $R$.
  (b) $\{(a, b) \mid a \in R \text{ and } b \in R, b \neq 0\}$ called **the set of imaginary numbers** and designated by $I$. If $a = 0$, the element $(0, b)$ is called a **pure imaginary number.**

**9.** The field properties of the set of complex numbers follow from the corresponding properties of real numbers (see pages 288 and 289).

**10.** If $z_1 \in C$ and $z_2 \in C$, then

$$z_1 - z_2 = z_1 + (-z_2),$$

where the number $z_1 - z_2$ is called the **difference** of $z_2$ subtracted from $z_1$;

$$\frac{z_1}{z_2} = z_1 \cdot \frac{1}{z_2}, \quad [z_2 \neq (0, 0)],$$

where the number $z_1/z_2$ is called the **quotient** of $z_1$ divided by $z_2$.

**11.** If $(a_1, b_1) \in C$, $(a_2, 0)C$, and $a_2 \neq 0$, then

$$\frac{(a_1, b_1)}{(a_2, 0)} = \left(\frac{a_1}{a_2}, \frac{b_1}{a_2}\right).$$

**12.** If $z = (a, b)$ and $z \in C$, then the conjugate of $z$ is defined to be

$$\bar{z} = (a, -b).$$

**13.** If $z_1, z_2, z_3, \in C$ and $z_2, z_3 \neq (0, 0)$, then

$$\frac{z_1}{z_2} = \frac{z_1 z_3}{z_2 z_3}.$$

**14.** In the set $C$,

$$(0, b) = bi,$$

where $i$, the imaginary unit, equals $\sqrt{-1}$ and $i^2 = -1$.

**15.** The ordered pair $(a, b)$, $a \in R$ and $b \in R$, can be represented as

$$a + bi \quad \text{or} \quad a + b\sqrt{-1}.$$

**16.** The quotient of two complex numbers can be expressed in the form $a + bi$ by first multiplying the numerator and the denominator by the unit $i$ or the conjugate of the denominator, as appropriate.

**17.** To avoid difficulty in rewriting products or quotients, *expressions of the form $\sqrt{-b}$, $b > 0$, should first be expressed in the form $\sqrt{b}i$.*

**18.** A plane on which complex numbers are represented is called a **complex plane**. The $x$ axis is called the **real axis** and the $y$ axis is called the **imaginary axis**.

**19.** Complex numbers can be added graphically by the parallelogram law applicable to geometric vectors.

**20.** The **absolute value** or **modulus** of the complex number $z = (a, b) = a + bi$ is given by

$$\rho = |z| = |(a, b)| = |a + bi| = \sqrt{a^2 + b^2}.$$

The modulus is the distance from the origin to the graph of $(a, b)$.

**21.** An **argument, or amplitude**, of the complex number $z = (a, b) = a + bi$, denoted by $\mathbf{arg}(a, b)$ is an angle $\theta$ with initial side the positive $x$ axis and terminal side the ray from the origin through the graph of $(a, b)$. Thus,

$$\cos \theta = \frac{a}{\sqrt{a^2 + b^2}} = \frac{a}{\rho} \quad (a^2 + b^2 \neq 0).$$

**22.** If $\theta$ is an argument of $a + bi$, then $(\theta + k \cdot 360)°$, $k \in J$ is also an argument of $a + bi$. The angle $\theta$ with the least positive measure is called the **principal argument**.

**23.** A nonzero complex number $(a, b)$ can be expressed in **trigonometric form** as

$$(a, b) = a + bi = \rho[\cos (\theta + k \cdot 360°) + i \sin (\theta + k \cdot 360°)], \quad k \in J$$

**24.** If $z_1, z_2 \in C$, $z_1 = \rho_1(\cos \theta_1 + i \sin \theta_1)$ and $z_2 = \rho_2(\cos \theta_2 + i \sin \theta_2)$, then

$$z_1 \cdot z_2 = \rho_1 \cdot \rho_2 \left( \cos \left[ \theta_1 + \theta_2 \right] + i \sin \left( \theta_1 + \theta_2 \right) \right),$$

and

$$\frac{z_1}{z_2} = \frac{\rho_1}{\rho_2} \left[ \cos \left( \theta_1 - \theta_2 \right) + i \sin \left( \theta_1 - \theta_2 \right) \right].$$

**25.** If $z \in C$, $z \neq 0 + 0i$, $n \in J$, and $z = \rho(\cos \theta + i \sin \theta)$, then

$$z^n = \rho^n(\cos n\theta + i \sin n\theta).$$

This is known as **De Moivre's Theorem**.

**26.** If $z \neq 0 + 0i$, then

$$z^0 = 1 + 0i, \quad \text{and} \quad z^{-n} = \frac{1}{z^n}, \quad n \in N.$$

**27.** For $z \in C$ and $n \in N$, $z^{1/n}$ is called the $n$th root of $z$, that is,

$$(z^{1/n})^n = z.$$

Furthermore,

$$z^{1/n} = \rho^{1/n} \left( \cos \frac{\theta}{n} + i \sin \frac{\theta}{n} \right).$$

## Chapter Review

### A.

Write each expression as an ordered pair.

**1.** $(3, 2) + (-6, -9)$.

**2.** $(3, -4) + (-7, 5)$.

**3.** $(-9, -6) \cdot (7, 9)$.

**4.** $(-2, 5) \cdot (8, 4)$.

**5.** $(-7, 2) - (8, 3)$.

**6.** $(7, -9) - (3, -7)$.

**7.** $\dfrac{(-6, -1)}{(3, 3)}$.

**8.** $\dfrac{(0, -9)}{(-7, 8)}$.

Write each complex number in the form $a + bi$.

**9.** $(-4, 1)$.

**10.** $(2\sqrt{3}, -2)$.

Write the conjugate of each complex number in the form $a + bi$.

**11.** $7 - 5i$.

**12.** $-8 + 8i$.

Find real numbers $x$ and $y$ for which each statement is true.

**13.** $9x - 6yi = -3 + 4i$.

**14.** $-8x - 5i = -2y - 5x^2 i$.

Write each expression in the form $a + bi$.

**15.** $(-4 - i) + (-4 + 9i)$.

**16.** $(-2 + 5i) \cdot (8 + 6i)$.

**17.** $i - (3 - 2i)$.

**18.** $(-6 + 8i)^2$.

**19.** $\dfrac{8 + 3i}{2 + 4i}$.

**20.** $\dfrac{-7 + i}{6 - 10i}$.

Write each expression as one of the complex numbers $i$, $-1$, $-i$, or $1$.

**21.** $i^{37}$.

**22.** $i^{-47}$.

Solve each quadratic equation.

**23.** $x^2 - x + 2 = 0$.

**24.** $4x^2 + x + 3 = 0$.

Graph each complex number, its negative, and its conjugate.

**▪25.** $(-2, -6)$.

**26.** $-3 + 5i$.

Represent each sum graphically and check the results by analytic methods.

**27.** $(-1, -8) + (6, 7)$.

**28.** $(-4 + 3i) + (8 - 2i)$.

Find the absolute value and the argument of each complex number.

**29.** $(-2, -5)$.

**30.** $4 + 2i$.

Represent each complex number in trigonometric form.

**31.** $-8i$.

**32.** $-3 + 3\sqrt{3}i$.

Graph each complex number and rewrite it in the form $a + bi$.

**▪33.** $7(\cos 120° + i \sin 120°)$.

**34.** $2[\cos(-315°) + i \sin(-315°)]$.

In Problems 35 to 40, rewrite each expression in the form $a + bi$.

**35.** $3(\cos 200° + i \sin 200°) \cdot \frac{2}{9}(\cos 25° + i \sin 25°)$.

**36.** $\sqrt{3}(\cos 50° + i \sin 50°) \cdot \sqrt{12}\,[\cos(-20°) + i \sin(-20°)]$.

**37.** $\dfrac{8(\cos 309° + i \sin 309°)}{9(\cos 129° + i \sin 129°)}$.

**38.** $\dfrac{12[\cos(-315°) + i \sin(-315°)]}{3(\cos 270° + i \sin 270°)}$.

**39.** $(4[\cos(-60°) + i \sin(-60°)])^3$.

**40.** $(-1 + 5i)^{-7}$.

**41.** Find all cube roots of $[125(\cos 330° + i \sin 330°)]$.

**42.** Solve the equation $x^5 = -5 - 5\sqrt{3}i$, $x \in C$.

**B.**

**43.** Find $(a, b)$ for which $(a, b)^2 = -2$ is true.

**C.**

**44.** Use Euler's formula to find a value for $e^{2i}$, expressing the result in the form $a + bi$.

# Logarithmic Functions

## A.1
### Properties

The exponential function

$$\{(x, y) \,|\, y = b^x, b > 0, b \neq 1\}, \tag{1}$$

with domain $\{x \,|\, x \in R\}$ and range $\{y \,|\, y > 0\}$ was discussed briefly in Section 1.6 as an example of a function that is a one-to-one function; there is one and only one $x$ associated with each $y$ and one and only one $y$ associated with each $x$. Therefore, its inverse,

$$\{(x, y) \,|\, x = b^y, b > 0, b \neq 1\}, \tag{2}$$

is also a function with domain and range the same as the range and domain of Function (1), respectively. Thus, the domain of function (2) is $\{x \,|\, x > 0\}$ and its range is $\{y \,|\, y \in R\}$. The graphs of the exponential function for $b = 10$ and its inverse are shown in Figure A-1 on page 270, where again we have assumed that the functions are continuous over the specified domains.

In order to express the variable $y$ in terms of the variable $x$ in the defining equation of Function (2), the notation

$$y = \log_b x \tag{3}$$

is used. This notation is read "$y$ equals the logarithm to the base $b$ of $x$," or "$y$ equals the logarithm of $x$ to the base $b$." Functions defined by equations of the form $y = \log_b x$ are called **logarithmic functions**.

Observe that

$$x = b^y \quad \text{and} \quad y = \log_b x \tag{4}$$

are simply *different forms of an equation that define the same function*, in the same

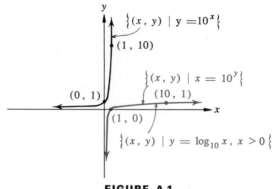

**FIGURE A.1**

way that $x - y = 5$ and $y = x - 5$ define the same function, and you may use whichever form that suits your purpose. Observe from Equations (4) that $\log_b x$ is an exponent given by the following.

**Definition A.1.**   *For all $b$, $x \in R$, $x > 0$, $b > 0$, and $b \neq 1$,*

$$\log_b x \text{ is the real number such that } x = b^{\log_b x}.$$

Observe that $\log_b x$ is the exponent on $b$ such that this power of $b$ is equal to $x$.

***Examples.***   Express each exponential statement in logarithmic form.

(a) $4^2 = 16$.  (b) $27^{1/3} = 3$.  (c) $5^{-2} = \frac{1}{25}$.

*Solutions.*   (a) $\log_4 16 = 2$.  (b) $\log_{27} 3 = \frac{1}{3}$.  (c) $\log_5 \frac{1}{25} = -2$.

***Examples.***   Express each logarithmic statement in exponential form.

(a) $\log_{10} 1{,}000 = 3$.  (b) $\log_{64} 2 = \frac{1}{6}$.  (c) $\log_2 \frac{1}{8} = -3$.

*Solutions.*   (a) $10^3 = 1{,}000$.  (b) $64^{1/6} = 2$.  (c) $2^{-3} = \frac{1}{8}$.

Certain relationships concerning logarithms follow from the fact that a logarithm is an exponent.

**Theorem A.1.**   *If $x_1, x_2, b \in R$, $x_1, x_2 > 0$, and $b > 0$, $b \neq 1$, then*

$$\log_b (x_1 x_2) = \log_b x_1 + \log_b x_2.$$

*Proof.*   Because, by Definition A.1,

$$x_1 = b^{\log_b x_1} \quad \text{and} \quad x_2 = b^{\log_b x_2},$$

therefore

$$x_1 x_2 = b^{\log_b x_2} \cdot b^{\log_b x_2},$$

and

$$x_1 x_2 = b^{\log_b x_1 + \log_b x_2}.$$

Thus from Definition A.1,

$$\log_b (x_1 x_2) = \log_b x_1 + \log_b x_2,$$

which proves the theorem.

Two other theorems will be useful. Their proofs will be left as exercises.

**Theorem A.2.** *If $x_1, x_2, b \in R$, $x_1, x_2 > 0$, and $b > 0$, $b \neq 1$, then*

$$\log_b \frac{x_2}{x_1} = \log_b x_2 - \log_b x_1.$$

*Examples.* Express each of the following as the sum or the difference of logarithmic expressions.

(a) $\log_b (4y)$.     (b) $\log_b (3x/z)$.

*Solutions.* (a) From Theorem A.1, $\log_b (4y) = \log_b 4 + \log_b y$.

(b) From Theorems A.2 and A-1, $\log_b \left( \dfrac{3x}{z} \right) = \log_b 3 + \log_b x - \log_b z$.

**Theorem A.3.** *If $x, b, m \in R$, $x > 0$, $b > 0$, $b \neq 1$, then*

$$\log_b x^m = m \log_b x .$$

*Examples.* Express each of the following as the sum or the difference of logarithmic expressions.

(a) $\log_b x^2 y^{1/3}$.     (b) $\log_b \dfrac{x^3}{y}$.

*Solutions.* (a) From Theorems A.1 and A.3,

$$\log_b x^2 y^{1/3} = \log_b x^2 + \log_b y^{1/3} = 2 \log_b x + \tfrac{1}{3} \log_b y.$$

(b) From Theorems A.2 and A.3,

$$\log_b \frac{x^3}{y} = \log_b x^3 - \log_b y = 3 \log_b x - \log_b y.$$

## EXERCISE A.1

A.

Express each exponential statement in logarithmic notation.

*Examples.* (a) $2^5 = 32$.       (b) $4^{-1} = \tfrac{1}{4}$.

*Solutions.* (a) $\log_2 32 = 5$.     (b) $\log_4 \tfrac{1}{4} = -1$.

1. $2^6 = 64$.                3. $32^{1/5} = 2$.              5. $5^{-2} = \tfrac{1}{25}$.
2. $4^3 = 64$.               4. $81^{1/4} = 3$.              6. $36^{-1/2} = \tfrac{1}{6}$.

Express each logarithmic statement in exponential notation.

7. $\log_2 16 = 4$.               9. $\log_9 3 = \tfrac{1}{2}$.               11. $\log_5 5 = 1$.
8. $\log_5 25 = 2$.              10. $\log_{27} 3 = \tfrac{1}{3}$.              12. $\log_5 1 = 0$.

Find the value of each of the following.

*Examples.* (a) $\log_2 8$.       (b) $\log_{10} 0.01$.

*Solutions.* (a) $\log_2 8 = 3$, because $\log_2 8$ is the exponent which when "placed" on 2 equals 8; that is, $2^{\log_2 8} = 8$.

*(Solution continued on the next page.)*

(b) $\log_{10} 0.01 = -2$, because $\log_{10} 0.01$ is the exponent which when "placed" on 10 equals 0.01; that is, $10^{\log_{10} 0.01} = 0.01$.

**13.** $\log_7 49$.

**14.** $\log_2 8$.

**15.** $\log_4 2$.

**16.** $\log_9 3$.

**17.** $\log_3 3$.

**18.** $\log_3 1$.

**19.** $\log_{10} 10$.

**20.** $\log_{10} 100$.

**21.** $\log_{10} 1$.

**22.** $\log_{10} 0.1$.

**23.** $\log_{10} 0.01$.

**24.** $\log_{10} 0.001$.

In Problems 25 to 32, express each of the following as the sum or the difference of logarithmic expressions.

**Example.**   $\log_{10} \sqrt{x^3 y/z}$.

**Solution.**   First write the radical expression in exponential form and then use Theorems A.1, A.2, and A.3 as appropriate.

$$\log_{10} \sqrt{\frac{x^3 y}{z}} = \log_{10} \left(\frac{x^3 y}{z}\right)^{1/2}.$$

$$= \tfrac{1}{2} \log_{10} \left(\frac{x^3 y}{z}\right) \qquad \text{(Theorem A.3)}$$

$$= \tfrac{1}{2} [\log_{10} x^3 y - \log_{10} z] \qquad \text{(Theorem A.2)}$$

$$= \tfrac{1}{2} [\log_{10} x^3 + \log_{10} y - \log_{10} z] \qquad \text{(Theorem A.1)}$$

$$= \tfrac{1}{2} [3 \log_{10} x + \log_{10} y - \log_{10} z]. \qquad \text{(Theorem A.3)}$$

**25.** $\log_{10} xyz$.

**26.** $\log_{10} \dfrac{xy}{z}$.

**27.** $\log_{10} x^2 y$.

**28.** $\log_{10} \dfrac{x^2}{y^3}$.

**29.** $\log_{10} \dfrac{x}{\sqrt{x-1}}$.

**30.** $\log_{10} \dfrac{\sqrt[3]{x+1}}{xy}$.

**31.** $\log_{10} 2\pi \sqrt{\dfrac{l}{g}}$.

**32.** $\log_{10} \sqrt{\dfrac{2s}{g}}$.

In Problems 33 to 38, express each of the following as a single logarithm.

**Example.**   $\tfrac{1}{2}(\log_{10} x - 3 \log_{10} y)$

**Solution.**   $\tfrac{1}{2}(\log_{10} x - 3 \log_{10} y) = \tfrac{1}{2}(\log_{10} x - \log_{10} y^3) \qquad \text{(Theorem A.3)}$

$$= \tfrac{1}{2} \log_{10} \frac{x}{y^3} \qquad \text{(Theorem A.2)}$$

$$= \log_{10} \left(\frac{x}{y^3}\right)^{1/2}. \qquad \text{(Theorem A.3)}$$

**33.** $\log_{10} x + \log_{10} y - \log_{10} z$.

**34.** $2 \log_{10} x - 3 \log_{10} y$.

**35.** $\tfrac{3}{4}(\log_{10} x - \tfrac{1}{2} \log_{10} y)$.

**36.** $\tfrac{1}{2}(\log_{10} x + \tfrac{1}{3} \log_{10} y)$.

**37.** $\log_{10} \pi + 2 \log_{10} r - \tfrac{1}{3} \log_{10} s$.

**38.** $\tfrac{1}{3}[\log_{10} (x+2) - \tfrac{1}{2} \log_{10} (x-2)]$.

**B.**

**39.** Show that $\frac{1}{3} \log_{10} 8 - \frac{1}{4} \log_{10} 16 = 0$.
**40.** Show that $2 \log_{10} 6 - \log_{10} 4 = 2 \log_{10} 3$.
**41.** Show that $10^{3 \log_{10} x} = x^3$.
**42.** Show that $\log_5 [\log_3 (\log_3 27)] = 0$.
**43.** Argue that $\log_{10} 23$ is between 1 and 2.
**44.** Argue that $\log_{10} 247$ is between 2 and 3.
**45.** Prove Theorem A.2.
**46.** Prove Theorem A.3.

# A.2

## Logarithms to the Base 10

A logarithmic function of special interest in mathematics is

$$\{(x, y) \,|\, y = \log_{10} x, \, x > 0\}.$$

See Figure A.1, page 270. Values for $\log_{10} x$ are called **common logarithms**. Since $\log_{10} x$ is an exponent defined by

$$10^{\log_{10} x} = x, \qquad x > 0,$$

you can find values of $\log_{10} x$ for each positive $x$ which is an integral power of 10. For example,

$$\log_{10} 10 = \log_{10} \phantom{}^1 = 1, \qquad \text{since } 10^1 = 10;$$

$$\log_{10} 100 = \log_{10} \phantom{}^2 = 2, \qquad \text{since } 10^2 = 100;$$

$$\log_{10} 1000 = \log_{10} \phantom{}^3 = 3, \qquad \text{since } 10^3 = 1000; \qquad \text{etc};$$

and

$$\log_{10} 1 = \log_{10} 10^0 = 0, \qquad \text{since } 10^0 = 1;$$

$$\log_{10} 0.1 = \log_{10} 10^{-1} = -1, \qquad \text{since } 10^{-1} = 0.1;$$

$$\log_{10} 0.01 = \log_{10} 10^{-2} = -2, \qquad \text{since } 10^{-2} = 0.01; \qquad \text{etc.}$$

Values for $\log_{10} x$ for each positive $x$ which is not an integral power of 10 are in general irrational numbers. Rational number approximations for some such values, $1 < x < 10$, have been obtained by using power series similar to those used to obtain approximations for circular function values. Values are available in Table II, pages 292 and 293, an excerpt of which is shown in Figure A.2, page 274. Each number in the column headed $x$ represents the first two significant digits of the numeral for $x$, while each number in the horizontal row containing $x$ represents the third significant digit of the numeral for $x$.

You can use Table II in conjunction with Theorem A.1 to find rational number approximations for values of $\log_{10} x$ for all $x \in R$ and $x > 1$.

**Examples.** Find a rational number approximation for each of the following.
(a) $\log_{10} 6.87$.     (b) $\log_{10} 68.7$.     (c) $\log_{10} 687.0$.

*Solutions.* (a) The value $\log_{10} 6.87$ is found in Table II, (or Figure A.2) at the intersection of the row containing 6.8 under $x$ and the column headed by 7 to the

*(Solutions continued on the next page.)*

| $x$ | 0 | 1 | 2 | 3 | 4 | 5 | 6 | 7 | 8 | 9 |
|-----|---|---|---|---|---|---|---|---|---|---|
| 6.5 | .8129 | .8136 | .8142 | .8149 | .8156 | .8162 | .8169 | .8176 | .8182 | .8189 |
| 6.6 | .8195 | .8202 | .8209 | .8215 | .8222 | .8228 | .8235 | .8241 | .8248 | .8254 |
| 6.7 | .8261 | .8267 | .8274 | .8280 | .8287 | .8293 | .8299 | .8306 | .8312 | .8319 |
| 6.8 | .8325 | .8331 | .8338 | .8344 | .8351 | .8357 | .8363 | .8370 | .8376 | .8382 |
| 6.9 | .8388 | .8395 | .8401 | .8407 | .8414 | .8420 | .8426 | .8432 | .8439 | .8445 |
| 7.0 | .8451 | .8457 | .8463 | .8470 | .8476 | .8482 | .8488 | .8494 | .8500 | .8506 |
| 7.1 | .8513 | .8519 | .8525 | .8531 | .8537 | .8543 | .8549 | .8555 | .8561 | .8567 |
| 7.2 | .8573 | .8579 | .8585 | .8591 | .8597 | .8603 | .8609 | .8615 | .8621 | .8627 |
| 7.3 | .8633 | .8639 | .8645 | .8651 | .8657 | .8663 | .8669 | .8675 | .8681 | .8686 |
| 7.4 | .8692 | .8698 | .8704 | .8710 | .8716 | .8722 | .8727 | .8733 | .8739 | .8745 |

**FIGURE A.2**

right of $x$. Thus, $\log_{10} 6.87 \approx 0.8370$. (0.8370 is the number such that $10^{0.8370} \approx 6.87$.)

(b) From Theorem A.1,

$$\log_{10} 68.7 = \log_{10} (6.87 \times 10^1) = \log 6.87 + \log_{10} 10^1,$$

and using the result in part (a),

$$\log_{10} 68.7 = \log_{10} 6.87 + \log_{10} 10^1 \approx 0.8370 + 1 = 1.8370.$$

(1.8370 is the number such that $10^{1.8370} \approx 68.7$.)

(c) From Theorem A.1,

$$\log_{10} 687.0 = \log_{10} (6.87 \times 10^2)$$
$$= \log_{10} 6.87 + \log_{10} 10^2 \approx 0.8370 + 2 = 2.8370.$$

(2.8370 is the number such that $10^{2.8370} \approx 687$.)

Observe in the foregoing examples that the decimal portion of the logarithm was always approximately equal to 0.8370 and the integral portion was the exponent on 10 when the number was written as a product of a number between 1 and 10 and a power of 10. The integral part of the logarithm is called the **characteristic** and the nonnegative decimal part is called the **mantissa**.

You can also use Table II and Theorem A.1 to find a value for $\log_{10} x$ where $x \in R$ and $0 < x < 1$. For example,

$$\log_{10} 0.00712 = \log_{10} (7.12 \times 10^{-3}) = \log_{10} 7.12 + \log_{10} 10^{-3}.$$

From Table II you can find $\log_{10} 7.12 \approx 0.8525$. Since $\log_{10} 10^{-3} = -3$, then

$$\log_{10} 0.00712 \approx 0.8525 + (-3) \tag{1}$$

$$= -2.1475.$$

Thus, $10^{-2.1475} \approx 0.00712$. In this form ($-2.1475$), the decimal portion of the logarithm ($-0.1475$) and the integral part ($-2$) are both negative. In order to further use Table II, whose entries are positive, *logarithms are generally written with positive mantissas*. Thus, the right-hand member of ·

$$\log_{10} 0.00712 \approx 0.8525 - 3,$$

is a useful representation for $\log_{10} 0.00712$.

Alternatively, the characteristic of the logarithm, $-3$ can be written as $(1-4)$, $(2-5)$, or $(7-10)$, etc., from which

$$\log_{10} 0.00712 \approx 0.8525 + (-3),$$

$$\log_{10} 0.00712 \approx 0.8525 + (1-4) = 1.8525 - 4,$$

$$\log_{10} 0.00712 \approx 0.8525 + (2-5) = 2.8525 - 5,$$

$$\log_{10} 0.00712 \approx 0.8525 + (7-10) = 7.8525 - 10, \qquad \text{etc.}$$

The latter form is frequently used and, in general, we show this form in the answers to examples and exercises.

**Examples.**   Find a rational number approximation for each of the following.
(a) $\log_{10} 0.0734$.        (b) $\log_{10} 0.000669$.

*Solutions.*   Values for each logarithm are obtained from Table II.
(a) $\log_{10} 0.0734 = \log_{10} 7.34 \times 10^{-2} \approx 0.8657 - 2 = 8.8657 - 10$.
(b) $\log_{10} 0.000669 = \log_{10} 6.69 \times 10^{-4} \approx 0.8254 - 4 = 6.8254 - 10$.

Given the value of $\log_{10} x$, you can also find a rational number approximation for $x$ by using Table II. The value $x$, in this case, is called the **antilogarithm** (abbreviated **antilog$_{10}$**) of the $\log_{10} x$. In general, antilog$_b$ $(\log_b x) = x$; that is, the element $x$ in the domain of the logarithmic function associated with the element $\log_b x$ in the range.

**Examples.**   Find a rational number approximation for each of the following.
(a) antilog$_{10}$ 0.8395.        (b) antilog$_{10}$ 2.8156.        (c) antilog$_{10}$ 8.8739 − 10.

*Solutions.*   (a) From Table II, antilog$_{10}$ 0.8395 $\approx$ 6.91.

(b) From Table II, antilog$_{10}$ 0.8156 $\approx$ 6.54. Since the characteristic of 2.8156 is 2,
$$\text{antilog}_{10}\ 2.8156 \approx 6.54 \times 10^2 = 654.$$

(c) From Table II, antilog$_{10}$ 0.8739 $\approx$ 7.48. Since the characteristic of 8.8739 − 10 is −2,
$$\text{antilog}_{10}\ 8.8739 - 10 \approx 7.48 \times 10^{-2} = 0.0748.$$

You can find approximations to logarithms and antilogarithms of numbers which are between those listed in Table II by the method of linear interpolation discussed in Section 2.7. Several examples are shown in the exercises.

## EXERCISE A.2

A.

Find an approximation for each of the following.

**Examples.**   (a) $\log_{10} 4720$.        (b) $\log_{10} 0.073$.

*(Solutions on the next page.)*

*Solutions.*   (a) $\log_{10} 4720 = \log_{10} (4.720 \times 10^3) = \log_{10} 4.720 + \log_{10} 10^3$

$$\approx 0.6739 + 3 = 3.6739.$$

(b) $\log_{10} 0.073 = \log_{10} (7.3 \times 10^{-2}) = \log_{10} 7.3 + \log_{10} 10^{-2}$

$$\approx 0.8633 + (-2)$$

$$= 0.8633 - 2 \qquad \text{or} \qquad 8.8633 - 10.$$

| | | |
|---|---|---|
| **1.** $\log_{10} 4.39$. | **5.** $\log_{10} 243$. | **9.** $\log_{10} 0.006$. |
| **2.** $\log_{10} 6.94$. | **6.** $\log_{10} 300$. | **10.** $\log_{10} 0.034$. |
| **3.** $\log_{10} 83.1$. | **7.** $\log_{10} 0.621$. | **11.** $\log_{10} 0.0005$. |
| **4.** $\log_{10} 10.5$. | **8.** $\log_{10} 0.74$. | **12.** $\log_{10} 0.00642$. |

*Example.*   Find $\log_{10} 6.833$.

*Solution.*   From Table II, $\log_{10} 6.830 \approx 0.8344$ and $\log_{10} 6.840 \approx 0.8351$. The data is arranged for linear interpolation in the following tabular form:

$$
\begin{array}{cc}
x & \log_{10} x \\
\hline
\end{array}
$$

$$
0.010 \left\{
\begin{array}{l}
0.003 \left\{
\begin{array}{l}
6.830 \approx 0.8344 \\
6.833 \approx \quad ?
\end{array}
\right\} d \\
\\
6.840 \approx 0.8351
\end{array}
\right\} 0.0007.
$$

Assuming that the differences of $\log_{10} x$ vary directly with the differences of $x$, it follows that

$$\frac{0.003}{0.010} \approx \frac{d}{0.0007},$$

from which

$$d \approx \frac{3}{10} (0.0007)$$

and correct to four decimal places, $d = 0.0002$. Thus,

$$\log_{10} 6.833 \approx 0.8344 + 0.0002 = 0.8346.$$

| | | |
|---|---|---|
| **13.** $\log_{10} 3.475$. | **15.** $\log_{10} 10.73$. | **17.** $\log_{10} 0.3742$. |
| **14.** $\log_{10} 5.428$. | **16.** $\log_{10} 242.6$. | **18.** $\log_{10} 0.007654$. |

Find an approximation for each of the following.

*Example.*   (a) antilog$_{10}$ 2.9304.        (b) antilog$_{10}$ 7.5502 − 10.

*Solution.*   (a) From Table II, antilog$_{10}$ 0.9304 $\approx 8.52$. Because the characteristic of 2.9304 is 2,

$$\text{antilog}_{10}\ 2.9304 \approx 8.52 \times 10^2 = 852.$$

(b) Using Table II, we have

$$\text{antilog}_{10}\ 7.5502 - 10 = \text{antilog}_{10}\ [0.5502 + (-3)] \approx 3.55 \times 10^{-3} = 0.00355.$$

| | |
|---|---|
| **19.** antilog$_{10}$ 0.8149. | **21.** antilog$_{10}$ 1.4048. |
| **20.** antilog$_{10}$ 0.9581. | **22.** antilog$_{10}$ 2.9868. |

**23.** antilog$_{10}$ 3.3010.

**24.** antilog$_{10}$ 4.7782.

**25.** antilog$_{10}$ 0.9335 − 1.

**26.** antilog$_{10}$ 0.9768 − 2.

**27.** antilog$_{10}$ 4.3927 − 6.

**28.** antilog$_{10}$ 7.4014 − 10.

*Example.*   antilog$_{10}$ 0.8731.

*Solution.*   From Table II, log$_{10}$ 7.46 ≈ 0.8727 and log$_{10}$ 7.47 ≈ 0.8733. The data is set up for linear interpolation as follows:

$$x \qquad \log_{10} x$$

$$0.010 \left\{ d \left\{ \begin{array}{l} 7.460 \approx 0.8727 \\ ? \quad \approx 0.8731 \\ 7.470 \approx 0.8733 \end{array} \right. \left. {}^{0.0004} \right\} 0.0006. \right.$$

Thus,

$$\frac{d}{0.010} \approx \frac{0.0004}{0.0006},$$

from which

$$d \approx (\tfrac{2}{3})0.010;$$

and correct to three decimal places, $d = 0.007$. Thus,

$$\text{antilog}_{10}\ 0.8731 \approx 7.460 + 0.007 = 7.467.$$

**29.** antilog$_{10}$ 0.8765.

**30.** antilog$_{10}$ 0.6467.

**31.** antilog$_{10}$ 3.9742.

**32.** antilog$_{10}$ 1.4539.

**33.** antilog$_{10}$ 8.1860 − 10.

**34.** antilog$_{10}$ 7.3809 − 10.

**B.**

In Problems 35 to 38, find an approximation for each power.   *Hint*: Find the antilogarithm of the exponent to the base 10.

**35.** $10^{0.4014}$.

**36.** $10^{2.8692}$.

**37.** $10^{1.8993}$.

**38.** $10^{3.7612}$.

**39.** Find antilog$_{10}$ (log$_{10}$ 62).

**40.** Find log$_{10}$ (antilog$_{10}$ 2.8519).

**41.** Find the characteristic of: (a) log$_2$ 74.   (b) log$_3$ 74.

**42.** Find the characteristic of: (a) log$_2 \frac{1}{7}$.   (b) log$_3 \frac{1}{7}$.

# A.3
## Computations; Solutions of Exponential Equations

In this section you will use logarithms to perform routine numerical computations and solve exponential equations such as $3^x = 5$. Certainly these computations can be done more efficiently with high speed computers, desk calculators, or slide rules. However, the setting up of the equations that are preliminary to the computations will strengthen your understanding of the properties of logarithms. For easy reference, we repeat Theorems A.1, A.2, and A.3 using 10 as the base.

If $x_1$, $x_2 \in R$, $x_1$, $x_2 > 0$, then

1. $$\log_{10}(x_1 x_2) = \log_{10} x_1 + \log_{10} x_2,$$

2. $$\log_{10} \frac{x_2}{x_1} = \log_{10} x_2 - \log_{10} x_1,$$

3. $$\log_{10}(x_1)^m = m \log_{10} x_1, \qquad m \in R.$$

Since the logarithmic functions are one-to-one functions, we shall make two assumptions before proceeding with examples of computations involving logarithms.

**Axiom 1.**   If $x_1 = x_2$, $(x_1, x_2 > 0)$, then $\log_{10} x_1 = \log_{10} x_2$.

**Axiom 2.**   If $\log_{10} x_1 = \log_{10} x_2$, then $x_1 = x_2$.

*Example.*   Using logarithms, find the value of $Q = \dfrac{(0.214)(9.02)^3}{(8.52)^{1/2}}$.

*Solution.*   From Axiom 1 and Theorems A.1, A.2, and A.3,

$$\log_{10} Q = \log_{10} 0.214 + 3 \log_{10} 9.02 - \tfrac{1}{2} \log_{10} 8.52.$$

Table II provides data for the logarithms involved here. Thus,

$$\log_{10} Q \approx (0.3304 - 1) + 3(0.9552) - \tfrac{1}{2}(0.9304)$$

$$= 0.3304 - 1 + 2.8656 - 0.4652,$$

$$\log_{10} Q = 1.7308.$$

From Axiom 2, it follows that

$$Q = \operatorname{antilog}_{10} 1.7308 \approx 53.8.$$

The properties of logarithms and tables of logarithmic functions available now enable you to find a solution to an equation such as

$$3^x = 5, \tag{1}$$

or more precisely, in most cases, a rational number approximation for the solution. For example, since $3^x > 0$ for all $x \in R$, Axiom 1 and Theorem A.3 can be applied to write

$$\log_{10} 3^x = \log_{10} 5$$

and then

$$x \log_{10} 3 = \log_{10} 5.$$

Multiplying each member of this equation by $1/(\log_{10} 3)$ yields

$$x = \frac{\log_{10} 5}{\log_{10} 3},$$

and the solution set of Equation (1) is given by $\left\{ \dfrac{\log_{10} 5}{\log_{10} 3} \right\}$. You can obtain a rational number approximation by substituting appropriate entries from Table II. Thus,

$$x = \frac{\log_{10} 5}{\log_{10} 3} \approx \frac{0.6990}{0.4771} \approx 1.47,$$

rounded off to three significant digits, and the solution set is $\{\approx 1.47\}$. Observe that the value of $x$ was obtained by *dividing* 0.6990 by 0.4771, *not by subtracting* one number from the other. The distinction to be made here is that the quotient of two logarithms $(\log_{10} 5)/(\log_{10} 3) \neq \log_{10} \frac{5}{3}$. To facilitate computations, as above, in which numbers with four digit numerals are added or subtracted, a systematic approach is desirable. The first example in the following exercise illustrates one such procedure.

## EXERCISE A.3

**A.**

Find an approximation for each of the following by means of logarithms. Use linear interpolation as necessary.

*Example.* $\quad \sqrt{\dfrac{(0.631)(47.1)}{15.6}}$ .

*Solution.* Let $R = \sqrt{\dfrac{(0.631)(47.1)}{15.6}}$ ,

then by Axiom 1, page 278 and Theorems A.1, A.2, and A.3,

$$\log_{10} R = \tfrac{1}{2}(\log_{10} 0.631 + \log 47.1 - \log 15.6).$$

The data can be arranged for convenient computation as follows:

$$\left.\begin{array}{l} \log_{10} 0.631 \approx \quad 9.8000 - 10 \\ \log_{10} 47.1 \approx \quad 1.6730 \end{array}\right\} \text{add}$$

$$\left.\begin{array}{l} \log_{10} (0.631)(47.1) \approx 11.4730 - 10 \\ \log_{10} 15.6 \approx \quad 1.1931 \end{array}\right\} \text{subtract}$$

$$\log_{10} \dfrac{(0.631)(47.1)}{15.6} \approx 10.2799 - 10 = 0.2799$$

$$\tfrac{1}{2} \log_{10} \dfrac{(0.631)(47.1)}{15.6} \approx \tfrac{1}{2}(0.2799) \approx 0.1399.$$

Thus,

$$R \approx \text{antilog}_{10}\, 0.1399 \approx 1.38.$$

**1.** $(34.1)(0.793)$.

**2.** $\dfrac{47.4}{1.73}$ .

**3.** $(1.5)^5$.

**4.** $\sqrt[3]{23.2}$.

**5.** $\sqrt[4]{68.1}$.

**6.** $(0.92)^7$.

**7.** $\sqrt[5]{0.0063}$.

**8.** $\dfrac{(24.3)(0.72)}{1.97}$ .

**9.** $\dfrac{(4.1)^3(2.71)}{\sqrt{13.3}}$ .

**10.** $\sqrt{\dfrac{(3.62)(0.00291)}{(0.041)^3}}$ .

**11.** $\sqrt[3]{\dfrac{(6.18)^2}{(0.04)(49.2)}}$ .

**12.** $\dfrac{\sqrt{47.1}\ \sqrt[3]{2.74}}{\sqrt[4]{37.1}}$ .

Solve each exponential equation.

***Example.***   $5^{x-3} = 29$.

*Solution.*   From Axiom 1, page 278 and Theorem A.3,

$$\log_{10} 5^{x-3} = \log_{10} 29,$$

$$(x - 3)\log_{10} 5 = \log_{10} 29,$$

$$x - 3 = \frac{\log_{10} 29}{\log_{10} 5},$$

$$x = \frac{\log_{10} 29}{\log_{10} 5} + 3. \tag{1}$$

Using appropriate entries in Table II you can obtain a numerical solution. Thus,

$$x \approx \frac{1.4624}{0.6990} + 3 \approx 2.09 + 3 = 5.09,$$

and the solution set is $\{\approx 5.09\}$.

Notice again that in seeking a numerical approximation for the right-hand member of Equation (1), the logarithms are divided, not subtracted.

13. $3^x = 5$.      15. $5^{x+1} = 290$      17. $3^{-x} = 21$.
14. $7^x = 12$.      16. $2^{x-1} = 17$.      18. $4^{5-x} = 17.1$.

If $P$ dollars are compounded $t$ times per year at a yearly rate of interest $r$, the amount $A$ present at the end of $n$ years is given by

$$A = P\left(1 + \frac{r}{t}\right)^{tn}.$$

Use this formula in Problems 19 to 22.

19. Find the amount present at the end of ten years if $10,000 is invested at 4%, compounded annually.
20. Find the amount present at the end of 20 years if $1,000 is invested at 3%, compounded annually. When compounded semi annually.
21. How many years would be necessary for an account of $1,500 to increase to $2,400 when compounded annually at 4%?
22. What is the rate of interest if an initial investment of $3,000 compounded quarterly amounts to $4,500 in ten years?

The number of bacteria present in a culture is related to time by the formula $N = N_0 e^{0.04t}$, where $N_0$ is the number of bacteria present at time $t = 0$, $t$ is the time in hours and $e \approx 2.718$. Use this information in Problems 23 to 26.

23. If 6,000 bacteria are present initially, how many are present 8 hours later?
24. If 20,000 bacteria are present 4 hours after the beginning of an experiment, how many were present at the beginning when $t = 0$?
25. If 3,000 bacteria are present at the beginning of an experiment, in how many hours will there be 15,000 bacteria?

**26.** If 7,500 bacteria are present 2 hours after the beginning of an experiment, how many are present 5 hours later?

The amount of radioactive element available at any time, $t$, is given by the formula $N = N_0 e^{-0.4t}$, where $t$ is in seconds and $N_0$ is the amount present at time $t = 0$ and $e \approx 2.718$. Use this information in Problems 27 to 30.

**27.** How many grams of a radioactive element would remain after four seconds if 60 grams were present initially?

**28.** If 45 grams of a radioactive element is present 6 seconds after the beginning of its decay, how many grams were present when $t = 0$?

**29.** If 20 grams of a radioactive element are present initially, in how many seconds will there be only 10 grams?

**30.** If 12 grams of a radioactive element remain 15 seconds after decay starts, how many grams will remain 32 seconds from the start of decay?

**B.**

**31.** Find an approximation for $\log_2 7$ using a table of logarithms to the base 10. *Hint*: Let $x = \log_2 7$; write the equation in the exponential form $2^x = 7$, and then solve for $x$.

**32.** Use the method suggested in Problem 31 to find an approximation for $\log_3 8$.

**33.** Use the method suggested in Problem 31 to show that $\log_b a = \log_{10} a / \log_{10} b$.

**34.** Show that $\log_b a = 1/\log_a b$.

# A.4

## Solution of Triangles

In many cases the use of logarithms simplifies the routine computations involved in the solution of triangles. This is particularly true when the Law of Sines is being used.

For example, it is possible to find a rational number approximation for the logarithm of a circular or trigonometric function value, say $\log_{10} \sin 31°$, by first finding the value of $\sin 31°$ ($\approx 0.515$) in Table IV and then finding the value of $\log_{10} 0.515$ ($\approx 0.7118 - 1$) in Table II. This two-step process is not necessary however, because a table has been prepared which gives the common logarithms of the cosine, sine, tangent, and cotangent function values directly. Table V, page 304 is used for trigonometric function values of angles in degree measure.

*Examples.* Find the value for:
(a) $\log_{10} \sin 31°$.      (b) $\log_{10} \tan 58° 40'$.

*Solutions.*  (a) In Table V, $\log_{10} \sin 31° \approx 9.7118 - 10$ is located at the intersection of the row labeled $31° 00'$ in the left-hand column and the column labeled "log sin $\theta$" at the top. (This is the value obtained through the two-step process above.)

(b) From Table V, $\log_{10} (\tan 58° 40') \approx 0.2155$ is located at the intersection of the row labeled $58° 40'$ in the right-hand column (note that the measures of the angles increase upward) and the column labeled "log tan $\theta$" at the bottom.

Now consider how logarithms can be used in the solution of triangles.

*Example.*   In a right triangle $ABC$, find $a$, if the measure of $\alpha$ is $37°$ and $c$ is 15.0 units.

*Solution.*   Since

$$\sin 37° = \frac{a}{c} = \frac{a}{15},$$

$$a = 15 \sin 37°.$$

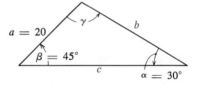

Thus,

$$\log_{10} a = \log_{10} 15 + \log_{10} \sin 37°$$

$$\approx 1.1761 + (9.7795 - 10) = 0.9556.$$

From Table II, with interpolation,

$$a = \text{antilog}_{10}\, 0.9556 \approx 9.03.$$

Although we have not discussed the use of the logarithmic slide rule for routine computations and do not intend to do so in any detail, we do note here that the slide rule is particularly easy to use in conjunction with the Law of Sines for solving triangles.

*Example.*   Solve the triangle where $a$ is 20 units, the measure of $\alpha$ is $30°$, and the measure of $\beta$ is $45°$.

*Solution.*

$$m°(\gamma) = 180° - (30° + 45°) = 105°. \qquad a = 20$$

From the Law of Sines,

$$\frac{20}{\sin 30°} = \frac{b}{\sin 45°} = \frac{c}{\sin 105°}.$$

Setting $30°$ on the $S$ (sine) scale opposite 20 on the $D$ scale, the values of $b$ and $c$ can be read on the $D$ scale directly opposite $45°$ and $75°$ ($180° - 105°$) on the $S$ scale, respectively, as shown in the figure:

$$b \approx 28.3 \qquad \text{and} \qquad c \approx 38.6.$$

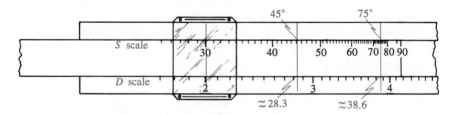

Solutions in Problems 25 to 42 in Exercise A.4 to be found using logarithms can be checked by using the slide rule. Also many problems in Chapter 6 can be used to provide additional practice in using the slide rule in cases where the Law of Sines is applicable.

Logarithms (or the slide rule) can be used efficiently in the solution of triangles in conjunction with the Law of Tangents discussed on page 179 or with a half-angle formula discussed on page 184. Consider two examples which were solved earlier without using logarithms.

**Example.**   Solve the triangle in which $a = 27$, $b = 23$, and $\gamma = 30°$.

*Solution.*   From the same example on page 179 we observe that $\alpha + \beta = 150°$ and that

$$\tan \tfrac{1}{2}(\alpha - \beta) = \frac{4 \tan 75°}{50}.$$

From Axiom 1 and Theorems A.1 and A.2,

$$\log_{10} \tan \tfrac{1}{2}(\alpha - \beta) = \log_{10} 4 + \log_{10} \tan 75° - \log_{10} 50.$$

Tables II and V provide data for the logarithms. The data is arranged for convenient computation as follows:

$$\left. \begin{array}{l} \log_{10} 4 \approx 0.6021 \\ \log_{10} \tan 75° \approx 0.5719 \end{array} \right\} \text{add}$$

$$\left. \begin{array}{l} \log_{10} 4 + \log_{10} \tan 75° \approx 1.1740 = 11.1740 - 10 \\ \log_{10} 50 \approx 1.6990 = \phantom{1}1.6990 \end{array} \right\} \text{subtract}$$

$$\log_{10} \tan \tfrac{1}{2}(\alpha - \beta) \approx \phantom{1}9.4750 - 10$$

Here we have written $1.1740$ as $11.1740 - 10$ in order to write the difference $9.4750 - 10$ with a positive mantissa. From Axiom 2, and Table V,

$$\tfrac{1}{2}(\alpha - \beta) \approx 16° \, 40',$$

and the solution can be completed as shown on page 180.

**Example.**   Find the measure of $\alpha$ for the triangle in which $a = 5$, $b = 6$, and $c = 7$.

*Solution.*   From the same example on page 184 we observe that

$$\cos \tfrac{1}{2}\alpha = \sqrt{\frac{6}{7}}.$$

From Axiom 1 and Theorems A.2 and A.3, we have

$$\log_{10} \cos \tfrac{1}{2}\alpha = \tfrac{1}{2}[\log_{10} 6 - \log_{10} 7].$$

Table II provides data for the logarithms as follows:

$$\left. \begin{array}{l} \log_{10} 6 \approx 0.7782 = 10.7782 - 10 \\ \log_{10} 7 \approx 0.8451 = \phantom{1}0.8451 \end{array} \right\} \text{subtract.}$$

$$\log_{10} 6 - \log_{10} 7 \approx 9.9331 - 10.$$

Here we have written $0.7782$ as $10.7782 - 10$ in order to write the difference $9.9331 - 10$ with a positive mantissa. Then

$$\log_{10} \tfrac{1}{2}\cos\alpha = \tfrac{1}{2}[\log_{10} 6 - \log_{10} 7] \approx 4.9665 - 5.$$

*Solution continued on the next page.*

From Axiom 2 and Table V,

$$\tfrac{1}{2}\alpha \approx 22°\ 10',$$

$$\alpha \approx 44°\ 20'.$$

### EXERCISE A.4

A.

Find each of the following from Table V. Interpolate when necessary.

1. $\log_{10} \sin 35°$.
2. $\log_{10} \tan 68°\ 10'$.
3. $\log_{10} \cos 50°\ 10'$.
4. $\log_{10} \sin 17°\ 30'$.
5. $\log_{10} \cot 68°\ 10'$.

6. $\log_{10} \cos 74°\ 10'$.
7. $\log_{10} \sin 18°\ 25'$.
8. $\log_{10} \cos 64°\ 37'$.
9. $\log_{10} \cos 16°\ 13'$.
10. $\log_{10} \sin 84°\ 06'$.

Find each least positive measure of $\alpha$ to the nearest ten minutes.

11. $\log_{10} \sin \alpha = 9.4223 - 10$.
12. $\log_{10} \cos \alpha = 9.9914 - 10$.
13. $\log_{10} \tan \alpha = 0.5147$.

14. $\log_{10} \cot \alpha = 0.0634$.
15. $\log_{10} \cos \alpha = 9.8414 - 10$.
16. $\log_{10} \tan \alpha = 9.7125 - 10$.

Find the value of each of the following, using logarithms whenever possible.

*Example.* $\dfrac{12.6 \sin 68°\ 10'}{\sin 23°\ 20'}$ .

*Solution.* Let $b$ equal the quotient. From Axiom 1 and Theorems A.1 and A.2,

$$\log_{10} b = \log_{10} 12.6 + \log \sin_{10} 68°\ 10' - \log_{10} \sin 23°\ 20'.$$

Tables II and V provide data for the logarithms. The data is arranged for convenient computation as follows:

$$
\left.
\begin{array}{l}
\log_{10} 12.6 \approx \quad 1.1004 \\
\log_{10} \sin 68°\ 10' \approx \quad 9.9677 - 10
\end{array}
\right\} \text{add}
$$

$$
\left.
\begin{array}{l}
\phantom{\log_{10} \sin 68°\ 10' \approx} 11.0681 - 10 \\
\log_{10} \sin 23°\ 20' \approx \quad 9.5978 - 10
\end{array}
\right\} \text{subtract}
$$

$$\log_{10} b \approx 1.4703$$

From Axiom 2 and Table II,

$$b = \text{antilog}_{10}\ 1.4703 \approx 2.95 \times 10^1 = 29.5.$$

17. $4.93 \sin 23°\ 30'$.

18. $62.1 \tan 68°\ 10'$.

19. $\dfrac{\cos 27°\ 10'}{0.712}$ .

20. $\dfrac{\sin 45°\ 50'}{2.46}$ .

21. $\dfrac{10.6 \tan 16°\ 20'}{8.63}$ .

22. $\dfrac{0.93 \cos 78°\ 10'}{0.62}$ .

23. $\dfrac{45.1 \sin 30°\ 10'}{\sin 48°\ 20'}$ .

24. $\dfrac{1.73 \sin 108°\ 10'}{\sin 20°\ 30'}$ .

Solve each right triangle in Problems 25 to 30 where in each case $\gamma = 90°$. Use logarithms in computations whenever possible.

**25.** $a = 9$, $b = 12$.                    **27.** $b = 9.6$, $\beta = 68° \ 10'$.
**26.** $a = 2.6$, $\beta = 34° \ 20'$.        **28.** $c = 4.75$, $\alpha = 8° \ 50'$.

Solve each triangle in Problems 29 to 34. Use logarithms in the computations whenever possible.

**29.** $a = 8$, $\alpha = 32°$, $\beta = 49°$.                **32.** $a = 0.024$, $c = 0.07$, $\gamma = 96° \ 10'$.
**30.** $a = 1.6$, $\alpha = 19°$, $\gamma = 57°$.            **33.** $c = 5.3$, $b = 3.11$, $\beta = 17° \ 10'$.
**31.** $a = 12.6$, $b = 9.4$, $\alpha = 22° \ 50'$.          **34.** $a = 0.29$, $c = 0.41$, $\gamma = 34° \ 20'$.

**B.**

In Problems 35 to 38, use the Law of Tangents and use logarithms in the computations whenever possible.

**35.** $a = 4$, $c = 5$, $\beta = 30°$.                      **37.** $a = 5.2$, $b = 8.9$, $\gamma = 51° \ 40'$.
**36.** $b = 3$, $c = 2$, $\alpha = 60°$.                     **38.** $a = 6.2$, $c = 9.7$, $\beta = 47° \ 20'$.

In Problems 39 to 42, use a half-angle formula and use logarithms in the computations whenever possible.

**39.** $a = 5$, $b = 3$, $c = 6$.                            **41.** $a = 9.2$, $b = 3.7$, $c = 8.1$.
**40.** $a = 7.9$, $b = 4.8$, $c = 6.1$.                      **42.** $a = 0.34$, $b = 0.52$, $c = 0.66$.

## Summary

**1.** The exponential function

$$\{(x, y) \mid y = b^x, b > 0, b \neq 1\}$$

and its inverse

$$\{(x, y) \mid x = b^y, b > 0, b \neq 1\}$$

are one-to-one functions. The domain, $\{x \mid x \in R\}$, and the range, $\{y \mid y > 0\}$, of the first are the range and domain of the second, respectively.

**2.** The equations

$$y = \log_b x \qquad \text{and} \qquad x = b^y$$

are equivalent forms of an equation that defines the **logarithmic function** in which $x$ is an element in the domain and $y$ is an element in the range.

**3.** The following relationships are consequences of the definition of a logarithm:
If $x_1, x_2, m \in R$, $x_1, x_2 > 0$, $b > 0$, $b \neq 1$, then

I.   $\log_b (x_1 x_2) = \log_b x_1 + \log_b x_2$,

II.  $\log_b \dfrac{x_1}{x_2} = \log_b x_1 - \log_b x_2$,

III. $\log_b x^m = m \log_b x$.

**4.** The integral part of a logarithm is called the **characteristic** and the nonnegative decimal part is called the **mantissa**.

5. Table II can be used to find approximations for $\log_{10} x$ and antilog$_{10}$ ($\log_{10} x$), $x \in R$, $x > 0$, for values of $x$ which are entries in the table.

6. Using linear interpolation in conjunction with Table II, approximations for $\log_{10} x$ and antilog$_{10}$ ($\log_{10} x$), $x \in R$, $x > 0$, can be found for values of $x$ which are not entries in the table.

7. Equation III in paragraph 3 above is helpful in solving exponential equations. Such equations are first rewritten equivalently without exponents and then solved by standard methods.

8. In some cases the use of logarithms simplifies routine computations including those involved in the solution of triangles.

## Review

### A.

Express in logarithmic notation.

**1.** $3^4 = 81$.

**2.** $25^{-1/2} = \frac{1}{5}$.

Express in exponential notation.

**3.** $\log_4 64 = 3$.

**4.** $\log_{125} 5 = \frac{1}{3}$.

Find the value of each of the following without using tables.

**5.** $\log_3 27$.

**7.** $\log_x x$, $(x \in R, x > 0)$.

**6.** $\log_{10} 0.001$.

**8.** $\log_a 1$ $(a \in R, a > 0)$.

Express each of the following as the sum or difference of simple logarithmic quantities.

**9.** $\log_{10} 2x^3 y$.

**10.** $\log_{10} \dfrac{x\sqrt{y}}{z}$.

Express each of the following as a single logarithm.

**11.** $2 \log_{10} x - \frac{1}{3} \log_{10} y$.

**12.** $\log_{10} \pi + \frac{1}{2} \log_{10} (x + 1) - 3 \log_{10} y$.

Find an approximation for each of the following.

**13.** $\log_{10} 7.13$.

**15.** antilog$_{10}$ $8.1234 - 10$.

**14.** $\log_{10} 0.00426$.

**16.** antilog$_{10}$ $4.9762$.

Compute by means of logarithms.

**17.** $(2.91)(13.8)$.

**18.** $\dfrac{6.62}{\sqrt{29.53}}$.

Solve each exponential equation.

**19.** $2^x = 14$.

**20.** $3^{x-5} = 28$.

Find each of the following. Interpolate when necessary.

**21.** $\log_{10} \sin 71° 40'$.

**23.** $\log_{10} \cos 76°24'$.

**22.** $\log_{10} \tan 23°30'$.

**24.** $\log_{10} \sec 156° 24'$.

Find $\alpha$ to the nearest ten minutes.

**25.** $\log_{10} \cot \alpha = 9.6401 - 10$.　　　**26.** $\log_{10} \sin \alpha = 9.8055 - 10$.

Use logarithms in the computations to solve the following triangles.

**27.** $b = 10$, $c = 13$, $\gamma = 90°$.　　　**29.** $b = 6.9$, $\beta = 119° \ 10'$, $\gamma = 28° \ 30'$.
**28.** $b = 6.1$, $\alpha = 79° \ 10'$, $\gamma = 90°$.　　　**30.** $a = 0.44$, $c = 5.7$, $\gamma = 67° \ 50'$.

**B.**

Find each power.

**31.** $10^{1.6064}$.　　　**32.** $10^{8.8463 - 10}$.

In Problems 33 and 34, find the characteristic of each logarithm.

**33.** $\log_3 214$.　　　**34.** $\log_4 179$.
**35.** For what values of $x$ will $\log_{10} x < 6$?　**36.** Find $\log_5 39$.

Without using tables show that each of the following statements is true.

**37.** $4 \log_{10} 3 - 2 \log_{10} 3 = \log_{10} 9$.　　　**39.** $b^{5 \log_b x} = x^5$.
**38.** $\frac{1}{4} \log_{10} 8 - \log_{10} 2 + \frac{1}{4} \log_{10} 2 = 0$.　　**40.** $\log_3 [\log_3 (\log_2 8)] = 0$.

# Axioms and Some Theorems in the Real Number System

## $a, b, c,$ and $d \in R$

**Equality Axioms**

| | | |
|---|---|---|
| E-1 | $a = a.$ | Reflexive Law |
| E-2 | If $a = b$, then $b = a.$ | Symmetric Law |
| E-3 | If $a = b$ and $b = c$, then $a = c.$ | Transitive Law |
| E-4 | A number may be substituted for its equal in an expression involving the addition or the multiplication operation. | Substitution Law |

**Axioms for Operations**

| | | |
|---|---|---|
| F-1 | $a + b \in R.$ | Closure Law for Addition |
| F-2 | $a + b = b + a.$ | Commutative Law of Addition |
| F-3 | $(a + b) + c = a + (b + c).$ | Associative Law of Addition |
| F-4 | $a + 0 = 0 + a = a.$ | Identity Law of Addition |
| F-5 | $a + (-a) = (-a) + a = 0.$ | Additive Inverse Law |
| F-6 | $a \cdot b \in R.$ | Closure Law for Multiplication |
| F-7 | $a \cdot b = b \cdot a.$ | Commutative Law of Multiplication. |
| F-8 | $(a \cdot b) \cdot c = a \cdot (b \cdot c).$ | Associative Law of Multiplication |
| F-9 | $a \cdot 1 = 1 \cdot a = a.$ | Identity Law of Multiplication |
| F-10 | $a \cdot \dfrac{1}{a} = \dfrac{1}{a} \cdot a = 1 \quad (a \neq 0).$ | Multiplicative Inverse Law |
| F-11 | $a \cdot (b + c) = a \cdot b + a \cdot c.$ | Distributive Law |

**Order Axioms**

| | | |
|---|---|---|
| O-1 | Exactly one of the following is true: $a < b$, $a = b$, or $a > b.$ | Trichotomy Law |
| O-2 | If $a < b$ and $b < c$, then $a < c.$ | Transitive Law |
| O-3 | If $a, b > 0$, then $a + b > 0$ and $a \cdot b > 0.$ | Closure for positive numbers |

## Completeness Axiom

If nonempty sets $A$ and $B$ are contained in set $R$, and $a < b$ for each $a \in A$ and each $b \in B$, then there is at least one real number $c$ with the property that $a < c < b$ for each $a \in A$ and for each $b \in B$.

## Some Theorems

**1.** $a \cdot 0 = 0$.

**2.** $-(-a) = a$.

**3.** If $a > 0$, then $-a < 0$.

**4.** If $a = b$, then $a + c = b + c$.

**5.** If $a + c = b + c$, then $a = b$.

**6.** If $a = b$, then $a \cdot c = b \cdot c$.

**7.** If $a \cdot c = b \cdot c$, then $a = b$, $(c \neq 0)$.

**8.** $a - b = a + (-b)$.

**9.** $\dfrac{a}{b} = \dfrac{c}{d}$ if and only if $a \cdot d = b \cdot c$,

$\qquad\qquad (b, d \neq 0)$.

**10.** $\dfrac{a}{b} = \dfrac{a \cdot c}{b \cdot c}$, $\quad (b, c \neq 0)$.

**11.** $\dfrac{a}{c} + \dfrac{b}{c} = \dfrac{a + b}{c}$, $\quad (c \neq 0)$.

**12.** $\dfrac{a}{b} \cdot \dfrac{c}{d} = \dfrac{a \cdot c}{b \cdot d}$, $\quad (b, d \neq 0)$.

**13.** $\dfrac{a}{b} = a \cdot \dfrac{1}{b}$, $\quad (b \neq 0)$.

**14.** $\dfrac{a}{b} \div \dfrac{c}{d} = \dfrac{a}{b} \cdot \dfrac{d}{c}$, $\quad (b, c, d \neq 0)$.

**15.** $\dfrac{a}{b} = \dfrac{-a}{-b} = -\dfrac{a}{-b} = -\dfrac{-a}{b}$ ;

$\qquad \dfrac{-a}{b} = \dfrac{a}{-b} = -\dfrac{a}{b} = -\dfrac{-a}{-b}$, $\quad (b \neq 0)$.

**16.** $ab = 0$ if and only if $a = 0$ or $b = 0$.

# Tables

TABLE I  Powers, Roots, and Reciprocals

I

| N | $N^2$ | $\sqrt{N}$ | $\sqrt{10N}$ | $1/N$ | N | $N^2$ | $\sqrt{N}$ | $\sqrt{10N}$ | $1/N$ |
|---|---|---|---|---|---|---|---|---|---|
| 1 | 1 | 1.0 00 | 3.1 62 | 1.00 00 | 51 | 2 601 | 7.1 41 | 22.5 83 | .019 61 |
| 2 | 4 | 1.4 14 | 4.4 72 | .500 00 | 52 | 2 704 | 7.2 11 | 22.8 04 | .019 23 |
| 3 | 9 | 1.732 | 5.4 77 | .333 33 | 53 | 2 809 | 7.2 80 | 23.0 22 | .018 87 |
| 4 | 16 | 2.0 00 | 6.3 25 | .250 00 | 54 | 2 916 | 7.3 48 | 23.2 38 | .018 52 |
| 5 | 25 | 2.2 36 | 7.0 71 | .200 00 | 55 | 3 025 | 7.4 16 | 23.4 52 | .018 18 |
| 6 | 36 | 2.4 49 | 7.7 46 | .166 67 | 56 | 3 136 | 7.4 83 | 23.6 64 | .017 86 |
| 7 | 49 | 2.6 46 | 8.3 67 | .142 86 | 57 | 3 249 | 7.5 50 | 23.8 75 | .017 54 |
| 8 | 64 | 2.8 28 | 8.9 44 | .125 00 | 58 | 3 364 | 7.6 16 | 24.0 83 | .017 24 |
| 9 | 81 | 3.0 00 | 9.4 87 | .111 11 | 59 | 3 481 | 7.6 81 | 24.2 90 | .016 95 |
| 10 | 100 | 3.1 62 | 10.0 00 | .100 00 | 60 | 3 600 | 7.7 46 | 24.4 95 | .016 67 |
| 11 | 121 | 3.3 17 | 10.4 88 | .090 91 | 61 | 3 721 | 7.8 10 | 24.6 98 | .016 39 |
| 12 | 144 | 3.4 64 | 10.9 54 | .083 33 | 62 | 3 844 | 7.8 74 | 24.9 00 | .016 13 |
| 13 | 169 | 3.6 06 | 11.4 02 | .076 92 | 63 | 3 969 | 7.9 37 | 25.1 00 | .015 87 |
| 14 | 196 | 3.7 42 | 11.8 32 | .071 43 | 64 | 4 096 | 8.0 00 | 25.2 98 | .015 62 |
| 15 | 225 | 3.8 73 | 12.2 47 | .066 67 | 65 | 4 225 | 8.0 62 | 25.4 95 | .015 38 |
| 16 | 256 | 4.0 00 | 12.6 49 | .062 50 | 66 | 4 356 | 8.1 24 | 25.6 90 | .015 15 |
| 17 | 289 | 4.1 23 | 13.0 38 | .058 82 | 67 | 4 489 | 8.1 85 | 25.8 84 | .014 93 |
| 18 | 324 | 4.2 43 | 13.4 16 | .055 56 | 68 | 4 624 | 8.2 46 | 26.0 77 | .014 71 |
| 19 | 361 | 4.3 59 | 13.7 84 | .052 63 | 69 | 4 761 | 8.3 07 | 26.2 68 | .014 49 |
| 20 | 400 | 4.4 72 | 14.1 42 | .050 00 | 70 | 4 900 | 8.3 67 | 26.4 58 | .014 29 |
| 21 | 441 | 4.5 83 | 14.4 91 | .047 62 | 71 | 5 041 | 8.4 26 | 26.6 46 | .014 08 |
| 22 | 484 | 4.6 90 | 14.8 32 | .045 45 | 72 | 5 184 | 8.4 85 | 26.8 33 | .013 89 |
| 23 | 529 | 4.7 96 | 15.1 66 | .043 48 | 73 | 5 329 | 8.5 44 | 27.0 19 | .013 70 |
| 24 | 576 | 4.8 99 | 15.4 92 | .041 67 | 74 | 5 476 | 8.6 02 | 27.2 03 | .013 51 |
| 25 | 625 | 5.0 00 | 15.8 11 | .040 00 | 75 | 5 625 | 8.6 60 | 27.3 86 | .013 33 |
| 26 | 676 | 5.0 99 | 16.1 25 | .038 46 | 76 | 5 776 | 8.7 18 | 27.5 68 | .013 16 |
| 27 | 729 | 5.1 96 | 16.4 32 | .037 04 | 77 | 5 929 | 8.7 75 | 27.7 49 | .012 99 |
| 28 | 784 | 5.2 92 | 16.7 33 | .035 71 | 78 | 6 084 | 8.8 32 | 27.9 28 | .012 82 |
| 29 | 841 | 5.3 85 | 17.0 29 | .034 48 | 79 | 6 241 | 8.8 88 | 28.1 07 | .012 66 |
| 30 | 900 | 5.4 77 | 17.3 21 | .033 33 | 80 | 6 400 | 8.9 44 | 28.2 84 | .012 50 |
| 31 | 961 | 5.5 68 | 17.6 07 | .032 26 | 81 | 6 561 | 9.0 00 | 28.4 60 | .012 35 |
| 32 | 1 024 | 5.6 57 | 17.8 89 | .031 25 | 82 | 6 724 | 9.0 55 | 28.6 36 | .012 20 |
| 33 | 1 089 | 5.7 45 | 18.1 66 | .030 30 | 83 | 6 889 | 9.1 10 | 28.8 10 | .012 05 |
| 34 | 1 156 | 5.8 31 | 18.4 39 | .029 41 | 84 | 7 056 | 9.1 65 | 28.9 83 | .011 90 |
| 35 | 1 225 | 5.9 16 | 18.7 08 | .028 57 | 85 | 7 225 | 9.2 20 | 29.1 55 | .011 76 |
| 36 | 1 296 | 6.0 00 | 18.9 74 | .027 78 | 86 | 7 396 | 9.2 74 | 29.3 26 | .011 63 |
| 37 | 1 369 | 6.0 83 | 19.2 35 | .027 03 | 87 | 7 569 | 9.3 27 | 29.4 96 | .011 49 |
| 38 | 1 444 | 6.1 64 | 19.4 94 | .026 32 | 88 | 7 744 | 9.3 81 | 29.6 65 | .011 36 |
| 39 | 1 521 | 6.2 45 | 19.7 48 | .025 64 | 89 | 7 921 | 9.4 34 | 29.8 33 | .011 24 |
| 40 | 1 600 | 6.3 25 | 20.0 00 | .025 00 | 90 | 8 100 | 9.4 87 | 30.0 00 | .011 11 |
| 41 | 1 681 | 6.4 03 | 20.2 48 | .024 39 | 91 | 8 281 | 9.5 39 | 30.1 66 | .010 99 |
| 42 | 1 764 | 6.4 81 | 20.4 94 | .023 81 | 92 | 8 464 | 9.5 92 | 30.3 32 | .010 87 |
| 43 | 1 849 | 6.5 57 | 20.7 36 | .023 26 | 93 | 8 649 | 9.6 44 | 30.4 96 | .010 75 |
| 44 | 1 936 | 6.6 33 | 20.9 76 | .022 73 | 94 | 8 836 | 9.6 95 | 30.6 59 | .010 64 |
| 45 | 2 025 | 6.7 08 | 21.2 13 | .022 22 | 95 | 9 025 | 9.7 47 | 30.8 22 | .010 53 |
| 46 | 2 116 | 6.7 82 | 21.4 48 | .021 74 | 96 | 9 216 | 9.7 98 | 30.9 84 | .010 42 |
| 47 | 2 209 | 6.8 56 | 21.6 79 | .021 28 | 97 | 9 409 | 9.8 49 | 31.1 45 | .010 31 |
| 48 | 2 304 | 6.9 28 | 21.9 09 | .020 83 | 98 | 9 604 | 9.8 99 | 31.3 05 | .010 20 |
| 49 | 2 401 | 7.0 00 | 22.1 36 | .020 41 | 99 | 9 801 | 9.9 50 | 31.4 64 | .010 10 |
| 50 | 2 500 | 7.0 71 | 22.3 61 | .020 00 | 100 | 10 000 | 10.0 00 | 31.6 23 | .010 00 |
| N | $N^2$ | $\sqrt{N}$ | $\sqrt{10N}$ | $1/N$ | N | $N^2$ | $\sqrt{N}$ | $\sqrt{10N}$ | $1/N$ |

**TABLE II** **Common Logarithms**

| x | 0 | 1 | 2 | 3 | 4 | 5 | 6 | 7 | 8 | 9 |
|---|---|---|---|---|---|---|---|---|---|---|
| 1.0 | .0000 | .0043 | .0086 | .0128 | .0170 | .0212 | .0253 | .0294 | .0334 | .0374 |
| 1.1 | .0414 | .0453 | .0492 | .0531 | .0569 | .0607 | .0645 | .0682 | .0719 | .0755 |
| 1.2 | .0792 | .0828 | .0864 | .0899 | .0934 | .0969 | .1004 | .1038 | .1072 | .1106 |
| 1.3 | .1139 | .1173 | .1206 | .1239 | .1271 | .1303 | .1335 | .1367 | .1399 | .1430 |
| 1.4 | .1461 | .1492 | .1523 | .1553 | .1584 | .1614 | .1644 | .1673 | .1703 | .1732 |
| 1.5 | .1761 | .1790 | .1818 | .1847 | .1875 | .1903 | .1931 | .1959 | .1987 | .2014 |
| 1.6 | .2041 | .2068 | .2095 | .2122 | .2148 | .2175 | .2201 | .2227 | .2253 | .2279 |
| 1.7 | .2304 | .2330 | .2355 | .2380 | .2405 | .2430 | .2455 | .2480 | .2504 | .2529 |
| 1.8 | .2553 | .2577 | .2601 | .2625 | .2648 | .2672 | .2695 | .2718 | .2742 | .2765 |
| 1.9 | .2788 | .2810 | .2833 | .2856 | .2878 | .2900 | .2923 | .2945 | .2967 | .2989 |
| 2.0 | .3010 | .3032 | .3054 | .3075 | .3096 | .3118 | .3139 | .3160 | .3181 | .3201 |
| 2.1 | .3222 | .3243 | .3263 | .3284 | .3304 | .3324 | .3345 | .3365 | .3385 | .3404 |
| 2.2 | .3424 | .3444 | .3464 | .3483 | .3502 | .3522 | .3541 | .3560 | .3579 | .3598 |
| 2.3 | .3617 | .3636 | .3655 | .3674 | .3692 | .3711 | .3729 | .3747 | .3766 | .3784 |
| 2.4 | .3802 | .3820 | .3838 | .3856 | .3874 | .3892 | .3909 | .3927 | .3945 | .3962 |
| 2.5 | .3979 | .3997 | .4014 | .4031 | .4048 | .4065 | .4082 | .4099 | .4116 | .4133 |
| 2.6 | .4150 | .4166 | .4183 | .4200 | .4216 | .4232 | .4249 | .4265 | .4281 | .4298 |
| 2.7 | .4314 | .4330 | .4346 | .4362 | .4378 | .4393 | .4409 | .4425 | .4440 | .4456 |
| 2.8 | .4472 | .4487 | .4502 | .4518 | .4533 | .4548 | .4564 | .4579 | .4594 | .4609 |
| 2.9 | .4624 | .4639 | .4654 | .4669 | .4683 | .4698 | .4713 | .4728 | .4742 | .4757 |
| 3.0 | .4771 | .4786 | .4800 | .4814 | .4829 | .4843 | .4857 | .4871 | .4886 | .4900 |
| 3.1 | .4914 | .4928 | .4942 | .4955 | .4969 | .4983 | .4997 | .5011 | .5024 | .5038 |
| 3.2 | .5051 | .5065 | .5079 | .5092 | .5105 | .5119 | .5132 | .5145 | .5159 | .5172 |
| 3.3 | .5185 | .5198 | .5211 | .5224 | .5237 | .5250 | .5263 | .5276 | .5289 | .5302 |
| 3.4 | .5315 | .5328 | .5340 | .5353 | .5366 | .5378 | .5391 | .5403 | .5416 | .5428 |
| 3.5 | .5441 | .5453 | .5465 | .5478 | .5490 | .5502 | .5514 | .5527 | .5539 | .5551 |
| 3.6 | .5563 | .5575 | .5587 | .5599 | .5611 | .5623 | .5635 | .5647 | .5658 | .5670 |
| 3.7 | .5682 | .5694 | .5705 | .5717 | .5729 | .5740 | .5752 | .5763 | .5775 | .5786 |
| 3.8 | .5798 | .5809 | .5821 | .5832 | .5843 | .5855 | .5866 | .5877 | .5888 | .5899 |
| 3.9 | .5911 | .5922 | .5933 | .5944 | .5955 | .5966 | .5977 | .5988 | .5999 | .6010 |
| 4.0 | .6021 | .6031 | .6042 | .6053 | .6064 | .6075 | .6085 | .6096 | .6107 | .6117 |
| 4.1 | .6128 | .6138 | .6149 | .6160 | .6170 | .6180 | .6191 | .6201 | .6212 | .6222 |
| 4.2 | .6232 | .6243 | .6253 | .6263 | .6274 | .6284 | .6294 | .6304 | .6314 | .6325 |
| 4.3 | .6335 | .6345 | .6355 | .6365 | .6375 | .6385 | .6395 | .6405 | .6415 | .6425 |
| 4.4 | .6435 | .6444 | .6454 | .6464 | .6474 | .6484 | .6493 | .6503 | .6513 | .6522 |
| 4.5 | .6532 | .6542 | .6551 | .6561 | .6571 | .6580 | .6590 | .6599 | .6609 | .6618 |
| 4.6 | .6628 | .6637 | .6646 | .6656 | .6665 | .6675 | .6684 | .6693 | .6702 | .6712 |
| 4.7 | .6721 | .6730 | .6739 | .6749 | .6758 | .6767 | .6776 | .6785 | .6794 | .6803 |
| 4.8 | .6812 | .6821 | .6830 | .6839 | .6848 | .6857 | .6866 | .6875 | .6884 | .6893 |
| 4.9 | .6902 | .6911 | .6920 | .6928 | .6937 | .6946 | .6955 | .6964 | .6972 | .6981 |
| 5.0 | .6990 | .6998 | .7007 | .7016 | .7024 | .7033 | .7042 | .7050 | .7059 | .7067 |
| 5.1 | .7076 | .7084 | .7093 | .7101 | .7110 | .7118 | .7126 | .7135 | .7143 | .7152 |
| 5.2 | .7160 | .7168 | .7177 | .7185 | .7193 | .7202 | .7210 | .7218 | .7226 | .7235 |
| 5.3 | .7243 | .7251 | .7259 | .7267 | .7275 | .7284 | .7292 | .7300 | .7308 | .7316 |
| 5.4 | .7324 | .7332 | .7340 | .7348 | .7356 | .7364 | .7372 | .7380 | .7388 | .7396 |
| x | 0 | 1 | 2 | 3 | 4 | 5 | 6 | 7 | 8 | 9 |

**TABLE II** (*continued*)

| x | 0 | 1 | 2 | 3 | 4 | 5 | 6 | 7 | 8 | 9 |
|---|---|---|---|---|---|---|---|---|---|---|
| 5.5 | .7404 | .7412 | .7419 | .7427 | .7435 | .7443 | .7451 | .7459 | .7466 | .7474 |
| 5.6 | .7482 | .7490 | .7497 | .7505 | .7513 | .7520 | .7528 | .7536 | .7543 | .7551 |
| 5.7 | .7559 | .7566 | .7574 | .7582 | .7589 | .7597 | .7604 | .7612 | .7619 | .7627 |
| 5.8 | .7634 | .7642 | .7649 | .7657 | .7664 | .7672 | .7679 | .7686 | .7694 | .7701 |
| 5.9 | .7709 | .7716 | .7723 | .7731 | .7738 | .7745 | .7752 | .7760 | .7767 | .7774 |
| 6.0 | .7782 | .7789 | .7796 | .7803 | .7810 | .7818 | .7825 | .7832 | .7839 | .7846 |
| 6.1 | .7853 | .7860 | .7868 | .7875 | .7882 | .7889 | .7896 | .7903 | .7910 | .7917 |
| 6.2 | .7924 | .7931 | .7938 | .7945 | .7952 | .7959 | .7966 | .7973 | .7980 | .7987 |
| 6.3 | .7993 | .8000 | .8007 | .8014 | .8021 | .8028 | .8035 | .8041 | .8048 | .8055 |
| 6.4 | .8062 | .8069 | .8075 | .8082 | .8089 | .8096 | .8102 | .8109 | .8116 | .8122 |
| 6.5 | .8129 | .8136 | .8142 | .8149 | .8156 | .8162 | .8169 | .8176 | .8182 | .8189 |
| 6.6 | .8195 | .8202 | .8209 | .8215 | .8222 | .8228 | .8235 | .8241 | .8248 | .8254 |
| 6.7 | .8261 | .8267 | .8274 | .8280 | .8287 | .8293 | .8299 | .8306 | .8312 | .8319 |
| 6.8 | .8325 | .8331 | .8338 | .8344 | .8351 | .8357 | .8363 | .8370 | .8376 | .8382 |
| 6.9 | .8388 | .8395 | .8401 | .8407 | .8414 | .8420 | .8426 | .8432 | .8439 | .8445 |
| 7.0 | .8451 | .8457 | .8463 | .8470 | .8476 | .8482 | .8488 | .8494 | .8500 | .8506 |
| 7.1 | .8513 | .8519 | .8525 | .8531 | .8537 | .8543 | .8549 | .8555 | .8561 | .8567 |
| 7.2 | .8573 | .8579 | .8585 | .8591 | .8597 | .8603 | .8609 | .8615 | .8621 | .8627 |
| 7.3 | .8633 | .8639 | .8645 | .8651 | .8657 | .8663 | .8669 | .8675 | .8681 | .8686 |
| 7.4 | .8692 | .8698 | .8704 | .8710 | .8716 | .8722 | .8727 | .8733 | .8739 | .8745 |
| 7.5 | .8751 | .8756 | .8762 | .8768 | .8774 | .8779 | .8785 | .8791 | .8797 | .8802 |
| 7.6 | .8808 | .8814 | .8820 | .8825 | .8831 | .8837 | .8842 | .8848 | .8854 | .8859 |
| 7.7 | .8865 | .8871 | .8876 | .8882 | .8887 | .8893 | .8899 | .8904 | .8910 | .8915 |
| 7.8 | .8921 | .8927 | .8932 | .8938 | .8943 | .8949 | .8954 | .8960 | .8965 | .8971 |
| 7.9 | .8976 | .8982 | .8987 | .8993 | .8998 | .9004 | .9009 | .9015 | .9020 | .9025 |
| 8.0 | .9031 | .9036 | .9042 | .9047 | .9053 | .9058 | .9063 | .9069 | .9074 | .9079 |
| 8.1 | .9085 | .9090 | .9096 | .9101 | .9106 | .9112 | .9117 | .9122 | .9128 | .9133 |
| 8.2 | .9138 | .9143 | .9149 | .9154 | .9159 | .9165 | .9170 | .9175 | .9180 | .9186 |
| 8.3 | .9191 | .9196 | .9201 | .9206 | .9212 | .9217 | .9222 | .9227 | .9232 | .9238 |
| 8.4 | .9243 | .9248 | .9253 | .9258 | .9263 | .9269 | .9274 | .9279 | .9284 | .9289 |
| 8.5 | .9294 | .9299 | .9304 | .9309 | .9315 | .9320 | .9325 | .9330 | .9335 | .9340 |
| 8.6 | .9345 | .9350 | .9355 | .9360 | .9365 | .9370 | .9375 | .9380 | .9385 | .9390 |
| 8.7 | .9395 | .9400 | .9405 | .9410 | .9415 | .9420 | .9425 | .9430 | .9435 | .9440 |
| 8.8 | .9445 | .9450 | .9455 | .9460 | .9465 | .9469 | .9474 | .9479 | .9484 | .9489 |
| 8.9 | .9494 | .9499 | .9504 | .9509 | .9513 | .9518 | .9523 | .9528 | .9533 | .9538 |
| 9.0 | .9542 | .9547 | .9552 | .9557 | .9562 | .9566 | .9571 | .9576 | .9581 | .9586 |
| 9.1 | .9590 | .9595 | .9600 | .9605 | .9609 | .9614 | .9619 | .9624 | .9628 | .9633 |
| 9.2 | .9638 | .9643 | .9647 | .9652 | .9657 | .9661 | .9666 | .9671 | .9675 | .9680 |
| 9.3 | .9685 | .9689 | .9694 | .9699 | .9703 | .9708 | .9713 | .9717 | .9722 | .9727 |
| 9.4 | .9731 | .9736 | .9741 | .9745 | .9750 | .9754 | .9759 | .9763 | .9768 | .9773 |
| 9.5 | .9777 | .9782 | .9786 | .9791 | .9795 | .9800 | .9805 | .9809 | .9814 | .9818 |
| 9.6 | .9823 | .9827 | .9832 | .9836 | .9841 | .9845 | .9850 | .9854 | .9859 | .9863 |
| 9.7 | .9868 | .9872 | .9877 | .9881 | .9886 | .9890 | .9894 | .9899 | .9903 | .9908 |
| 9.8 | .9912 | .9917 | .9921 | .9926 | .9930 | .9934 | .9939 | .9943 | .9948 | .9952 |
| 9.9 | .9956 | .9961 | .9965 | .9969 | .9974 | .9978 | .9983 | .9987 | .9991 | .9996 |
| x | 0 | 1 | 2 | 3 | 4 | 5 | 6 | 7 | 8 | 9 |

## TABLE III  Values of Circular and Trigonometric Functions

| Real Number $x$ or $\theta$ radians | $\theta$ degrees | $\cos x$ or $\cos \theta$ | $\sin x$ or $\sin \theta$ | $\tan x$ or $\tan \theta$ | $\sec x$ or $\sec \theta$ | $\csc x$ or $\csc \theta$ | $\cot x$ or $\cot \theta$ |
|---|---|---|---|---|---|---|---|
| 0.00 | 0° 00′ | 1.000 | 0.0000 | 0.0000 | 1.000 | No value | No value |
| .01 | 0° 34′ | 1.000 | .0100 | .0100 | 1.000 | 100.0 | 100.0 |
| .02 | 1° 09′ | 0.9998 | .0200 | .0200 | 1.000 | 50.00 | 49.99 |
| .03 | 1° 43′ | 0.9996 | .0300 | .0300 | 1.000 | 33.34 | 33.32 |
| .04 | 2° 18′ | 0.9992 | .0400 | .0400 | 1.001 | 25.01 | 24.99 |
| 0.05 | 2° 52′ | 0.9988 | 0.0500 | 0.0500 | 1.001 | 20.01 | 19.98 |
| .06 | 3° 26′ | .9982 | .0600 | .0601 | 1.002 | 16.68 | 16.65 |
| .07 | 4° 01′ | .9976 | .0699 | .0701 | 1.002 | 14.30 | 14.26 |
| .08 | 4° 35′ | .9968 | .0799 | .0802 | 1.003 | 12.51 | 12.47 |
| .09 | 5° 09′ | .9960 | .0899 | .0902 | 1.004 | 11.13 | 11.08 |
| 0.10 | 5° 44′ | 0.9950 | 0.0998 | 0.1003 | 1.005 | 10.02 | 9.967 |
| .11 | 6° 18′ | .9940 | .1098 | .1104 | 1.006 | 9.109 | 9.054 |
| .12 | 6° 53′ | .9928 | .1197 | .1206 | 1.007 | 8.353 | 8.293 |
| .13 | 7° 27′ | .9916 | .1296 | .1307 | 1.009 | 7.714 | 7.649 |
| .14 | 8° 01′ | .9902 | .1395 | .1409 | 1.010 | 7.166 | 7.096 |
| 0.15 | 8° 36′ | 0.9888 | 0.1494 | 0.1511 | 1.011 | 6.692 | 6.617 |
| .16 | 9° 10′ | .9872 | .1593 | .1614 | 1.013 | 6.277 | 6.197 |
| .17 | 9° 44′ | .9856 | .1692 | .1717 | 1.015 | 5.911 | 5.826 |
| .18 | 10° 19′ | .9838 | .1790 | .1820 | 1.016 | 5.586 | 5.495 |
| .19 | 10° 53′ | .9820 | .1889 | .1923 | 1.018 | 5.295 | 5.200 |
| 0.20 | 11° 28′ | 0.9801 | 0.1987 | 0.2027 | 1.020 | 5.033 | 4.933 |
| .21 | 12° 02′ | .9780 | .2085 | .2131 | 1.022 | 4.797 | 4.692 |
| .22 | 12° 36′ | .9759 | .2182 | .2236 | 1.025 | 4.582 | 4.472 |
| .23 | 13° 11′ | .9737 | .2280 | .2341 | 1.027 | 4.386 | 4.271 |
| .24 | 13° 45′ | .9713 | .2377 | .2447 | 1.030 | 4.207 | 4.086 |
| 0.25 | 14° 19′ | 0.9689 | 0.2474 | 0.2553 | 1.032 | 4.042 | 3.916 |
| .26 | 14° 54′ | .9664 | .2571 | .2660 | 1.035 | 3.890 | 3.759 |
| .27 | 15° 28′ | .9638 | .2667 | .2768 | 1.038 | 3.749 | 3.613 |
| .28 | 16° 03′ | .9611 | .2764 | .2876 | 1.041 | 3.619 | 3.478 |
| .29 | 16° 37′ | .9582 | .2860 | .2984 | 1.044 | 3.497 | 3.351 |
| 0.30 | 17° 11′ | 0.9553 | 0.2955 | 0.3093 | 1.047 | 3.384 | 3.233 |
| .31 | 17° 46′ | .9523 | .3051 | .3203 | 1.050 | 3.278 | 3.122 |
| .32 | 18° 20′ | .9492 | .3146 | .3314 | 1.053 | 3.179 | 3.018 |
| .33 | 18° 54′ | .9460 | .3240 | .3425 | 1.057 | 3.086 | 2.920 |
| .34 | 19° 29′ | .9428 | .3335 | .3537 | 1.061 | 2.999 | 2.827 |
| 0.35 | 20° 03′ | 0.9394 | 0.3429 | 0.3650 | 1.065 | 2.916 | 2.740 |
| .36 | 20° 38′ | .9359 | .3523 | .3764 | 1.068 | 2.839 | 2.657 |
| .37 | 21° 12′ | .9323 | .3616 | .3879 | 1.073 | 2.765 | 2.578 |
| .38 | 21° 46′ | .9287 | .3709 | .3994 | 1.077 | 2.696 | 2.504 |
| .39 | 22° 21′ | .9249 | .3802 | .4111 | 1.081 | 2.630 | 2.433 |
| 0.40 | 22° 55′ | 0.9211 | 0.3894 | 0.4228 | 1.086 | 2.568 | 2.365 |
| .41 | 23° 29′ | .9171 | .3986 | .4346 | 1.090 | 2.509 | 2.301 |
| .42 | 24° 04′ | .9131 | .4078 | .4466 | 1.095 | 2.452 | 2.239 |
| .43 | 24° 38′ | .9090 | .4169 | .4586 | 1.100 | 2.399 | 2.180 |
| .44 | 25° 13′ | .9048 | .4259 | .4708 | 1.105 | 2.348 | 2.124 |
| 0.45 | 25° 47′ | 0.9004 | 0.4350 | 0.4831 | 1.111 | 2.299 | 2.070 |

**TABLE III** (*continued*)

| Real Number $x$ or $\theta$ radians | $\theta$ degrees | $\cos x$ or $\cos \theta$ | $\sin x$ or $\sin \theta$ | $\tan x$ or $\tan \theta$ | $\sec x$ or $\sec \theta$ | $\csc x$ or $\csc \theta$ | $\cot x$ or $\cot \theta$ |
|---|---|---|---|---|---|---|---|
| 0.45 | 25° 47′ | 0.9004 | 0.4350 | 0.4831 | 1.111 | 2.299 | 2.070 |
| .46 | 26° 21′ | .8961 | .4439 | .4954 | 1.116 | 2.253 | 2.018 |
| .47 | 26° 56′ | .8916 | .4529 | .5080 | 1.122 | 2.208 | 1.969 |
| .48 | 27° 30′ | .8870 | .4618 | .5206 | 1.127 | 2.166 | 1.921 |
| .49 | 28° 04′ | .8823 | .4706 | .5334 | 1.133 | 2.125 | 1.875 |
| 0.50 | 28° 39′ | 0.8776 | 0.4794 | 0.5463 | 1.139 | 2.086 | 1.830 |
| .51 | 29° 13′ | .8727 | .4882 | .5594 | 1.146 | 2.048 | 1.788 |
| .52 | 29° 48′ | .8678 | .4969 | .5726 | 1.152 | 2.013 | 1.747 |
| .53 | 30° 22′ | .8628 | .5055 | .5859 | 1.159 | 1.978 | 1.707 |
| .54 | 30° 56′ | .8577 | .5141 | .5994 | 1.166 | 1.945 | 1.668 |
| 0.55 | 31° 31′ | 0.8525 | 0.5227 | 0.6131 | 1.173 | 1.913 | 1.631 |
| .56 | 32° 05′ | .8473 | .5312 | .6269 | 1.180 | 1.883 | 1.595 |
| .57 | 32° 40 | .8419 | .5396 | .6410 | 1.188 | 1.853 | 1.560 |
| .58 | 33° 14′ | .8365 | .5480 | .6552 | 1.196 | 1.825 | 1.526 |
| .59 | 33° 48′ | .8309 | .5564 | .6696 | 1.203 | 1.797 | 1.494 |
| 0.60 | 34° 23′ | 0.8253 | 0.5646 | 0.6841 | 1.212 | 1.771 | 1.462 |
| .61 | 34° 57′ | .8196 | .5729 | .6989 | 1.220 | 1.746 | 1.431 |
| .62 | 35° 31′ | .8139 | .5810 | .7139 | 1.229 | 1.721 | 1.401 |
| .63 | 36° 06′ | .8080 | .5891 | .7291 | 1.238 | 1.697 | 1.372 |
| .64 | 36° 40′ | .8021 | .5972 | .7445 | 1.247 | 1.674 | 1.343 |
| 0.65 | 37° 15′ | 0.7961 | 0.6052 | 0.7602 | 1.256 | 1.652 | 1.315 |
| .66 | 37° 49′ | .7900 | .6131 | .7761 | 1.266 | 1.631 | 1.288 |
| .67 | 38° 23′ | .7838 | .6210 | .7923 | 1.276 | 1.610 | 1.262 |
| .68 | 38° 58′ | .7776 | .6288 | .8087 | 1.286 | 1.590 | 1.237 |
| .69 | 39° 32′ | .7712 | .6365 | .8253 | 1.297 | 1.571 | 1.212 |
| 0.70 | 40° 06′ | 0.7648 | 0.6442 | 0.8423 | 1.307 | 1.552 | 1.187 |
| .71 | 40° 41′ | .7584 | .6518 | .8595 | 1.319 | 1.534 | 1.163 |
| .72 | 41° 15′ | .7518 | .6594 | .8771 | 1.330 | 1.517 | 1.140 |
| .73 | 41° 50′ | .7452 | .6669 | .8949 | 1.342 | 1.500 | 1.117 |
| .74 | 42° 24′ | .7385 | .6743 | .9131 | 1.354 | 1.483 | 1.095 |
| 0.75 | 42° 58′ | 0.7317 | 0.6816 | 0.9316 | 1.367 | 1.467 | 1.073 |
| .76 | 43° 33′ | .7248 | .6889 | .9505 | 1.380 | 1.452 | 1.052 |
| .77 | 44′ 07′ | .7179 | .6961 | .9697 | 1.393 | 1.436 | 1.031 |
| .78 | 44° 41′ | .7109 | .7033 | .9893 | 1.407 | 1.422 | 1.011 |
| .79 | 45° 16′ | .7038 | .7104 | 1.009 | 1.421 | 1.408 | 0.9908 |
| 0.80 | 45° 50′ | 0.6967 | 0.7174 | 1.030 | 1.435 | 1.394 | 0.9712 |
| .81 | 46° 25′ | .6895 | .7243 | 1.050 | 1.450 | 1.381 | .9520 |
| .82 | 46° 59′ | .6822 | .7311 | 1.072 | 1.466 | 1.368 | .9331 |
| .83 | 47° 33′ | .6749 | .7379 | 1.093 | 1.482 | 1.355 | .9146 |
| .84 | 48° 08′ | .6675 | .7446 | 1.116 | 1.498 | 1.343 | .8964 |
| 0.85 | 48° 42′ | 0.6600 | 0.7513 | 1.138 | 1.515 | 1.331 | 0.8785 |
| .86 | 49° 16′ | .6524 | .7578 | 1.162 | 1.533 | 1.320 | .8609 |
| .87 | 49° 51′ | .6448 | .7643 | 1.185 | 1.551 | 1.308 | .8437 |
| .88 | 50° 25′ | .6372 | .7707 | 1.210 | 1.569 | 1.297 | .8267 |
| .89 | 51° 00′ | .6294 | .7771 | 1.235 | 1.589 | 1.287 | .8100 |
| 0.90 | 51° 34′ | 0.6216 | 0.7833 | 1.260 | 1.609 | 1.277 | 0.7936 |
| .91 | 52° 08′ | .6137 | .7895 | 1.286 | 1.629 | 1.267 | .7774 |
| .92 | 52° 43′ | .6058 | .7956 | 1.313 | 1.651 | 1.257 | .7615 |
| .93 | 53° 17′ | .5978 | .8016 | 1.341 | 1.673 | 1.247 | .7458 |
| .94 | 53° 51′ | .5898 | .8076 | 1.369 | 1.696 | 1.238 | .7303 |
| 0.95 | 54° 26′ | 0.5817 | 0.8134 | 1.398 | 1.719 | 1.229 | 0.7151 |

TABLE III  (*continued*)

| Real Number x or θ radians | θ degrees | cos x or cos θ | sin x or sin θ | tan x or tan θ | sec x or sec θ | csc x or csc θ | cot x or cot θ |
|---|---|---|---|---|---|---|---|
| 0.95 | 54° 26′ | 0.5817 | 0.8134 | 1.398 | 1.719 | 1.229 | 0.7151 |
| .96 | 55° 00′ | .5735 | .8192 | 1.428 | 1.744 | 1.221 | .7001 |
| .97 | 55° 35′ | .5653 | .8249 | 1.459 | 1.769 | 1.212 | .6853 |
| .98 | 56° 09′ | .5570 | .8305 | 1.491 | 1.795 | 1.204 | .6707 |
| .99 | 56° 43′ | .5487 | .8360 | 1.524 | 1.823 | 1.196 | .6563 |
| 1.00 | 57° 18′ | 0.5403 | 0.8415 | 1.557 | 1.851 | 1.188 | 0.6421 |
| 1.01 | 57° 52′ | .5319 | .8468 | 1.592 | 1.880 | 1.181 | .6281 |
| 1.02 | 58° 27′ | .5234 | .8521 | 1.628 | 1.911 | 1.174 | .6142 |
| 1.03 | 59° 01′ | .5148 | .8573 | 1.665 | 1.942 | 1.166 | .6005 |
| 1.04 | 59° 35′ | .5062 | .8624 | 1.704 | 1.975 | 1.160 | .5870 |
| 1.05 | 60° 10′ | 0.4976 | 0.8674 | 1.743 | 2.010 | 1.153 | 0.5736 |
| 1.06 | 60° 44′ | .4889 | .8724 | 1.784 | 2.046 | 1.146 | .5604 |
| 1.07 | 61° 18′ | .4801 | .8772 | 1.827 | 2.083 | 1.140 | .5473 |
| 1.08 | 61° 53′ | .4713 | .8820 | 1.871 | 2.122 | 1.134 | .5344 |
| 1.09 | 62° 27′ | .4625 | .8866 | 1.917 | 2.162 | 1.128 | .5216 |
| 1.10 | 63° 02′ | 0.4536 | 0.8912 | 1.965 | 2.205 | 1.122 | 0.5090 |
| 1.11 | 63° 36′ | .4447 | .8957 | 2.014 | 2.249 | 1.116 | .4964 |
| 1.12 | 64° 10′ | .4357 | .9001 | 2.066 | 2.295 | 1.111 | .4840 |
| 1.13 | 64° 45′ | .4267 | .9044 | 2.120 | 2.344 | 1.106 | .4718 |
| 1.14 | 65° 19′ | .4176 | .9086 | 2.176 | 2.395 | 1.101 | .4596 |
| 1.15 | 65° 53′ | 0.4085 | 0.9128 | 2.234 | 2.448 | 1.096 | 0.4475 |
| 1.16 | 66° 28′ | .3993 | .9168 | 2.296 | 2.504 | 1.091 | .4356 |
| 1.17 | 67′ 02′ | .3902 | .9208 | 2.360 | 2.563 | 1.086 | .4237 |
| 1.18 | 67° 37′ | .3809 | .9246 | 2.427 | 2.625 | 1.082 | .4120 |
| 1.19 | 68° 11′ | .3717 | .9284 | 2.498 | 2.691 | 1.077 | .4003 |
| 1.20 | 68° 45′ | 0.3624 | .09320 | 2.572 | 2.760 | 1.073 | 0.3888 |
| 1.21 | 69° 20′ | .3530 | .9356 | 2.650 | 2.833 | 1.069 | .3773 |
| 1.22 | 69° 54′ | .3436 | .9391 | 2.733 | 2.910 | 1.065 | .3659 |
| 1.23 | 70° 28′ | .3342 | .9425 | 2.820 | 2.992 | 1.061 | .3546 |
| 1.24 | 71° 03′ | .3248 | .9458 | 2.912 | 3.079 | 1.057 | .3434 |
| 1.25 | 71° 37′ | 0.3153 | 0.9490 | 3.010 | 3.171 | 1.054 | 0.3323 |
| 1.26 | 72° 12′ | .3058 | .9521 | 3.113 | 3.270 | 1.050 | .3212 |
| 1.27 | 72° 46′ | .2963 | .9551 | 3.224 | 3.375 | 1.047 | .3102 |
| 1.28 | 73° 20′ | .2867 | .9580 | 3.341 | 1.488 | 1.044 | .2993 |
| 1.29 | 73° 55′ | .2771 | .9608 | 3.467 | 3.609 | 1.041 | .2884 |
| 1.30 | 74° 29′ | 0.2675 | 0.9636 | 3.602 | 3.738 | 1.038 | 0.2776 |
| 1.31 | 75° 03′ | .2579 | .9662 | 3.747 | 3.878 | 1.035 | .2669 |
| 1.32 | 75° 38′ | .2482 | .9687 | 3.903 | 4.029 | 1.032 | .2562 |
| 1.33 | 76° 12° | .2385 | .9711 | 4.072 | 4.193 | 1.030 | .2456 |
| 1.34 | 76° 47′ | .2288 | .9735 | 4.256 | 4.372 | 1.027 | .2350 |
| 1.35 | 77° 21′ | 0.2190 | 0.9757 | 4.455 | 4.566 | 1.025 | 0.2245 |
| 1.36 | 77° 55′ | .2092 | .9779 | 4.673 | 4.779 | 1.023 | .2140 |
| 1.37 | 78° 30′ | .1994 | .9799 | 4.913 | 5.014 | 1.021 | .2035 |
| 1.38 | 79° 04′ | .1896 | .9819 | 5.177 | 5.273 | 1.018 | .1931 |
| 1.39 | 79° 38′ | .1798 | .9837 | 5.471 | 5.561 | 1.017 | .1828 |
| 1.40 | 80° 13′ | 0.1700 | 0.9854 | 5.798 | 5.883 | 1.015 | 0.1725 |
| 1.41 | 80° 47′ | .1601 | .9871 | 6.165 | 6.246 | 1.013 | .1622 |
| 1.42 | 81° 22′ | .1502 | .9887 | 6.581 | 6.657 | 1.011 | .1519 |
| 1.43 | 81° 56′ | .1403 | .9901 | 7.055 | 7.126 | 1.010 | .1417 |
| 1.44 | 82° 30′ | .1304 | .9915 | 7.602 | 7.667 | 1.009 | .1315 |
| 1.45 | 83° 05′ | 0.1205 | 0.9927 | 8.238 | 8.299 | 1.007 | 0.1214 |

**TABLE III** (*continued*)

| Real Number $x$ or $\theta$ radians | $\theta$ degrees | $\cos x$ or $\cos \theta$ | $\sin x$ or $\sin \theta$ | $\tan x$ or $\tan \theta$ | $\sec x$ or $\sec \theta$ | $\csc x$ or $\csc \theta$ | $\cot x$ or $\cot \theta$ |
|---|---|---|---|---|---|---|---|
| 1.45 | 83° 05′ | 0.1205 | 0.9927 | 8.238 | 8.299 | 1.007 | 0.1214 |
| 1.46 | 83° 39′ | .1106 | .9939 | 8.989 | 9.044 | 1.006 | .1113 |
| 1.47 | 84° 13′ | .1006 | .9949 | 9.887 | 9.938 | 1.005 | .1001 |
| 1.48 | 84° 48′ | .0907 | .9959 | 10.98 | 11.03 | 1.004 | .0910 |
| 1.49 | 85° 22′ | .0807 | .9967 | 12.35 | 12.39 | 1.003 | .0810 |
| 1.50 | 85° 57′ | 0.0707 | 0.9975 | 14.10 | 14.14 | 1.003 | 0.0709 |
| 1.51 | 86° 31′ | .0608 | .9982 | 16.43 | 16.46 | 1.002 | .0609 |
| 1.52 | 87° 05′ | .0508 | .9987 | 19.67 | 19.69 | 1.001 | .0508 |
| 1.53 | 87° 40′ | .0408 | .9992 | 24.50 | 24.52 | 1.001 | .0408 |
| 1.54 | 88° 14′ | .0308 | .9995 | 32.46 | 32.48 | 1.000 | .0308 |
| 1.55 | 88° 49′ | 0.0208 | 0.9998 | 48.08 | 48.09 | 1.000 | 0.0208 |
| 1.56 | 89° 23′ | .0108 | .9999 | 92.62 | 92.63 | 1.000 | .0108 |
| 1.57 | 89° 57′ | .0008 | 1.000 | 1256 | 1256 | 1.000 | .0008 |

# TABLE IV  Values of Trigonometric Functions

| Angle θ Degrees | Radians | cos θ | sin θ | tan θ | sec θ | csc θ | cot θ | | |
|---|---|---|---|---|---|---|---|---|---|
| 0° 00′ | .0000 | 1.0000 | .0000 | .0000 | 1.000 | No value | No value | 1.5708 | 90° 00′ |
| 10 | 029 | 000 | 029 | 029 | 000 | 343.8 | 343.8 | 679 | 50 |
| 20 | 058 | 000 | 058 | 058 | 000 | 171.9 | 171.9 | 650 | 40 |
| 30 | 087 | 1.0000 | 087 | 087 | 000 | 114.6 | 114.6 | 621 | 30 |
| 40 | 116 | .9999 | 116 | 116 | 000 | 85.95 | 85.94 | 592 | 20 |
| 50 | 145 | 999 | 145 | 145 | 000 | 68.76 | 68.75 | 563 | 10 |
| 1° 00′ | .0175 | .9998 | .0175 | .0175 | 1.000 | 57.30 | 57.29 | 1.5533 | 89° 00′ |
| 10 | 204 | 998 | 204 | 204 | 000 | 49.11 | 49.10 | 504 | 50 |
| 20 | 233 | 997 | 233 | 233 | 000 | 42.98 | 42.96 | 475 | 40 |
| 30 | 262 | 997 | 262 | 262 | 000 | 38.20 | 38.19 | 446 | 30 |
| 40 | 291 | 996 | 291 | 291 | 000 | 34.38 | 34.37 | 417 | 20 |
| 50 | 320 | 995 | 320 | 320 | 001 | 31.26 | 31.24 | 388 | 10 |
| 2° 00′ | .0349 | .9994 | .0349 | .0349 | 1.001 | 28.65 | 28.64 | 1.5359 | 88° 00′ |
| 10 | 378 | 993 | 378 | 378 | 001 | 26.45 | 26.43 | 330 | 50 |
| 20 | 407 | 992 | 407 | 407 | 001 | 24.56 | 24.54 | 301 | 40 |
| 30 | 436 | 990 | 436 | 437 | 001 | 22.93 | 22.90 | 272 | 30 |
| 40 | 465 | 989 | 465 | 466 | 001 | 21.49 | 21.47 | 243 | 20 |
| 50 | 495 | 988 | 494 | 495 | 001 | 20.23 | 20.21 | 213 | 10 |
| 3° 00′ | .0524 | .9986 | .0523 | .0524 | 1.001 | 19.11 | 19.08 | 1.5184 | 87° 00′ |
| 10 | 553 | 985 | 552 | 553 | 002 | 18.10 | 18.07 | 155 | 50 |
| 20 | 582 | 983 | 581 | 582 | 002 | 17.20 | 17.17 | 126 | 40 |
| 30 | 611 | 981 | 610 | 612 | 002 | 16.38 | 16.35 | 097 | 30 |
| 40 | 640 | 980 | 640 | 641 | 002 | 15.64 | 15.60 | 068 | 20 |
| 50 | 669 | 978 | 669 | 670 | 002 | 14.96 | 14.92 | 039 | 10 |
| 4° 00′ | .0698 | .9976 | .0698 | .0699 | 1.002 | 14.34 | 14.30 | 1.5010 | 86° 00′ |
| 10 | 727 | 974 | 727 | 729 | 003 | 13.76 | 13.73 | 981 | 50 |
| 20 | 765 | 971 | 756 | 758 | 003 | 13.23 | 13.20 | 952 | 40 |
| 30 | 785 | 969 | 785 | 787 | 003 | 12.75 | 12.71 | 923 | 30 |
| 40 | 814 | 967 | 814 | 816 | 003 | 12.29 | 12.25 | 893 | 20 |
| 50 | 844 | 964 | 843 | 846 | 004 | 11.87 | 11.83 | 864 | 10 |
| 5° 00′ | .0873 | .9962 | .0872 | .0875 | 1.004 | 11.47 | 11.43 | 1.4835 | 85° 00′ |
| 10 | 902 | 959 | 901 | 904 | 004 | 11.10 | 11.06 | 806 | 50 |
| 20 | 931 | 957 | 929 | 934 | 004 | 10.76 | 10.71 | 777 | 40 |
| 30 | 960 | 954 | 958 | 963 | 005 | 10.43 | 10.39 | 748 | 30 |
| 40 | .0989 | 951 | .0987 | .0992 | 005 | 10.13 | 10.08 | 719 | 20 |
| 50 | .1018 | 948 | .1016 | .1022 | 005 | 9.839 | 9.788 | 690 | 10 |
| 6° 00′ | .1047 | .9945 | .1045 | .1051 | 1.006 | 9.567 | 9.514 | 1.4661 | 84° 00′ |
| 10 | 076 | 942 | 074 | 080 | 006 | 9.309 | 9.255 | 632 | 50 |
| 20 | 105 | 939 | 103 | 110 | 006 | 9.065 | 9.010 | 603 | 40 |
| 30 | 134 | 936 | 132 | 139 | 006 | 8.834 | 8.777 | 573 | 30 |
| 40 | 164 | 932 | 161 | 169 | 007 | 8.614 | 8.556 | 544 | 20 |
| 50 | 193 | 929 | 190 | 198 | 007 | 8.405 | 8.345 | 515 | 10 |
| 7° 00′ | .1222 | .9925 | .1219 | .1228 | 1.008 | 8.206 | 8.144 | 1.4486 | 83° 00′ |
| 10 | 251 | 922 | 248 | 257 | 008 | 8.016 | 7.953 | 457 | 50 |
| 20 | 280 | 918 | 276 | 287 | 008 | 7.834 | 7.770 | 428 | 40 |
| 30 | 309 | 914 | 305 | 317 | 009 | 7.661 | 7.596 | 399 | 30 |
| 40 | 338 | 911 | 334 | 346 | 009 | 7.496 | 7.429 | 370 | 20 |
| 50 | 367 | 907 | 363 | 376 | 009 | 7.337 | 7.269 | 341 | 10 |
| 8° 00′ | .1396 | .9903 | .1392 | .1405 | 1.010 | 7.185 | 7.115 | 1.4312 | 82° 00′ |
| | | sin θ | cos θ | cot θ | csc θ | sec θ | tan θ | Radians | Degrees |
| | | | | | | | | Angle θ | |

**TABLE IV**  (*continued*)

| Angle θ Degrees | Angle θ Radians | cos θ | sin θ | tan θ | sec θ | csc θ | cot θ | | |
|---|---|---|---|---|---|---|---|---|---|
| 8° 00′ | .1396 | .9903 | .1392 | .1405 | 1.010 | 7.185 | 7.115 | 1.4312 | 82° 00′ |
| 10 | 425 | 899 | 421 | 435 | 010 | 7.040 | 6.968 | 283 | 50 |
| 20 | 454 | 894 | 449 | 465 | 011 | 6.900 | 827 | 254 | 40 |
| 30 | 484 | 890 | 478 | 495 | 011 | 765 | 691 | 224 | 30 |
| 40 | 513 | 886 | 507 | 524 | 012 | 636 | 561 | 195 | 20 |
| 50 | 542 | 881 | 536 | 554 | 012 | 512 | 435 | 166 | 10 |
| 9° 00′ | .1571 | .9877 | .1564 | .1584 | 1.012 | 6.392 | 6.314 | 1.4137 | 81° 00′ |
| 10 | 600 | 872 | 593 | 614 | 013 | 277 | 197 | 108 | 50 |
| 20 | 629 | 868 | 622 | 644 | 013 | 166 | 6.084 | 079 | 40 |
| 30 | 658 | 863 | 650 | 673 | 014 | 6.059 | 5.976 | 050 | 30 |
| 40 | 687 | 858 | 679 | 703 | 014 | 5.955 | 871 | 1.4021 | 20 |
| 50 | 716 | 853 | 708 | 733 | 015 | 855 | 769 | 1.3992 | 10 |
| 10° 00′ | .1745 | .9848 | .1736 | .1763 | 1.015 | 5.759 | 5.671 | 1.3963 | 80° 00′ |
| 10 | 774 | 843 | 765 | 793 | 016 | 665 | 576 | 934 | 50 |
| 20 | 804 | 838 | 794 | 823 | 016 | 575 | 485 | 904 | 40 |
| 30 | 833 | 833 | 822 | 853 | 017 | 487 | 396 | 875 | 30 |
| 40 | 862 | 827 | 851 | 883 | 018 | 403 | 309 | 846 | 20 |
| 50 | 891 | 822 | 880 | 914 | 018 | 320 | 226 | 817 | 10 |
| 11° 00′ | .1920 | .9816 | .1908 | .1944 | 1.019 | 5.241 | 5.145 | 1.3788 | 79° 00′ |
| 10 | 949 | 811 | 937 | .1974 | 019 | 164 | 5.066 | 759 | 50 |
| 20 | .1978 | 805 | 965 | .2004 | 020 | 089 | 4.989 | 730 | 40 |
| 30 | .2007 | 799 | .1994 | 035 | 020 | 5.016 | 915 | 701 | 30 |
| 40 | 036 | 793 | .2022 | 065 | 021 | 4.945 | 843 | 672 | 20 |
| 50 | 065 | 787 | 051 | 095 | 022 | 876 | 773 | 643 | 10 |
| 12° 00′ | .2094 | .9781 | .2079 | .2126 | 1.022 | 4.810 | 4.705 | 1.3614 | 78° 00′ |
| 10 | 123 | 775 | 108 | 156 | 023 | 745 | 638 | 584 | 50 |
| 20 | 153 | 769 | 136 | 186 | 024 | 682 | 574 | 555 | 40 |
| 30 | 182 | 763 | 164 | 217 | 024 | 620 | 511 | 526 | 30 |
| 40 | 211 | 757 | 193 | 247 | 025 | 560 | 449 | 497 | 20 |
| 50 | 240 | 750 | 221 | 278 | 026 | 502 | 390 | 468 | 10 |
| 13° 00′ | .2269 | .9744 | .2250 | .2309 | 1.026 | 4.445 | 4.331 | 1.3439 | 77° 00′ |
| 10 | 298 | 737 | 278 | 339 | 027 | 390 | 275 | 410 | 50 |
| 20 | 327 | 730 | 306 | 370 | 028 | 336 | 219 | 381 | 40 |
| 30 | 356 | 724 | 334 | 401 | 028 | 284 | 165 | 352 | 30 |
| 40 | 385 | 717 | 363 | 432 | 029 | 232 | 113 | 323 | 20 |
| 50 | 414 | 710 | 391 | 462 | 030 | 182 | 061 | 294 | 10 |
| 14° 00′ | .2443 | .9703 | .2419 | .2493 | 1.031 | 4.134 | 4.011 | 1.3265 | 76° 00′ |
| 10 | 473 | 696 | 447 | 524 | 031 | 086 | 3.962 | 235 | 50 |
| 20 | 502 | 689 | 476 | 555 | 032 | 4.039 | 914 | 206 | 40 |
| 30 | 531 | 681 | 504 | 586 | 033 | 3.994 | 867 | 177 | 30 |
| 40 | 560 | 674 | 532 | 617 | 034 | 950 | 821 | 148 | 20 |
| 50 | 589 | 667 | 560 | 648 | 034 | 906 | 776 | 119 | 10 |
| 15° 00′ | .2618 | .9659 | .2588 | .2679 | 1.035 | 3.864 | 3.732 | 1.3090 | 75° 00′ |
| 10 | 647 | 652 | 616 | 711 | 036 | 822 | 689 | 061 | 50 |
| 20 | 676 | 644 | 644 | 742 | 037 | 782 | 647 | 032 | 40 |
| 30 | 705 | 636 | 672 | 773 | 038 | 742 | 606 | 1.3003 | 30 |
| 40 | 734 | 628 | 700 | 805 | 039 | 703 | 566 | 1.2974 | 20 |
| 50 | 763 | 621 | 728 | 836 | 039 | 665 | 526 | 945 | 10 |
| 16° 00′ | .2793 | .9613 | .2756 | .2867 | 1.040 | 3.628 | 3.487 | 1.2915 | 74° 00′ |
| | | sin θ | cos θ | cot θ | csc θ | sec θ | tan θ | Radians | Degrees |
| | | | | | | | | Angle θ | |

**IV**

**TABLE IV** (*continued*)

| Angle Degrees Radians | | cos θ | sin θ | tan θ | sec θ | csc θ | cot θ | | |
|---|---|---|---|---|---|---|---|---|---|
| 16° 00′ | .2793 | .9613 | .2756 | .2867 | 1.040 | 3.628 | 3.487 | 1.2915 | 74° 00′ |
| 10 | 822 | 605 | 784 | 899 | 041 | 592 | 450 | 886 | 50 |
| 20 | 851 | 596 | 812 | 931 | 042 | 556 | 412 | 857 | 40 |
| 30 | 880 | 588 | 840 | 962 | 043 | 521 | 376 | 828 | 30 |
| 40 | 909 | 580 | 868 | .2944 | 044 | 487 | 340 | 799 | 20 |
| 50 | 938 | 572 | 896 | .3026 | 045 | 453 | 305 | 770 | 10 |
| 17° 00′ | .2967 | .9563 | .2924 | .3057 | 1.046 | 3.420 | 3.271 | 1.2741 | 73° 00′ |
| 10 | .2996 | 555 | 952 | 089 | 047 | 388 | 237 | 712 | 50 |
| 20 | .3025 | 546 | .2979 | 121 | 048 | 357 | 204 | 683 | 40 |
| 30 | 054 | 537 | .3007 | 153 | 048 | 326 | 172 | 654 | 30 |
| 40 | 083 | 528 | 035 | 185 | 049 | 295 | 140 | 625 | 20 |
| 50 | 113 | 520 | 062 | 217 | 050 | 265 | 108 | 595 | 10 |
| 18° 00′ | .3142 | .9511 | .3090 | .3249 | 1.051 | 3.236 | 3.078 | 1.2566 | 72° 00′ |
| 10 | 171 | 502 | 118 | 281 | 052 | 207 | 047 | 537 | 50 |
| 20 | 200 | 492 | 145 | 314 | 053 | 179 | 3.018 | 508 | 40 |
| 30 | 229 | 483 | 173 | 346 | 054 | 152 | 2.989 | 479 | 30 |
| 40 | 258 | 474 | 201 | 378 | 056 | 124 | 960 | 450 | 20 |
| 50 | 287 | 465 | 228 | 411 | 057 | 098 | 932 | 421 | 10 |
| 19° 00′ | .3316 | .9455 | .3256 | .3443 | 1.058 | 3.072 | 2.904 | 1.2392 | 71° 00′ |
| 10 | 345 | 446 | 283 | 476 | 059 | 046 | 877 | 363 | 50 |
| 20 | 374 | 436 | 311 | 508 | 060 | 3.021 | 850 | 334 | 40 |
| 30 | 403 | 426 | 338 | 541 | 061 | 2.996 | 824 | 305 | 30 |
| 40 | 432 | 417 | 365 | 574 | 062 | 971 | 798 | 275 | 20 |
| 50 | 462 | 407 | 393 | 607 | 063 | 947 | 773 | 246 | 10 |
| 20° 00′ | .3491 | .9397 | .3420 | .3640 | .1064 | 2.924 | 2.747 | 1.2217 | 70° 00′ |
| 10 | 520 | 387 | 448 | 673 | 065 | 901 | 723 | 188 | 50 |
| 20 | 549 | 377 | 475 | 706 | 066 | 878 | 699 | 159 | 40 |
| 30 | 578 | 367 | 502 | 739 | 068 | 855 | 675 | 130 | 30 |
| 40 | 607 | 356 | 529 | 772 | 069 | 833 | 651 | 101 | 20 |
| 50 | 636 | 346 | 557 | 805 | 070 | 812 | 628 | 072 | 10 |
| 21° 00′ | .3665 | .9336 | .3584 | .3839 | 1.071 | 2.790 | 2.605 | 1.2043 | 69° 00′ |
| 10 | 694 | 325 | 611 | 872 | 072 | 769 | 583 | 1.2014 | 50 |
| 20 | 723 | 315 | 638 | 906 | 074 | 749 | 560 | 1.1985 | 40 |
| 30 | 752 | 304 | 665 | 939 | 075 | 729 | 539 | 956 | 30 |
| 40 | 782 | 293 | 692 | .3973 | 076 | 709 | 517 | 926 | 20 |
| 50 | 811 | 283 | 719 | .4006 | 077 | 689 | 496 | 897 | 10 |
| 22° 00′ | .3840 | .9272 | .3746 | .4040 | 1.079 | 2.669 | 2.475 | 1.1868 | 68° 00′ |
| 10 | 869 | 261 | 773 | 074 | 080 | 650 | 455 | 839 | 50 |
| 20 | 898 | 250 | 800 | 108 | 081 | 632 | 434 | 810 | 40 |
| 30 | 927 | 239 | 827 | 142 | 082 | 613 | 414 | 781 | 30 |
| 40 | 956 | 228 | 854 | 176 | 084 | 595 | 394 | 752 | 20 |
| 50 | 985 | 216 | 881 | 210 | 085 | 577 | 375 | 723 | 10 |
| 23° 00′ | .4014 | .9205 | .3907 | .4245 | 1.086 | 2.559 | 2.356 | 1.1694 | 67° 00′ |
| 10 | 043 | 194 | 934 | 279 | 088 | 542 | 337 | 665 | 50 |
| 20 | 072 | 182 | 961 | 314 | 089 | 525 | 318 | 636 | 40 |
| 30 | 102 | 171 | .3987 | 348 | 090 | 508 | 300 | 606 | 30 |
| 40 | 131 | 159 | .4014 | 383 | 092 | 491 | 282 | 577 | 20 |
| 50 | 160 | 147 | 041 | 417 | 093 | 475 | 264 | 548 | 10 |
| 24° 00′ | .4189 | .9135 | .4067 | .4452 | 1.095 | 2.459 | 2.246 | 1.1519 | 66° 00′ |
| | | sin θ | cos θ | cot θ | csc θ | sec θ | tan θ | Radians Degrees | |
| | | | | | | | | Angle θ | |

TABLE IV   (continued)

| Angle θ Degrees | Angle θ Radians | cos θ | sin θ | tan θ | sec θ | csc θ | cot θ | Radians | Degrees |
|---|---|---|---|---|---|---|---|---|---|
| 24° 00′ | .4189 | .9135 | .4067 | .4452 | 1.095 | 2.459 | 2.246 | 1.1519 | 66° 00′ |
| 10 | 218 | 124 | 094 | 487 | 096 | 443 | 229 | 490 | 50 |
| 20 | 247 | 112 | 120 | 522 | 097 | 427 | 211 | 461 | 40 |
| 30 | 276 | 100 | 147 | 557 | 099 | 411 | 194 | 432 | 30 |
| 40 | 30 5 | 088 | 173 | 592 | 100 | 396 | 177 | 403 | 20 |
| 50 | 334 | 075 | 200 | 628 | 102 | 381 | 161 | 374 | 10 |
| 25° 00′ | .4363 | .9063 | .4226 | .4663 | 1.103 | 2.366 | 2.145 | 1.1345 | 65° 00′ |
| 10 | 392 | 051 | 253 | 699 | 105 | 352 | 128 | 316 | 50 |
| 20 | 422 | 038 | 279 | 734 | 106 | 337 | 112 | 286 | 40 |
| 30 | 451 | 026 | 305 | 770 | 108 | 323 | 097 | 257 | 30 |
| 40 | 480 | 013 | 331 | 806 | 109 | 309 | 081 | 228 | 20 |
| 50 | 509 | .9001 | 358 | 841 | 111 | 295 | 066 | 199 | 10 |
| 26° 00′ | .4538 | .8988 | .4384 | .4877 | 1.113 | 2.281 | 2.050 | 1.1170 | 64° 00′ |
| 10 | 567 | 975 | 410 | 913 | 114 | 268 | 035 | 141 | 50 |
| 20 | 596 | 962 | 436 | 950 | 116 | 254 | 020 | 112 | 40 |
| 30 | 625 | 949 | 462 | .4986 | 117 | 241 | 2.006 | 083 | 30 |
| 40 | 654 | 936 | 488 | .5022 | 119 | 228 | 1.991 | 054 | 20 |
| 50 | 683 | 923 | 514 | 059 | 121 | 215 | 977 | 1.1025 | 10 |
| 27° 00′ | .4712 | .8910 | .4540 | .5095 | 1.122 | 2.203 | 1.963 | 1.0996 | 63° 00′ |
| 10 | 741 | 897 | 566 | 132 | 124 | 190 | 949 | 966 | 50 |
| 20 | 771 | 884 | 592 | 169 | 126 | 178 | 935 | 937 | 40 |
| 30 | 800 | 870 | 617 | 206 | 127 | 166 | 921 | 908 | 30 |
| 40 | 829 | 857 | 643 | 243 | 129 | 154 | 907 | 879 | 20 |
| 50 | 858 | 843 | 669 | 280 | 131 | 142 | 894 | 850 | 10 |
| 28° 00′ | .4887 | .8829 | .4695 | .5317 | 1.133 | 2.130 | 1.881 | 1.0821 | 62° 00′ |
| 10 | 916 | 816 | 720 | 354 | 134 | 118 | 868 | 792 | 50 |
| 20 | 945 | 802 | 746 | 392 | 136 | 107 | 855 | 763 | 40 |
| 30 | .4974 | 788 | 772 | 430 | 138 | 096 | 842 | 734 | 30 |
| 40 | .5003 | 774 | 797 | 467 | 140 | 085 | 829 | 705 | 20 |
| 50 | 032 | 760 | 823 | 505 | 142 | 074 | 816 | 676 | 10 |
| 29° 00′ | .5061 | .8746 | .4848 | .5543 | 1.143 | 2.063 | 1.804 | 1.0647 | 61° 00′ |
| 10 | 091 | 732 | 874 | 581 | 145 | 052 | 792 | 617 | 50 |
| 20 | 120 | 718 | 899 | 619 | 147 | 041 | 780 | 588 | 40 |
| 30 | 149 | 704 | 924 | 658 | 149 | 031 | 767 | 559 | 30 |
| 40 | 178 | 689 | 950 | 696 | 151 | 020 | 756 | 530 | 20 |
| 50 | 207 | 675 | .4975 | 735 | 153 | 010 | 744 | 501 | 10 |
| 30° 00′ | .5236 | .8660 | .5000 | .5774 | 1.155 | 2.000 | 1.732 | 1.0472 | 60° 00′ |
| 10 | 265 | 646 | 025 | 812 | 157 | 1.990 | 720 | 443 | 50 |
| 20 | 294 | 631 | 050 | 851 | 159 | 980 | 709 | 414 | 40 |
| 30 | 323 | 616 | 075 | 890 | 161 | 970 | 698 | 385 | 30 |
| 40 | 352 | 601 | 100 | 930 | 163 | 961 | 686 | 356 | 20 |
| 50 | 381 | 587 | 125 | .5969 | 165 | 951 | 675 | 327 | 10 |
| 31° 00′ | .5411 | .8572 | .5150 | .6009 | 1.167 | 1.942 | 1.664 | 1.0297 | 59° 00′ |
| 10 | 440 | 557 | 175 | 048 | 169 | 932 | 653 | 268 | 50 |
| 20 | 469 | 542 | 200 | 088 | 171 | 923 | 643 | 239 | 40 |
| 30 | 498 | 526 | 225 | 128 | 173 | 914 | 632 | 210 | 30 |
| 40 | 527 | 511 | 250 | 168 | 175 | 905 | 621 | 181 | 20 |
| 50 | 556 | 496 | 275 | 208 | 177 | 896 | 611 | 152 | 10 |
| 32° 00′ | .5585 | .8480 | .5299 | .6249 | 1.179 | 1.887 | 1.600 | 1.0123 | 58° 00′ |

| | | sin θ | cos θ | cot θ | csc θ | sec θ | tan θ | Radians | Degrees |
| | | | | | | | | Angle θ | |

**TABLE IV** (*continued*)

| Angle $\theta$ | | $\cos \theta$ | $\sin \theta$ | $\tan \theta$ | $\sec \theta$ | $\csc \theta$ | $\cot \theta$ | | |
|---|---|---|---|---|---|---|---|---|---|
| Degrees | Radians | | | | | | | | |
| 32° 00′ | .5585 | .8480 | .5299 | .6249 | 1.179 | 1.887 | 1.600 | 1.0123 | 58° 00′ |
| 10 | 614 | 465 | 324 | 289 | 181 | 878 | 590 | 094 | 50 |
| 20 | 643 | 450 | 348 | 330 | 184 | 870 | 580 | 065 | 40 |
| 30 | 672 | 434 | 373 | 371 | 186 | 861 | 570 | 036 | 30 |
| 40 | 701 | 418 | 398 | 412 | 188 | 853 | 560 | 1.0007 | 20 |
| 50 | 730 | 403 | 422 | 453 | 190 | 844 | 550 | .9977 | 10 |
| 33° 00′ | .5760 | .8387 | .5446 | .6494 | 1.192 | 1.836 | 1.540 | .9948 | 57° 00′ |
| 10 | 789 | 371 | 471 | 536 | 195 | 828 | 530 | 919 | 50 |
| 20 | 818 | 355 | 495 | 577 | 197 | 820 | 520 | 890 | 40 |
| 30 | 847 | 339 | 519 | 619 | 199 | 812 | 511 | 861 | 30 |
| 40 | 876 | 323 | 544 | 661 | 202 | 804 | 501 | 832 | 20 |
| 50 | 905 | 307 | 568 | 703 | 204 | 796 | 492 | 803 | 10 |
| 34° 00′ | .5934 | .8290 | .5592 | .6745 | 1.206 | 1.788 | 1.483 | .9774 | 56° 00′ |
| 10 | 963 | 274 | 616 | 787 | 209 | 781 | 473 | 745 | 50 |
| 20 | .5992 | 258 | 640 | 830 | 211 | 773 | 464 | 716 | 40 |
| 30 | .6021 | 241 | 664 | 873 | 213 | 766 | 455 | 687 | 30 |
| 40 | 050 | 225 | 688 | 916 | 216 | 758 | 446 | 657 | 20 |
| 50 | 080 | 208 | 712 | .6959 | 218 | 751 | 437 | 628 | 10 |
| 35° 00′ | .6109 | .8192 | .5736 | .7002 | 1.221 | 1.743 | 1.428 | .9599 | 55° 00′ |
| 10 | 138 | 175 | 760 | 046 | 223 | 736 | 419 | 570 | 50 |
| 20 | 167 | 158 | 783 | 089 | 226 | 729 | 411 | 541 | 40 |
| 30 | 196 | 141 | 807 | 133 | 228 | 722 | 402 | 512 | 30 |
| 40 | 225 | 124 | 831 | 177 | 231 | 715 | 393 | 483 | 20 |
| 50 | 254 | 107 | 854 | 221 | 233 | 708 | 385 | 454 | 10 |
| 36° 00′ | .6283 | .8090 | .5878 | .7265 | 1.236 | 1.701 | 1.376 | .9425 | 54° 00′ |
| 10 | 312 | 073 | 901 | 310 | 239 | 695 | 368 | 396 | 50 |
| 20 | 341 | 056 | 925 | 355 | 241 | 688 | 360 | 367 | 40 |
| 30 | 370 | 039 | 948 | 400 | 244 | 681 | 351 | 338 | 30 |
| 40 | 400 | 021 | 972 | 445 | 247 | 675 | 343 | 308 | 20 |
| 50 | 429 | .8004 | .5995 | 490 | 249 | 668 | 335 | 279 | 10 |
| 37° 00′ | .6458 | .7986 | .6018 | .7536 | 1.252 | 1.662 | 1.327 | .9250 | 53° 00′ |
| 10 | 487 | 966 | 041 | 581 | 255 | 655 | 319 | 221 | 50 |
| 20 | 516 | 951 | 065 | 627 | 258 | 649 | 311 | 192 | 40 |
| 30 | 545 | 934 | 088 | 673 | 260 | 643 | 303 | 163 | 30 |
| 40 | 574 | 916 | 111 | 720 | 263 | 636 | 295 | 134 | 20 |
| 50 | 603 | 898 | 134 | 766 | 266 | 630 | 288 | 105 | 10 |
| 38° 00′ | .6632 | .7880 | .6157 | .7813 | 1.269 | 1.624 | 1.280 | .9076 | 52° 00′ |
| 10 | 661 | 862 | 180 | 860 | 272 | 618 | 272 | 047 | 50 |
| 20 | 690 | 844 | 202 | 907 | 275 | 612 | 265 | .9018 | 40 |
| 30 | 720 | 826 | 225 | .7954 | 278 | 606 | 257 | .8988 | 30 |
| 40 | 749 | 808 | 248 | .8002 | 281 | 601 | 250 | 959 | 20 |
| 50 | 778 | 790 | 271 | 050 | 284 | 595 | 242 | 930 | 10 |
| 39° 00′ | .6807 | .7771 | .6293 | .8098 | 1.287 | 1.589 | 1.235 | .8901 | 51° 00′ |
| 10 | 836 | 753 | 316 | 146 | 290 | 583 | 228 | 872 | 50 |
| 20 | 865 | 735 | 338 | 195 | 293 | 578 | 220 | 843 | 40 |
| 30 | 894 | 716 | 361 | 243 | 296 | 572 | 213 | 814 | 30 |
| 40 | 923 | 698 | 383 | 292 | 299 | 567 | 206 | 785 | 20 |
| 50 | 952 | 679 | 406 | 342 | 302 | 561 | 199 | 756 | 10 |
| 40° 00′ | .6981 | .7660 | .6428 | .8391 | 1.305 | 1.556 | 1.192 | .8727 | 50° 00′ |
| | | $\sin \theta$ | $\cos \theta$ | $\cot \theta$ | $\csc \theta$ | $\sec \theta$ | $\tan \theta$ | Radians | Degrees |
| | | | | | | | | Angle $\theta$ | |

**TABLE IV** (*continued*)

| Angle θ Degrees | Radians | cos θ | sin θ | tan θ | sec θ | csc θ | cot θ | | |
|---|---|---|---|---|---|---|---|---|---|
| 40° 00′ | .6981 | .7660 | .6428 | .8391 | 1.305 | 1.556 | 1.192 | .8727 | 50° 00′ |
| 10 | .7010 | 642 | 450 | 441 | 309 | 550 | 185 | 698 | 50 |
| 20 | 039 | 623 | 472 | 491 | 312 | 545 | 178 | 668 | 40 |
| 30 | 069 | 604 | 494 | 541 | 315 | 540 | 171 | 639 | 30 |
| 40 | 098 | 585 | 517 | 591 | 318 | 535 | 164 | 610 | 20 |
| 50 | 127 | 566 | 539 | 642 | 322 | 529 | 157 | 581 | 10 |
| 41° 00′ | .7156 | .7547 | .6561 | .8693 | 1.325 | 1.524 | 1.150 | .8552 | 49° 00′ |
| 10 | 185 | 528 | 583 | 744 | 328 | 519 | 144 | 523 | 50 |
| 20 | 214 | 509 | 604 | 796 | 332 | 514 | 137 | 494 | 40 |
| 30 | 243 | 490 | 626 | 847 | 335 | 509 | 130 | 465 | 30 |
| 40 | 272 | 470 | 648 | 899 | 339 | 504 | 124 | 436 | 20 |
| 50 | 301 | 451 | 670 | .8952 | 342 | 499 | 117 | 407 | 10 |
| 42° 00′ | .7330 | .7431 | .6691 | .9004 | 1.346 | 1.494 | 1.111 | .8378 | 48° 00′ |
| 10 | 359 | 412 | 713 | 057 | 349 | 490 | 104 | 348 | 50 |
| 20 | 389 | 392 | 734 | 110 | 353 | 485 | 098 | 319 | 40 |
| 30 | 418 | 373 | 756 | 163 | 356 | 480 | 091 | 290 | 30 |
| 40 | 447 | 353 | 777 | 217 | 360 | 476 | 085 | 261 | 20 |
| 50 | 476 | 333 | 799 | 271 | 364 | 471 | 079 | 232 | 10 |
| 43° 00′ | .7505 | .7314 | .6820 | .9325 | 1.367 | 1.466 | 1.072 | .8203 | 47° 00′ |
| 10 | 534 | 294 | 841 | 380 | 371 | 462 | 066 | 174 | 50 |
| 20 | 563 | 274 | 862 | 435 | 375 | 457 | 060 | 145 | 40 |
| 30 | 592 | 254 | 884 | 490 | 379 | 453 | 054 | 116 | 30 |
| 40 | 621 | 234 | 905 | 545 | 382 | 448 | 048 | 087 | 20 |
| 50 | 650 | 214 | 926 | 601 | 386 | 444 | 042 | 058 | 10 |
| 44° 00′ | .7679 | .7193 | .6947 | .9657 | 1.390 | 1.440 | 1.036 | .8029 | 46° 00′ |
| 10 | 709 | 173 | 967 | 713 | 394 | 435 | 030 | .7999 | 50 |
| 20 | 738 | 153 | .6988 | 770 | 398 | 431 | 024 | 970 | 40 |
| 30 | 767 | 133 | .7009 | 827 | 402 | 427 | 018 | 941 | 30 |
| 40 | 796 | 112 | 030 | 884 | 406 | 423 | 012 | 912 | 20 |
| 50 | 825 | 092 | 050 | .9942 | 410 | 418 | 006 | 883 | 10 |
| 45° 00′ | .7854 | .7071 | .7071 | 1.000 | 1.414 | 1.414 | 1.000 | .7854 | 45° 00′ |
| | | sin θ | cos θ | cot θ | csc θ | sec θ | tan θ | Radians | Degrees |
| | | | | | | | | Angle θ | |

TABLE V    0°—Logarithms* of Trigonometric Function Values—9°

| Angle θ | log cos θ | log sin θ | log tan θ | log cot θ | |
|---|---|---|---|---|---|
| **0°  0′** | 0.0000 | No value | No value | No value | **90°  0′** |
| 0° 10′ | 0.0000 | 7.4637−10 | 7.4637−10 | 2.5363 | 89° 50′ |
| 0° 20′ | 0.0000 | 7.7648−10 | 7.7648−10 | 2.2352 | 89° 40′ |
| 0° 30′ | 0.0000 | 7.9408−10 | 7.9409−10 | 2.0591 | 89° 30′ |
| 0° 40′ | 0.0000 | 8.0658−10 | 8.0658−10 | 1.9342 | 89° 20′ |
| 0° 50′ | 0.0000 | 8.1627−10 | 8.1627−10 | 1.8373 | 89° 10′ |
| **1°  0′** | 9.9999−10 | 8.2419−10 | 8.2419−10 | 1.7581 | **89°  0′** |
| 1° 10′ | 9.9999−10 | 8.3088−10 | 8.3089−10 | 1.6911 | 88° 50′ |
| 1° 20′ | 9.9999−10 | 8.3668−10 | 8.3669−10 | 1.6331 | 88° 40′ |
| 1° 30′ | 9.9999−10 | 8.4179−10 | 8.4181−10 | 1.5819 | 88° 30′ |
| 1° 40′ | 9.9998−10 | 8.4637−10 | 8.4638−10 | 1.5362 | 88° 20′ |
| 1° 50′ | 9.9998−10 | 8.5050−10 | 8.5053−10 | 1.4947 | 88° 10′ |
| **2°  0′** | 9.9997−10 | 8.5428−10 | 8.5431−10 | 1.4569 | **88°  0′** |
| 2° 10′ | 9.9997−10 | 8.5776−10 | 8.5779−10 | 1.4221 | 87° 50′ |
| 2° 20′ | 9.9996−10 | 8.6097−10 | 8.6101−10 | 1.3899 | 87° 40′ |
| 2° 30′ | 9.9996−10 | 8.6397−10 | 8.6401−10 | 1.3599 | 87° 30′ |
| 2° 40′ | 9.9995−10 | 8.6677−10 | 8.6682−10 | 1.3318 | 87° 20′ |
| 2° 50 | 9.9995−10 | 8.6940−10 | 8.6945−10 | 1.3055 | 87° 10′ |
| **3°  0′** | 9.9994−10 | 8.7188−10 | 8.7194−10 | 1.2806 | **87°  0′** |
| 3° 10′ | 9.9993−10 | 8.7423−10 | 8.7429−10 | 1.2571 | 86° 50′ |
| 3° 20′ | 9.9993−10 | 8.7645−10 | 8.7652−10 | 1.2348 | 86° 40′ |
| 3° 30′ | 9.9992−10 | 8.7857−10 | 8.7865−10 | 1.2135 | 86° 30′ |
| 3° 40′ | 9.9991−10 | 8.8059−10 | 8.8067−10 | 1.1933 | 86° 20′ |
| 3° 50′ | 9.9990−10 | 8.8251−10 | 8.8261−10 | 1.1739 | 86° 10′ |
| **4°  0′** | 9.9989−10 | 8.8436−10 | 8.8446−10 | 1.1554 | **86°  0′** |
| 4° 10′ | 9.9989−10 | 8.8613−10 | 8.8624−10 | 1.1376 | 85° 50′ |
| 4° 20′ | 9.9988−10 | 8.8783−10 | 8.8795−10 | 1.1205 | 85° 40′ |
| 4° 30′ | 9.9987−10 | 8.8946−10 | 8.8960−10 | 1.1040 | 85° 30′ |
| 4° 40′ | 9.9986−10 | 8.9104−10 | 8.9118−10 | 1.0882 | 85° 20′ |
| 4° 50′ | 9.9985−10 | 8.9256−10 | 8.9272−10 | 1.0728 | 85° 10′ |
| **5°  0′** | 9.9983−10 | 8.9403−10 | 8.9420−10 | 1.0580 | **85°  0′** |
| 5° 10′ | 9.9982−10 | 8.9545−10 | 8.9563−10 | 1.0437 | 84° 50′ |
| 5° 20′ | 9.9981−10 | 8.9682−10 | 8.9701−10 | 1.0299 | 84° 40′ |
| 5° 30′ | 9.9980−10 | 8.9816−10 | 8.9836−10 | 1.0164 | 84° 30′ |
| 5° 40′ | 9.9979−10 | 8.9945−10 | 8.9966−10 | 1.0034 | 84° 20′ |
| 5° 50′ | 9.9977−10 | 9.0070−10 | 9.0093−10 | 0.9907 | 84° 10′ |
| **6°  0′** | 9.9976−10 | 9.0192−10 | 9.0216−10 | 0.9784 | **84°  0′** |
| 6° 10′ | 9.9975−10 | 9.0311−10 | 9.0336−10 | 0.9664 | 83° 50′ |
| 6° 20′ | 9.9973−10 | 9.0426−10 | 9.0453−10 | 0.9547 | 83° 40′ |
| 6° 30′ | 9.9972−10 | 9.0539−10 | 9.0567−10 | 0.9433 | 83° 30′ |
| 6° 40′ | 9.9971−10 | 9.0648−10 | 9.0678−10 | 0.9322 | 83° 20′ |
| 6° 50′ | 9.9969−10 | 9.0755−10 | 9.0786−10 | 0.9214 | 83° 10′ |
| **7°  0′** | 9.9968−10 | 9.0859−10 | 9.0891−10 | 0.9109 | **83°  0′** |
| 7° 10′ | 9.9966−10 | 9.0961−10 | 9.0995−10 | 0.9005 | 82° 50′ |
| 7° 20′ | 9.9964−10 | 9.1060−10 | 9.1096−10 | 0.8904 | 82° 40″ |
| 7° 30′ | 9.9963−10 | 9.1157−10 | 9.1194−10 | 0.8806 | 82° 30′ |
| 7° 40′ | 9.9961−10 | 9.1252−10 | 9.1291−10 | 0.8709 | 82° 20′ |
| 7° 50′ | 9.9959−10 | 9.1345−10 | 9.1385−10 | 0.8615 | 82° 10′ |
| **8°  0′** | 9.9958−10 | 9.1436−10 | 9.1478−10 | 0.8522 | **82°  0′** |
| 8° 10′ | 9.9956−10 | 9.1525−10 | 9.1569−10 | 0.8431 | 81° 50′ |
| 8° 20′ | 9.9954−10 | 9.1612−10 | 9.1658−10 | 0.8342 | 81° 40′ |
| 8° 30′ | 9.9952−10 | 9.1697−10 | 9.1745−10 | 0.8255 | 81° 30′ |
| 8° 40′ | 9.9950−10 | 9.1781−10 | 9.1831−10 | 0.8169 | 81° 20′ |
| 8° 50′ | 9.9948−10 | 9.1863−10 | 9.1915−10 | 0.8085 | 81° 10′ |
| **9°  0′** | 9.9946−10 | 9.1943−10 | 9.1997−10 | 0.8003 | **81°  0′** |
| | log sin θ | log cos θ | log cot θ | log tan θ | Angle θ |

* Logarithms are to the base 10.

TABLE V  (*continued*)

## 9°—Logarithms of Trigonometric Function Values—18°

| Angle θ | log cos θ | log sin θ | log tan θ | log cot θ | |
|---------|-----------|-----------|-----------|-----------|---|
| 9° 0′   | 9.9946–10 | 9.1943–10 | 9.1997–10 | 0.8003 | 81° 0′ |
| 9° 10′  | 9.9944–10 | 9.2022–10 | 9.2078–10 | 0.7922 | 80° 50′ |
| 9° 20′  | 9.9942–10 | 9.2100–10 | 9.2158–10 | 0.7842 | 80° 40′ |
| 9° 30′  | 9.9940–10 | 9.2176–10 | 9.2236–10 | 0.7764 | 80° 30′ |
| 9° 40′  | 9.9938–10 | 9.2251–10 | 9.2313–10 | 0.7687 | 80° 20′ |
| 9° 50′  | 9.9936–10 | 9.2324–10 | 9.2389–10 | 0.7611 | 80° 10′ |
| 10° 0′  | 9.9934–10 | 9.2397–10 | 9.2463–10 | 0.7537 | 80° 0′ |
| 10° 10′ | 9.9931–10 | 9.2468–10 | 9.2536–10 | 0.7464 | 79° 50′ |
| 10° 20′ | 9.9929–10 | 9.2538–10 | 9.2609–10 | 0.7391 | 79° 40′ |
| 10° 30′ | 9.9927–10 | 9.2606–10 | 9.2680–10 | 0.7320 | 79° 30′ |
| 10° 40′ | 9.9924–10 | 9.2674–10 | 9.2750–10 | 0.7250 | 79° 20′ |
| 10° 50′ | 9.9922–10 | 9.2740–10 | 9.2819–10 | 0.7181 | 79° 10′ |
| 11° 0′  | 9.9919–10 | 9.2806–10 | 9.2887–10 | 0.7113 | 79° 0′ |
| 11° 10′ | 9.9917–10 | 9.2870–10 | 9.2953–10 | 0.7047 | 78° 50′ |
| 11° 20′ | 9.9914–10 | 9.2934–10 | 9.3020–10 | 0.6980 | 78° 40′ |
| 11° 30′ | 9.9912–10 | 9.2997–10 | 9.3085–10 | 0.6915 | 78° 30′ |
| 11° 40′ | 9.9909–10 | 9.3058–10 | 9.3149–10 | 0.6851 | 78° 20′ |
| 11° 50′ | 9.9907–10 | 9.3119–10 | 9.3212–10 | 0.6788 | 78° 10′ |
| 12° 0′  | 9.9904–10 | 9.3179–10 | 9.3275–10 | 0.6725 | 78° 0′ |
| 12° 10′ | 9.9901–10 | 9.3238–10 | 9.3336–10 | 0.6664 | 77° 50′ |
| 12° 20′ | 9.9899–10 | 9.3296–10 | 9.3397–10 | 0.6603 | 77° 40′ |
| 12° 30′ | 9.9896–10 | 9.3353–10 | 9.3458–10 | 0.6542 | 77° 30′ |
| 12° 40′ | 9.9893–10 | 9.3410–10 | 9.3517–10 | 0.6483 | 77° 20′ |
| 12° 50′ | 9.9890–10 | 9.3466–10 | 9.3576–10 | 0.6424 | 77° 10′ |
| 13° 0′  | 9.9887–10 | 9.3521–10 | 9.3634–10 | 0.6366 | 77° 0′ |
| 13° 10′ | 9.9884–10 | 9.3575–10 | 9.3691–10 | 0.6309 | 76° 50′ |
| 13° 20′ | 9.9881–10 | 9.3629–10 | 9.3748–10 | 0.6252 | 76° 40′ |
| 13° 30′ | 9.9878–10 | 9.3682–10 | 9.3804–10 | 0.6196 | 76° 30′ |
| 13° 40′ | 9.9875–10 | 9.3734–10 | 9.3859–10 | 0.6141 | 76° 20′ |
| 13° 50′ | 9.9872–10 | 9.3786–10 | 9.3914–10 | 0.6086 | 76° 10′ |
| 14° 0′  | 9.9869–10 | 9.3837–10 | 9.3968–10 | 0.6032 | 76° 0′ |
| 14° 10′ | 9.9866–10 | 9.3887–10 | 9.4021–10 | 0.5979 | 75° 50′ |
| 14° 20′ | 9.9863–10 | 9.3937–10 | 9.4074–10 | 0.5926 | 75° 40′ |
| 14° 30′ | 9.9859–10 | 9.3986–10 | 9.4127–10 | 0.5873 | 75° 30′ |
| 14° 40′ | 9.9856–10 | 9.4035–10 | 9.4178–10 | 0.5822 | 75° 20′ |
| 14° 50′ | 9.9853–10 | 9.4083–10 | 9.4230–10 | 0.5770 | 75° 10′ |
| 15° 0′  | 9.9849–10 | 9.4130–10 | 9.4281–10 | 0.5719 | 75° 0′ |
| 15° 10′ | 9.9846–10 | 9.4177–10 | 9.4331–10 | 0.5669 | 74° 50′ |
| 15° 20′ | 9.9843–10 | 9.4223–10 | 9.4381–10 | 0.5619 | 74° 40′ |
| 15° 30′ | 9.9839–10 | 9.4269–10 | 9.4430–10 | 0.5570 | 74° 30′ |
| 15° 40′ | 9.9836–10 | 9.4314–10 | 9.4479–10 | 0.5521 | 74° 20′ |
| 15° 50′ | 9.9832–10 | 9.4359–10 | 9.4427–10 | 0.5473 | 74° 10′ |
| 16° 0′  | 9.9828–10 | 9.4403–10 | 9.4575–10 | 0.5425 | 74° 0′ |
| 16° 10′ | 9.9825–10 | 9.4447–10 | 9.4622–10 | 0.5378 | 73° 50′ |
| 16° 20′ | 9.9821–10 | 9.4491–10 | 9.4669–10 | 0.5331 | 73° 40′ |
| 16° 30′ | 9.9817–10 | 9.4533–10 | 9.4716–10 | 0.5284 | 73° 30′ |
| 16° 40′ | 9.9814–10 | 9.4576–10 | 9.4762–10 | 0.5238 | 73° 20′ |
| 16° 50′ | 9.9810–10 | 9.4618–10 | 9.4808–10 | 0.5192 | 73° 10′ |
| 17° 0′  | 9.9806–10 | 9.4659–10 | 9.4853–10 | 0.5147 | 73° 0′ |
| 17° 10′ | 9.9802–10 | 9.4700–10 | 9.4898–10 | 0.5102 | 72° 50′ |
| 17° 20′ | 9.9798–10 | 9.4741–10 | 9.4943–10 | 0.5057 | 72° 40′ |
| 17° 30′ | 9.9794–10 | 9.4781–10 | 9.4987–10 | 0.5013 | 72° 30′ |
| 17° 40′ | 9.9790–10 | 9.4821–10 | 9.5031–10 | 0.4969 | 72° 20′ |
| 17° 50′ | 9.9786–10 | 9.4861–10 | 9.5075–10 | 0.4925 | 72° 10′ |
| 18° 0′  | 9.9782–10 | 9.4900–10 | 9.5118–10 | 0.4882 | 72° 0′ |
| | log sin θ | log cos θ | log cot θ | log tan θ | Angle θ |

TABLE V  (*continued*)
## 18°—Logarithms of Trigonometric Function Values—27°

| Angle θ | log cos θ | log sin θ | log tan θ | log cot θ | |
|---|---|---|---|---|---|
| **18° 0′** | 9.9782–10 | 9.4900–10 | 9.5118–10 | 0.4882 | **72° 0′** |
| 18° 10′ | 9.9778–10 | 9.4939–10 | 9.5161–10 | 0.4839 | 71° 50′ |
| 18° 20′ | 9.9774–10 | 9.4977–10 | 9.5203–10 | 0.4797 | 71° 40′ |
| 18° 30′ | 9.9770–10 | 9.5015–10 | 9.5245–10 | 0.4755 | 71° 30′ |
| 18° 40′ | 9.9765–10 | 9.5052–10 | 9.5287–10 | 0.4713 | 71° 20′ |
| 18° 50′ | 9.9761–10 | 9.5090–10 | 9.5329–10 | 0.4671 | 71° 10′ |
| **19° 0′** | 9.9757–10 | 9.5126–10 | 9.5370–10 | 0.4630 | **71° 0′** |
| 19° 10′ | 9.9752–10 | 9.5163–10 | 0.5411–10 | 0.4589 | 70° 50′ |
| 19° 20′ | 9.9748–10 | 9.5199–10 | 9.5451–10 | 0.4549 | 70° 40′ |
| 19° 30′ | 9.9743–10 | 9.5235–10 | 9.5491–10 | 0.4509 | 70° 30′ |
| 19° 40′ | 9.9739–10 | 9.5270–10 | 9.5531–10 | 0.4469 | 70° 20′ |
| 19° 50′ | 9.9734–10 | 9.5306–10 | 9.5571–10 | 0.4429 | 70° 10′ |
| **20° 0′** | 9.9730–10 | 9.5341–10 | 9.5611–10 | 0.4389 | **70° 0′** |
| 20° 10′ | 9.9725–10 | 9.5375–10 | 9.5650–10 | 0.4350 | 69° 50′ |
| 20° 20′ | 9.9721–10 | 9.5409–10 | 9.5689–10 | 0.4311 | 69° 40′ |
| 20° 30′ | 9.9716–10 | 9.5443–10 | 9.5727–10 | 0.4273 | 69° 30′ |
| 20° 40′ | 9.9711–10 | 9.5477–10 | 9.5766–10 | 0.4234 | 69° 20′ |
| 20° 50′ | 9.9706–10 | 9.5510–10 | 9.5804–10 | 0.4196 | 69° 10′ |
| **21° 0′** | 9.9702–10 | 9.5543–10 | 9.5842–10 | 0.4158 | **69° 0′** |
| 21° 10′ | 9.9697–10 | 9.5576–10 | 9.5879–10 | 0.4121 | 68° 50′ |
| 21° 20′ | 9.9692–10 | 9.5609–10 | 9.5917–10 | 0.4083 | 68° 40′ |
| 21° 30′ | 9.9687–10 | 9.5641–10 | 9.5954–10 | 0.4046 | 68° 30′ |
| 21° 40′ | 9.9682–10 | 9.5673–10 | 9.5991–10 | 0.4009 | 68° 20′ |
| 21° 50′ | 9.9677–10 | 9.5704–10 | 9.6028–10 | 0.3972 | 68° 10′ |
| **22° 0′** | 9.9672–10 | 9.5736–10 | 9.6064–10 | 0.3936 | **68° 0′** |
| 22° 10′ | 9.9667–10 | 9.5767–10 | 9.6100–10 | 0.3900 | 67° 50′ |
| 22° 20′ | 9.9661–10 | 9.5798–10 | 9.6136–10 | 0.3864 | 67° 40′ |
| 22° 30′ | 9.9656–10 | 9.5828–10 | 9.6172–10 | 0.3828 | 67° 30′ |
| 22° 40′ | 9.9651–10 | 9.5859–10 | 9.6208–10 | 0.3792 | 67° 20′ |
| 22° 50′ | 9.9646–10 | 9.5889–10 | 9.6243–10 | 0.3757 | 67° 10′ |
| **23° 0′** | 9.9640–10 | 9.5919–10 | 9.6279–10 | 0.3721 | **67° 0′** |
| 23° 10′ | 9.9635–10 | 9.5948–10 | 9.6314–10 | 0.3686 | 66° 50′ |
| 23° 20′ | 9.9629–10 | 9.5978–10 | 9.6348–10 | 0.3652 | 66° 40′ |
| 23° 30° | 9.9624–10 | 9.6007–10 | 9.6383–10 | 0.3617 | 66° 30′ |
| 23° 40′ | 9.9618–10 | 9.6036–10 | 9.6417–10 | 0.3583 | 66° 20′ |
| 23° 50′ | 9.9613–10 | 9.6065–10 | 9.6452–10 | 0.3548 | 66° 10′ |
| **24° 0′** | 9.9607–10 | 9.6093–10 | 9.6486–10 | 0.3514 | **66° 0′** |
| 24° 10′ | 9.9602–10 | 9.6121–10 | 9.6520–10 | 0.3480 | 65° 50′ |
| 24° 20′ | 9.9596–10 | 9.6149–10 | 9.6553–10 | 0.3447 | 65° 40′ |
| 24° 30′ | 9.9590–10 | 9.6177–10 | 9.6587–10 | 0.3413 | 65° 30′ |
| 24° 40′ | 9.9584–10 | 9.6205–10 | 9.6620–10 | 0.3380 | 65° 20′ |
| 24° 50′ | 9.9579–10 | 9.6232–10 | 9.6654–10 | 0.3346 | 65° 10′ |
| **25° 0′** | 9.9573–10 | 9.6259–10 | 9.6687–10 | 0.3313 | **65° 0′** |
| 25° 10′ | 9.9567–10 | 9.6286–10 | 9.6720–10 | 0.3280 | 64° 50′ |
| 25° 20′ | 9.9561–10 | 9.6313–10 | 9.6752–10 | 0.3248 | 64° 40′ |
| 25° 30′ | 9.9555–10 | 9.6340–10 | 9.6785–10 | 0.3215 | 64° 30′ |
| 25° 40′ | 9.9549–10 | 9.6366–10 | 9.6817–10 | 0.3183 | 64° 20′ |
| 25° 50′ | 9.9543–10 | 9.6392–10 | 9.6850–10 | 0.3150 | 64° 10′ |
| **26° 0′** | 9.9537–10 | 9.6418–10 | 9.6882–10 | 0.3118 | **64° 0′** |
| 26° 10′ | 9.9530–10 | 9.6444–10 | 9.6914–10 | 0.3086 | 63° 50′ |
| 26° 20′ | 9.9524–10 | 9.6470–10 | 9.6946–10 | 0.3054 | 63° 40′ |
| 26° 30′ | 9.9518–10 | 9.6495–10 | 9.6977–10 | 0.3023 | 63° 30′ |
| 26° 40′ | 9.9512–10 | 9.6521–10 | 9.7009–10 | 0.2991 | 63° 20′ |
| 26° 50′ | 9.9505–10 | 9.6546–10 | 9.7040–10 | 0.2960 | 63° 10′ |
| **27° 0′** | 9.9499–10 | 9.6570–10 | 9.7072–10 | 0.2928 | **63° 0′** |
| | log sin θ | log cos θ | log cot θ | log tan θ | Angle θ |

## 63°—Logarithms of Trigonometric Function Values—72°

TABLE V  (*continued*)

### 27°—Logarithms of Trigonometric Function Values—36°

| Angle θ | log cos θ | log sin θ | log tan θ | log cot θ | |
|---|---|---|---|---|---|
| 27° 0′ | 9.9499–10 | 9.6570–10 | 9.7072–10 | 0.2928 | 63° 0′ |
| 27° 10′ | 9.9492–10 | 9.6595–10 | 9.7103–10 | 0.2897 | 62° 50′ |
| 27° 20′ | 9.9486–10 | 9.6620–10 | 9.7134–10 | 0.2866 | 62° 40′ |
| 27° 30′ | 9.9479–10 | 9.6644–10 | 9.7165–10 | 0.2835 | 62° 30′ |
| 27° 40′ | 9.9473–10 | 9.6668–10 | 9.7196–10 | 0.2804 | 62° 20′ |
| 27° 50′ | 9.9466–10 | 9.6692–10 | 9.7226–10 | 0.2774 | 62° 10′ |
| 28° 0′ | 9.9459–10 | 9.6716–10 | 9.7257–10 | 0.2743 | 62° 0′ |
| 28° 10′ | 9.9453–10 | 9.6740–10 | 9.7287–10 | 0.2713 | 61° 50′ |
| 28° 20′ | 9.9446–10 | 9.6763–10 | 9.7317–10 | 0.2683 | 61° 40′ |
| 28° 30′ | 9.9439–10 | 9.6787–10 | 9.7348–10 | 0.2652 | 61° 30′ |
| 28° 40′ | 9.9432–10 | 9.6810–10 | 9.7378–10 | 0.2622 | 61° 20′ |
| 28° 50′ | 9.9425–10 | 9.6833–10 | 9.7408–10 | 0.2592 | 61° 10′ |
| 29° 0′ | 9.9418–10 | 9.6856–10 | 9.7438–10 | 0.2562 | 61° 0′ |
| 29° 10′ | 9.9411–10 | 9.6878–10 | 9.7467–10 | 0.2533 | 60° 50′ |
| 29° 20′ | 9.9404–10 | 9.6901–10 | 9.7497–10 | 0.2503 | 60° 40′ |
| 29° 30′ | 9.9397–10 | 9.6923–10 | 9.7526–10 | 0.2474 | 60° 30′ |
| 29° 40′ | 9.9390–10 | 9.6946–10 | 9.7556–10 | 0.2444 | 60° 20′ |
| 29° 50′ | 9.9383–10 | 9.6968–10 | 9.7585–10 | 0.2415 | 60° 10′ |
| 30° 0′ | 9.9375–10 | 9.6990–10 | 9.7614–10 | 0.2386 | 60° 0′ |
| 30° 10′ | 9.9368–10 | 9.7012–10 | 9.7644–10 | 0.2356 | 59° 50′ |
| 30° 20′ | 9.9361–10 | 9.7033–10 | 9.7673–10 | 0.2327 | 59° 40′ |
| 30° 30′ | 9.9353–10 | 9.7055–10 | 9.7701–10 | 0.2299 | 59° 30′ |
| 30° 40′ | 9.9346–10 | 9.7076–16 | 9.7730–10 | 0.2270 | 59° 20′ |
| 30° 50′ | 9.9338–10 | 9.7097–10 | 9.7759–10 | 0.2241 | 59° 10′ |
| 31° 0′ | 9.9331–10 | 9.7118–10 | 9.7788–10 | 0.2212 | 59° 0′ |
| 31° 10′ | 9.9323–10 | 9.7139–10 | 9.7816–10 | 0.2184 | 58° 50′ |
| 31° 20′ | 9.9315–10 | 9.7160–10 | 9.7845–10 | 0.2155 | 58° 40′ |
| 31° 30′ | 9.9308–10 | 9.7181–10 | 9.7873–10 | 0.2127 | 58° 30′ |
| 31° 40′ | 9.9300–10 | 9.7201–10 | 9.7902–10 | 0.2098 | 58° 20′ |
| 31° 50′ | 9.9292–10 | 9.7222–10 | 9.7930–10 | 0.2070 | 58° 10′ |
| 32° 0′ | 9.9284–10 | 9.7242–10 | 9.7958–10 | 0.2042 | 58° 0′ |
| 32° 10′ | 9.9276–10 | 9.7262–10 | 9.7986–10 | 0.2014 | 57° 50′ |
| 32° 20′ | 9.9268–10 | 9.7282–10 | 9.8014–10 | 0.1986 | 57° 40′ |
| 32° 30′ | 9.9260–10 | 9.7302–10 | 9.8042–10 | 0.1958 | 57° 30′ |
| 32° 40′ | 9.9252–10 | 9.7322–10 | 9.8070–10 | 0.1930 | 57° 20′ |
| 32° 50′ | 9.9244–10 | 9.7342–10 | 9.8097–10 | 0.1903 | 57° 10′ |
| 33° 0′ | 9.9236–10 | 9.7361–10 | 9.8125–10 | 0.1875 | 57° 0′ |
| 33° 10′ | 9.9228–10 | 9.7380–10 | 9.8153–10 | 0.1847 | 56° 50′ |
| 33° 20′ | 9.9219–10 | 9.7400–10 | 9.8180–10 | 0.1820 | 56° 40′ |
| 33° 30′ | 9.9211–10 | 9.7419–10 | 9.8208–10 | 0.1792 | 56° 30′ |
| 33° 40′ | 9.9203–10 | 9.7438–10 | 9.8235–10 | 0.1765 | 56° 20′ |
| 33° 50′ | 9.9194–10 | 9.7457–10 | 9.8263–10 | 0.1737 | 56° 10′ |
| 34° 0′ | 9.9186–10 | 9.7465–10 | 9.8290–10 | 0.1710 | 56° 0′ |
| 34° 10′ | 9.9177–10 | 9.7494–10 | 9.8317–10 | 0.1683 | 55° 50′ |
| 34° 20′ | 9.9169–10 | 9.7513–10 | 9.8344–10 | 0.1656 | 55° 40′ |
| 34° 30′ | 9.9160–10 | 9.7531–10 | 9.8371–10 | 0.1629 | 55° 30′ |
| 34° 40′ | 9.9151–10 | 9.7550–10 | 9.8398–10 | 0.1602 | 55° 20′ |
| 34° 50′ | 9.9142–10 | 9.7568–10 | 9.8425–10 | 0.1575 | 55° 10′ |
| 35° 0′ | 9.9134–10 | 9.7586–10 | 9.8452–10 | 0.1548 | 55° 0′ |
| 35° 10′ | 9.9125–10 | 9.7604–10 | 9.8479–10 | 0.1521 | 54° 50′ |
| 35° 20′ | 9.9116–10 | 9.7622–10 | 9.8506–10 | 0.1494 | 54° 40′ |
| 35° 30′ | 9.9107–10 | 9.7640–10 | 9.8533–10 | 0.1467 | 54° 30′ |
| 35° 40′ | 9.9098–10 | 9.7657–10 | 9.8559–10 | 0.1441 | 54° 20′ |
| 35° 50′ | 9.9089–10 | 9.7675–10 | 9.8586–10 | 0.1414 | 54° 10′ |
| 36° 0′ | 9.9080–10 | 9.7692–10 | 9.8613–10 | 0.1387 | 54° 0′ |
| | log sin θ | log cos θ | log cot θ | log tan θ | Angle θ |

### 54°—Logarithms of Trigonometric Function Values—63°

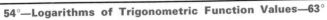

**TABLE V**  (*continued*)

## 36°—Logarithms of Trigonometric Function Values—45°

| Angle θ | log cos θ | log sin θ | log tan θ | log cot θ | |
|---------|-----------|-----------|-----------|-----------|---|
| **36° 0′** | 9.9080–10 | 9.7692–10 | 9.8613–10 | 0.1387 | **54° 0′** |
| 36° 10′ | 9.9070–10 | 9.7710–10 | 9.8639–10 | 0.1361 | 53° 50′ |
| 36° 20′ | 9.9061–10 | 9.7727–10 | 9.8666–10 | 0.1334 | 53° 40′ |
| 36° 30′ | 9.9052–10 | 9.7744–10 | 9.8692–10 | 0.1308 | 53° 30′ |
| 36° 40′ | 9.9042–10 | 9.7761–10 | 9.8718–10 | 0.1282 | 53° 20′ |
| 36° 50′ | 9.9033–10 | 9.7778–10 | 9.8745–10 | 0.1255 | 53° 10′ |
| **37° 0′** | 9.9023–10 | 9.7795–10 | 9.8771–10 | 0.1229 | **53° 0′** |
| 37° 10′ | 9.9014–10 | 9.7811–10 | 9.8797–10 | 0.1203 | 52° 50′ |
| 37° 20′ | 9.9004–10 | 9.7828–10 | 9.8824–10 | 0.1176 | 52° 40′ |
| 37° 30′ | 9.8995–10 | 9.7844–10 | 9.8850–10 | 0.1150 | 52° 30′ |
| 37° 40′ | 9.8985–10 | 9.7861–10 | 9.8876–10 | 0.1124 | 52° 20′ |
| 37° 50′ | 9.8975–10 | 9.7877–10 | 9.8902–10 | 0.1098 | 52° 10′ |
| **38° 0′** | 9.8965–10 | 9.7893–10 | 9.8928–10 | 0.1072 | **52° 0′** |
| 38° 10′ | 9.8955–10 | 9.7910–10 | 9.8954–10 | 0.1046 | 51° 50′ |
| 38° 20′ | 9.8945–10 | 9.7926–10 | 9.8980–10 | 0.1020 | 51° 40′ |
| 38° 30′ | 9.8935–10 | 9.7941–10 | 9.9006–10 | 0.0994 | 51° 30′ |
| 38° 40′ | 9.8925–10 | 9.7957–10 | 9.9032–10 | 0.0968 | 51° 20′ |
| 38° 50′ | 9.8915–10 | 9.7973–10 | 9.9058–10 | 0.0942 | 51° 10′ |
| **39° 0′** | 9.8905–10 | 9.7989–10 | 9.9084–10 | 0.0916 | **51° 0′** |
| 39° 10′ | 9.8895–10 | 9.8004–10 | 9.9110–10 | 0.0890 | 50° 50′ |
| 39° 20′ | 9.8884–10 | 9.8020–10 | 9.9135–10 | 0.0865 | 50° 40′ |
| 39° 30′ | 9.8874–10 | 9.8035–10 | 9.9161–10 | 0.0839 | 50° 30′ |
| 39° 40′ | 9.8864–10 | 9.8050–10 | 9.9187–10 | 0.0813 | 50° 20′ |
| 39° 50′ | 9.8853–10 | 9.8066–10 | 9.9212–10 | 0.0788 | 50° 10′ |
| **40° 0′** | 9.8843–10 | 9.8081–10 | 9.9238–10 | 0.0762 | **50° 0′** |
| 40° 10′ | 9.8832–10 | 9.8096–10 | 9.9264–10 | 0.0736 | 49° 50′ |
| 40° 20′ | 9.8821–10 | 9.8111–10 | 9.9289–10 | 0.0711 | 49° 40′ |
| 40° 30′ | 9.8810–10 | 9.8125–10 | 9.9315–10 | 0.0685 | 49° 30′ |
| 40° 40′ | 9.8800–10 | 9.8140–10 | 9.9341–10 | 0.0659 | 49° 20′ |
| 40° 50′ | 9.8789–10 | 9.8155–10 | 9.9366–10 | 0.0634 | 49° 10′ |
| **41° 0′** | 9.8778–10 | 9.8169–10 | 9.9392–10 | 0.0608 | **49° 0′** |
| 41° 10′ | 9.8767–10 | 9.8184–10 | 9.9417–10 | 0.0583 | 48° 50′ |
| 41° 20′ | 9.8756–10 | 9.8198–10 | 9.9443–10 | 0.0557 | 48° 40′ |
| 41° 30′ | 9.8745–10 | 9.8213–10 | 9.9468–10 | 0.0532 | 48° 30′ |
| 41° 40′ | 9.8733–10 | 9.8227–10 | 9.9494–10 | 0.0506 | 48° 20′ |
| 41° 50′ | 9.8722–10 | 9.8241–10 | 9.9519–10 | 0.0481 | 48° 10′ |
| **42° 0′** | 9.8711–10 | 9.8255–10 | 9.9544–10 | 0.0456 | **48° 0′** |
| 42° 10′ | 9.8699–10 | 9.8269–10 | 9.9570–10 | 0.0430 | 47° 50′ |
| 42° 20′ | 9.8688–10 | 9.8283–10 | 9.9595–10 | 0.0405 | 47° 40′ |
| 42° 30′ | 9.8676–10 | 9.8297–10 | 9.9621–10 | 0.0379 | 47° 30′ |
| 42° 40′ | 9.8665–10 | 9.8311–10 | 9.9646–10 | 0.0354 | 47° 20′ |
| 42° 50′ | 9.8653–10 | 9.8324–10 | 9.9671–10 | 0.0329 | 47° 10° |
| **43° 0′** | 9.8641–10 | 9.8338–10 | 9.9697–10 | 0.0303 | **47° 0′** |
| 43° 10″ | 9.8629–10 | 9.8351–10 | 9.9722–10 | 0.0278 | 46° 50′ |
| 43° 20′ | 9.8618–10 | 9.8365–10 | 9.9747–10 | 0.0253 | 46° 40′ |
| 43° 30′ | 9.8606–10 | 9.8378–10 | 9.9772–10 | 0.0228 | 46° 30′ |
| 43° 40′ | 9.8594–10 | 9.8391–10 | 9.9798–10 | 0.0202 | 46° 20′ |
| 43° 50′ | 9.8582–10 | 9.8405–10 | 9.9823–10 | 0.0177 | 46° 10′ |
| **44° 0′** | 9.8569–10 | 9.8418–10 | 9.9848–10 | 0.0152 | **46° 0′** |
| 44° 10′ | 9.8557–10 | 9.8431–10 | 9.9874–10 | 0.0126 | 45° 50′ |
| 44° 20′ | 9.8545–10 | 9.8444–10 | 9.9899–10 | 0.0101 | 45° 40′ |
| 44° 30′ | 9.8532–10 | 9.8457–10 | 9.9924–10 | 0.0076 | 45° 30′ |
| 44° 40′ | 9.8520–10 | 9.8469–10 | 9.9949–10 | 0.0051 | 45° 20′ |
| 44° 50′ | 9.8507–10 | 9.8482–10 | 9.9975–10 | 0.0025 | 45° 10′ |
| **45° 0′** | 9.8495–10 | 9.8495–10 | 0.0000 | 0.0000 | **45° 0′** |
| | log sin θ | log cos θ | log cot θ | log tan θ | Angle θ |

## 45°—Logarithms of Trigonometric Function Values—54°

# Answers to Odd-Numbered Problems

**Chapter 1**

**Exercise 1.1** (page 3)

1. $b \in P$.
3. $T \neq V$.
5. $c \notin W$.
7. $\{j, k, l, m\}$.
9. $\{j, k, m, n, o\}$.
11. $\phi$.
13. $\{k\}$.

15. $\{j, k, m\}$.
17. $\{6, 7\}$.
19. $\{1, 3, 5, 6, 7, 8, 9\}$.
21. $\{1, 2, 3, 4, 5, 6, 7, 8\}$.
23. $\{6, 8\}$.
25. $\{5, 7\}$.

27. $\{2, 4, 6, 7, 8, 9\}$.
29. $\phi$.
31. $\{1, 3, 5, 7\}$.
33. True.
35. True.
37. True.

**Exercise 1.2** (page 8)

1. $\{3, -1, 0, \sqrt{25}, 4\}$.
3. $\{-\sqrt{7}, \sqrt{5}\}$.

5. $5 < 7$.
7. $|-4| = 4$.

9. $|-4| < |-7|$.
11. $\dfrac{-4}{5} < \left| \dfrac{3}{-4} \right|$.

13.

17.

21.

**25.**

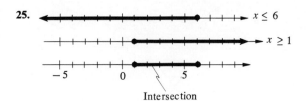

Intersection

**27.** By Definition 1.7,

$$b + (a - b) = a,$$
$$b + (a - b) + (-b) = a + (-b),$$
$$[(b + (-b)] + (a - b) = a + (-b),$$
$$0 + (a - b) = a + (-b),$$
$$a - b = a + (-b).$$

**Exercise 1.3** (page 13)

1. $\{(1, 4), (1, 5), (2, 4), (2, 5), (3, 4), (3, 5)\}$.
3. $\{(0, 1), (2, 1), (4, 1)\}$.
5. $\{(x, 0), (x, 1), (x, 2), (x, 3)\}$.
7. $\{(2, 1), (2, 2), (2, 3), \ldots\}$.
9. Domain: $\{0, 3\}$; Range: $\{2, 4\}$; not a function.
11. Domain: $\{2, 4, 8, 16, 32\}$; Range: $\{0, 1\}$; a function.
13. Domain: $\{0, 1, 2, 3, 4, 5\}$; Range: $\{-1, -2, -3, -4, -5, -6\}$; a function.
25. (a) Domain: $\{-2, 0, 2\}$.  (b) Rule: $y = 2x + 1$.  (c) Range: $\{-3, 1, 5\}$.
    (d) Relation: $\{(-2, -3), (0, 1), (2, 5)\}$.
27. (a) Domain: $\{2, 4, 6, 8\}$.  (b) Rule: $2y + x = 6$.  (c) Range: $\{2, 1, 0, -1\}$.
    (d) Relation: $\{(2, 2), (4, 1), (6, 0), (8, -1)\}$.
29. (a) Domain: $\{0, \sqrt{3}, \sqrt{8}, \sqrt{15}\}$.  (b) Rule: $y = 2\sqrt{x^2 + 1}$.  (c) Range: $\{2, 4, 6, 8\}$.
    (d) Relation: $\{(0, 2), (\sqrt{3}, 4), (\sqrt{8}, 6), (\sqrt{15}, 8)\}$.
31. (a) Domain: $\{-4, -2, 0, 2, 4\}$.  (b) Rule: $y = |x|$.  (c) Range: $\{0, 2, 4\}$.
    (d) Relation: $\{(-4, 4), (-2, 2), (0, 0), (2, 2), (4, 4)\}$.
33. $x = 2$.           35. Yes.            37. No.              39. $-2$.
41. $a^2 - 2$.         43. 5.              45. $2x^2 - 1$.      47. $x$.

**Exercise 1.4** (page 22)

1. $(0, 8); (-8, 0); (-5, 3)$.                    3. $(0, 5/2); (5/4, 0); (-1/2, 7/2)$.

**5.**

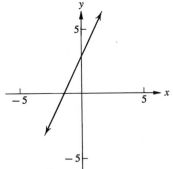

**9.**

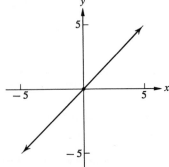

**13.** $-6$.                      **15.** 6.                              **17.** 5.

**19.** $d = \sqrt{13}$; slope: $\frac{3}{2}$.              **21.** $d = 4\sqrt{10}$; slope: 3.

**23.**

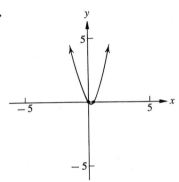

**27.**

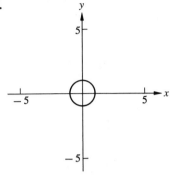

**31.**

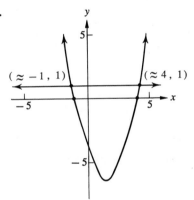

**35.**

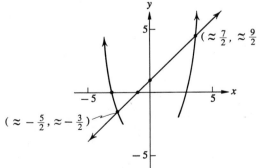

**39.**

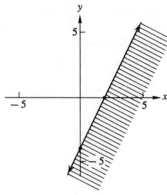

**43.**

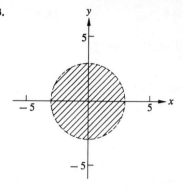

**47.**

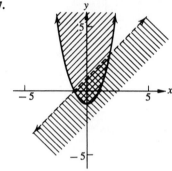

**Exercise 1.5** (page 28)

**1.** {1}.     **3.** {−1}.     **5.** {$\frac{5}{4}$}.     **7.** {−2}.     **9.** {$\sqrt{2}, -\sqrt{2}$}.     **11.** {1, 6}.

**13.**

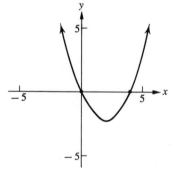

**15.** {−2, −1}.                 **19.** {−1, 7}.                 **23.** $\left(\dfrac{-5-\sqrt{33}}{2}, \dfrac{-5+\sqrt{33}}{2}\right)$.

**17.** {−$\sqrt{7}$, $\sqrt{7}$}.             **21.** {−1, 6}.                 **25.** {−$\frac{3}{2}$, 3}.

**27.**

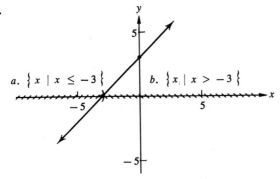

*a.* $\{x \mid x \leq -3\}$   *b.* $\{x_i \mid x > -3\}$

**31.**
$$ax^2 + bx + c = 0, \qquad (a \neq 0),$$

$$x^2 + \frac{b}{a}x = -\frac{c}{a},$$

$$x^2 + \frac{b}{a}x + \frac{b^2}{4a^2} = -\frac{c}{a} + \frac{b^2}{4a^2},$$

$$\left(x + \frac{b}{2a}\right)^2 = \frac{b^2 - 4ac}{4a^2},$$

$$x + \frac{b}{2a} = \frac{\pm\sqrt{b^2 - 4ac}}{2a},$$

$$\left\{\frac{-b - \sqrt{b^2 - 4ac}}{2a}, \quad \frac{-b + \sqrt{b^2 - 4ac}}{2a}\right\}.$$

**Exercise 1.6** (page 33)

**1.** $\{(1, 0), (3, 1), (5, 2), (7, 3)\}$.

**3.** $\{(1, 0), (\frac{1}{2}, \pi/6), (\sqrt{3}/2, \pi/3), (0, \pi/2)\}$.

**5.**

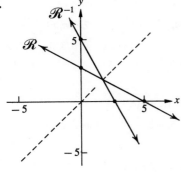

$\mathscr{R}^{-1}$: $\{(x, y) \mid y + 2x = 5\}$ or $\{(x, y) \mid y = 5 - 2x\}$.

**7.** $\mathscr{R}^{-1}$: $\{(x, y) \mid x = 6 - 2y\}$ or $\{(x, y) \mid y = (6 - x)/2\}$.

**9.**

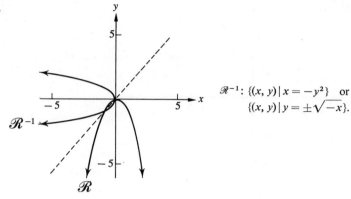

$\mathcal{R}^{-1}$: $\{(x, y) \mid x = -y^2\}$  or  $\{(x, y) \mid y = \pm\sqrt{-x}\}$.

**11.** $\mathcal{R}^{-1}$: $\{(x, y) \mid x = 2y^2 - 2\}$  or  $\{(x, y) \mid y = \pm\sqrt{(x+2)/2}\}$.

**13.**

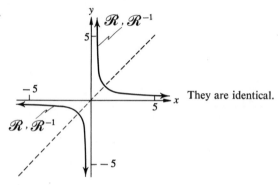

They are identical.

**15.** $\mathcal{F}^{-1}$ is defined by $y = \frac{1}{2}x$.

**17.** $\mathcal{F}^{-1}$ is defined by $y = (x + 1)/2$.

**19.** $\mathcal{F}^{-1}$ is defined by $y = (x + 6)/2$.

*Chapter 1 Review* (page 36)

**1.** $p \notin D$.          **3.** $D = A$.          **5.** $\{l, m, n, p\}$.          **7.** $\{m, n\}$.

**9.** $\{m, n, p\}$.                    **11.** $\{5, -3, 0, 2\}$.                    **13.** $\{5, -3, 7/8, 0, -1/5, 2\}$.

**15.**

**17.** $\{(a, c), (a, d), (a, e), (b, c), (b, d), (b, e), (c, c), (c, d), (c, e)\}$.

**19.** Domain: $\{1, 2, 3, 4, 5\}$; Range: $\{-2, -4, -6, -8, -10\}$; a function.

**21.** Conditional equation.

**23.** (a) $\{\frac{5}{3}, \frac{7}{3}, 3, \frac{11}{3}\}$.                    (b) $y = -3x + 4$.
    (c) $\{-1, -3, -5, -7\}$.                    (d) $\{(\frac{5}{3}, -1), (\frac{7}{3}, -3), (3, -5), (\frac{11}{3}, -7)\}$.

**25.** $x \in R$,     $(x \neq 1, -3)$.          **27.** 4.          **29.** 7.          **31.** $-11$.          **33.** $d = 10\sqrt{2}$; slope: 1.

**35.**

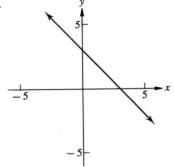

**39.** $\{-1, 3\}$.

**41.** $\{-\sqrt{5/2}, \sqrt{5/2}\}$.

**43.** $\mathcal{R}^{-1}$: $\{(x, y) \mid 5y + 2x = 10\}$

    or   $\{(x, y) \mid y = (10 - 2x)/5\}$.

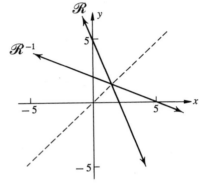

**49.**

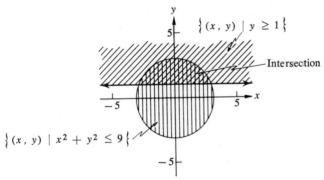

$\{(x, y) \mid y \geq 1\}$

Intersection

$\{(x, y) \mid x^2 + y^2 \leq 9\}$

**51.** $\{(x, y) \mid y = x^2 - 1\}$

$\{x \mid x \leq -1 \text{ or } x \geq 1\}$

**Chapter 2**

**Exercise 2.1** (page 44)

**1.** Domain: $\{s \mid s \in R\}$; Range: $\{(x, y) \mid x^2 + y^2 = 1,\ x,\ y \in R\}$.

**3.**

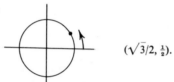

| $s$ | $x$ | $y$ |
|---|---|---|
| $0$ | $1$ | $0$ |
| $\dfrac{\pi}{2}$ | $0$ | $1$ |
| $\pi$ | $-1$ | $0$ |
| $\dfrac{3\pi}{2}$ | $0$ | $-1$ |

**5.**

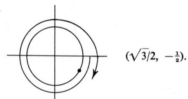

$(\sqrt{3}/2,\ \frac{1}{2})$.

**7.** $(1/\sqrt{2},\ 1/\sqrt{2})$.

**9.**

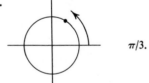

$(\sqrt{3}/2,\ -\frac{1}{2})$.

**11.** $\pi/4$.

**13.**

$\pi/3$.

**15.** 0.

**17.** II.            **19.** III.            **21.** III.

**23.**

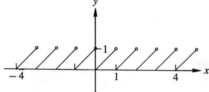

Period is 1.

**Exercise 2.2** (page 49)

1. $\sqrt{3}/2$.         7. $-1/\sqrt{2}$.         13. $\pi/4$.         23. $\frac{4}{5}$.

3. $1/\sqrt{2}$.         9. $\frac{1}{2}$.         15. 0.         25. $-\frac{12}{13}$.

5. $\frac{1}{2}$.         11. 1.         17. $\pi/3$.

27. (a) $\dfrac{\sqrt{3}}{2}$.     (b) $\dfrac{1}{\sqrt{3}}$.     (c) 2.     (d) $\dfrac{2}{\sqrt{3}}$.     (e) $\sqrt{3}$.

29. 0 inch;    2 inches;   0 inch;   $-2$ inches;   0 inch.         **31.** 30 inches.

**Exercise 2.3** (page 56)

1. Negative.         5. Positive.         9. Positive.

3. Positive.         7. Negative.         11. Negative.

13.
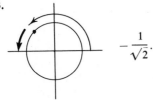
$-\dfrac{1}{\sqrt{2}}$.

15. $\dfrac{1}{2}$

17.
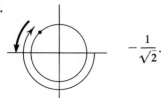
$-\dfrac{1}{\sqrt{2}}$.

19. $\dfrac{\sqrt{3}}{2}$.

21.
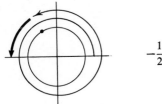
$-\dfrac{1}{2}$.

23. $\dfrac{\sqrt{3}}{2}$.

25.
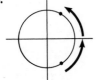
$\dfrac{\pi}{3}$; $\dfrac{\pi}{3}$, $\dfrac{5\pi}{3}$.

**27.** $\dfrac{3\pi}{4}$; $\dfrac{3\pi}{4}$, $\dfrac{5\pi}{4}$.

**29.**

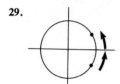     $\dfrac{\pi}{6}$; $\dfrac{\pi}{6}$, $\dfrac{11\pi}{6}$.

**31.**

**33.** $\dfrac{1+\sqrt{3}}{2\sqrt{2}}$.

**37.** $\{s \mid s = \dfrac{\pi}{3} + k \cdot 2\pi\} \cup \{s \mid s = \dfrac{5\pi}{3} + k \cdot 2\pi\}, \quad k \in J.$

**Exercise 2.4** (page 61)

**1.** Positive.          **5.** Positive.          **9.** Negative.

**3.** Negative.          **7.** Positive.          **11.** Negative.

**13.**

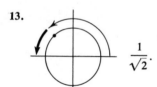     $\dfrac{1}{\sqrt{2}}$.                    **15.** $-\dfrac{1}{\sqrt{2}}$.

**17.**

     $\dfrac{1}{\sqrt{2}}$.                    **19.** $\dfrac{\sqrt{3}}{2}$

**21.**

     $-\dfrac{\sqrt{3}}{2}$.                 **23.** $\dfrac{1}{2}$.

**25.**

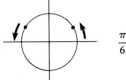

$\dfrac{\pi}{6}$; $\dfrac{\pi}{6}$, $\dfrac{5\pi}{6}$.

**27.** $\dfrac{\pi}{4}$; $\dfrac{\pi}{4}$, $\dfrac{3\pi}{4}$.

**29.**

$\dfrac{4\pi}{3}$; $\dfrac{4\pi}{3}$, $\dfrac{5\pi}{3}$.

**31.** $\dfrac{\pi}{2}$; $\dfrac{\pi}{2}$.

**33.**

| $\sin s$ | $\dfrac{1}{\sqrt{2}}$ | $1$ | $\dfrac{1}{\sqrt{2}}$ | $0$ | $-\dfrac{1}{\sqrt{2}}$ | $-1$ | $-\dfrac{1}{\sqrt{2}}$ | $0$ |
|---|---|---|---|---|---|---|---|---|

**35.** $\dfrac{\sqrt{3}-1}{2\sqrt{2}}$.

**39.** $\left\{s \,\middle|\, s = \dfrac{\pi}{6} + k \cdot 2\pi\right\} \cup \left\{s \,\middle|\, s = \dfrac{5\pi}{6} + k \cdot 2\pi\right\}, \quad k \in J.$

**Exercise 2.5** (page 67)

**1.** Negative.

**3.** Positive.

**5.** Negative.

**7.** Negative.

**9.** Positive.

**11.** Negative.

**13.** $-\dfrac{1}{\sqrt{3}}$.

**15.** $-\sqrt{3}.$

**17.**

1.

**19.** $-\sqrt{3}.$

**21.**

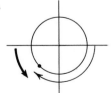

1.

**23.** $\dfrac{1}{\sqrt{3}}.$

**25.**

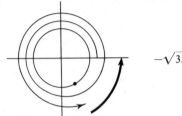

$-\sqrt{3}.$

**27.** $-\dfrac{1}{\sqrt{3}}.$

**29.**

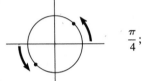

$\dfrac{\pi}{4}; \dfrac{\pi}{4}, \dfrac{5\pi}{4}.$

**31.** $\dfrac{5\pi}{6}; \dfrac{5\pi}{6}, \dfrac{11\pi}{6}.$

**33.**

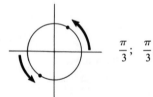

$\dfrac{\pi}{3}; \dfrac{\pi}{3}, \dfrac{4\pi}{3}.$

**35.** $\overline{\tan s \;\|\; 1 \;|\; \text{undef.} \;|\; -1 \;|\; 0 \;|\; 1 \;|\; \text{undef.} \;|\; -1 \;|\; 0}$

**37.** $\dfrac{\sqrt{3}-1}{1+\sqrt{3}}.$              **39.** $\left\{s \,\middle|\, s = \dfrac{\pi}{4} + k\pi\right\}, \qquad k \in J.$

**Exercise 2.6** (page 73)

**1.** Negative.              **5.** Positive.              **9.** Negative.
**3.** Positive.              **7.** Negative.              **11.** Negative.

**13.**

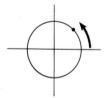

1.

**15.** $-\dfrac{2}{\sqrt{3}}.$

**17.**

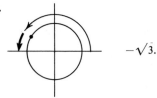

$-\sqrt{3}.$

**19.** $-\dfrac{2}{\sqrt{3}}.$

**21.**

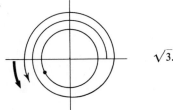

$\sqrt{3}.$

**23.** $-\sqrt{2}.$

**25.**

$\dfrac{\pi}{4}; \ \dfrac{\pi}{4}, \ \dfrac{7\pi}{4}.$

**27.** $\dfrac{7\pi}{6}; \ \dfrac{7\pi}{6}, \ \dfrac{11\pi}{6}.$

**29.**

$\dfrac{3\pi}{4}; \ \dfrac{3\pi}{4}, \ \dfrac{7\pi}{4}.$

**33.** $\left\{s \mid s = \dfrac{\pi}{4} + k \cdot 2\pi\right\} \cup \left\{s \mid s = \dfrac{7\pi}{4} + k \cdot 2\pi\right\}, \quad k \in J.$

**45.** Odd.      **47.** Even.      **49.** Even.      **51.** Odd.

**Exercise 2.7** (page 78)

| | | |
|---|---|---|
| **1.** 0.0998. | **15.** 1.10. | **29.** 0.292. |
| **3.** 0.3994. | **17.** 0.2523. | **31.** 0.946. |
| **5.** 1.421. | **19.** 1.617. | **33.** 1.435. |
| **7.** 0.4357. | **21.** 1.124. | **35.** 0.0100. |
| **9.** 0.12. | **23.** 0.3830. | **37.** 0.8776. |
| **11.** 0.40. | **25.** 0.3639. | **39.** 0.8415. |
| **13.** 0.81. | **27.** 1.036. | |

**Exercise 2.8** (page 82)

**1.** $\sin x$.    **3.** $\tan x$.    **5.** $-\csc x$.    **7.** $\cos x$.    **9.** $\sec x$.    **11.** $-\cot x$.

**13.**

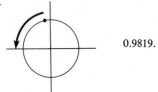

0.9819.

**15.** $-4.913$.

**17.**

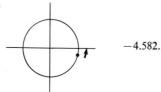

$-4.582$.

**19.** 0.9664.

**21.**

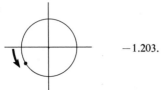

$-1.203$.

**23.** 1.163.

**25.**

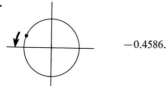

$-0.4586$.

**27.** 1.050.

**29.**

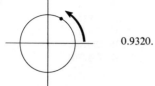

0.9320.

**31.** 0.9884.

**33.**

 $-0.6827.$

**35.** $-3.996.$

**37.**

 0.23; 0.23, 2.91.

**39.** 0.95; 0.95, 4.09.

**41.**

 1.11; 1.11, 2.03.

**43.** 1.80; 1.80, 4.48.

**45.**

2.29; 2.29, 5.43.

**47.** $\{x \mid x \approx 0.23 + k \cdot 2\pi\} \cup \{x \mid x \approx 2.91 + k \cdot 2\pi\}, \qquad k \in J.$

**49.** $\{x \mid x \approx 0.95 + k \cdot \pi\}, \qquad k \in J.$

**51.** Decreases.      **55.** Increases.      **59.** 0.242 feet.

**53.** Increases.      **57.** Decreases.

*Chapter 2 Review*  (page 86)

**1.** $\{s \mid s \in R\}; \quad \{(x, y) \mid x^2 + y^2 = 1; \quad x, y \in R\}.$

**3.** (a) $\left(\dfrac{1}{2}, \dfrac{\sqrt{3}}{2}\right).$      (b) $\left(\dfrac{\sqrt{3}}{2}, -\dfrac{1}{2}\right).$

**5.** (a) Quadrant II.      (b) Quadrant IV.

**7.** (a) $\dfrac{\pi}{2}$.    (b) $\dfrac{\pi}{3}$.

**9.** (a) $-3/5$.    (b) $\dfrac{4}{3}$.    (c) $-\dfrac{5}{3}$.

**11.** (a) $\pi$.    (b) $\dfrac{7\pi}{6}$.    (c) $\dfrac{5\pi}{6}$.

**13.** (a) 0.7385.    (b) 0.9128.    (c) 2.414.

**15.** (a) 0.3342.    (b) $-0.6594$.    (c) 0.2079.

**17.** $\{x \mid x \approx 2.17 + k\pi\}$,    $k \in J$.

**19.** 0.9992.

## Chapter 3

**Exercise 3.1** (page 92)

**1.**

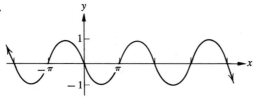

Zeros: $k\pi$, $k \in J$.

**3.** Zeros: $k\pi$, $k \in J$.

**5.**

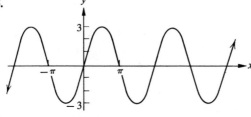

Zeros: $k\pi$, $k \in J$.

**7.** Zeros: $(\pi/2) + k\pi$, $k \in J$.

**9.**

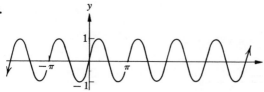

Zeros: $k\pi/2$, $k \in J$.

**11.** Zeros: $\pi + 2k\pi$, $k \in J$.

**13.**

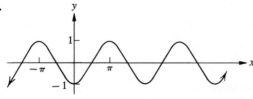

Zeros: $(\pi/2) + k\pi$, $k \in J$.

**17.**

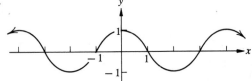

**19.** Less.            **21.** Greater.            **23.** Greater.

**3**

**Exercise 3.2** (page 97)

**1.**

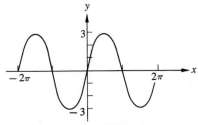

**5.**

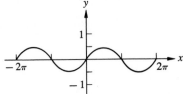

**9.**

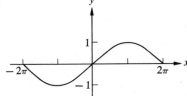

**13.**

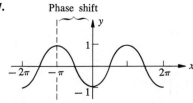

**17.**    Phase shift

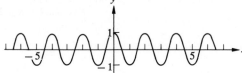

**21.**

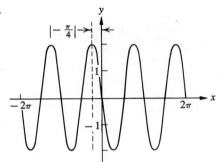

**27.** {(0, 1), (6.3, 1)}.

**29.**

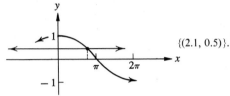                    {(2.1, 0.5)}.

**31.** {0, 1.5, 3.1, 5.0}.

**33.**

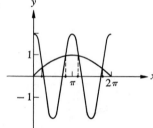                    {0.6, 2.7, 3.5, 5.7}.

**35.** $\{x \mid \pi < x < 2\pi\}$.

**37.**

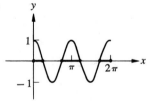          $\{x \mid 0 \leq x \leq \pi/4\} \cup \{x \mid 3\pi/4 \leq x \leq 5\pi/4\}$
                                           $\cup \{x \mid 7\pi/4 \leq x \leq 2\pi\}$.

**39.** $\{x \mid 0 \leq x < 3\pi/4\} \cup \{x \mid 7\pi/4 < x \leq 2\pi\}$.

**41.** Using Theorem 2.3,

$$\cos\left(x - \frac{\pi}{2}\right) = \cos x \cos \frac{\pi}{2} + \sin x \sin \frac{\pi}{2} = \sin x,$$

and

$$-\cos\left(x + \frac{\pi}{2}\right) = -\left[\cos x \cos \frac{\pi}{2} - \sin x \sin \frac{\pi}{2}\right] = \sin x.$$

**Exercise 3.3** (page 105)

**1.**

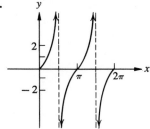

(a) $\{y\,|\,y \in R\}$.          (b) $x = (\pi/2) + k\pi,\ k \in J$.          (c) $\{x\,|\,x = k\pi,\ k \in J\}$.

**3.** (a) $\{y\,|\,|y| \geq 2\}$.          (b) $x = (\pi/2) + k\pi,\ k \in J$.          (c) $\phi$.

**5.**

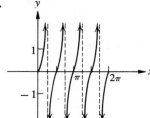

(a) $\{y\,|\,y \in R\}$.          (b) $x = (\pi/4) + (k\pi/2),\ k \in J$.          (c) $\{x\,|\,x = k\pi/2,\ k \in J\}$.

**7.** (a) $\{y\,|\,y \in R\}$.          (b) $x = (3\pi/4) + k\pi,\ k \in J$.          (c) $(x\,|\,x = (\pi/4) + k\pi,\ k \in J\}$.

**9.**

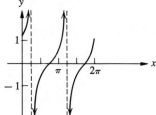

(a) $\{y\,|\,y \in R\}$.          (b) $x = (\pi/3) + k\pi,\ k \in J$.          (c) $\{x\,|\,x = (5\pi/6) + k\pi,\ k \in J\}$.

**13.**

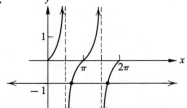

$\{(3\pi/4,\ -1),\ (7\pi/4,\ -1)\}$.

**15.** $\{(\pi/3, 6), (5\pi/3, 6)\}$.

**17.**

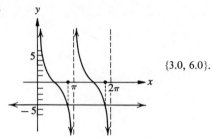

$\{3.0, 6.0\}$.

**19.** $\{1.0, 5.3\}$.

**21.**

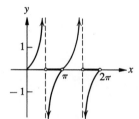

$\{x \mid \pi/2 < x < \pi\} \cup \{x \mid 3\pi/2 < x < 2\pi\}$.

**23.** $\{x \mid 0 < x < 2\pi\}$.

**25.**

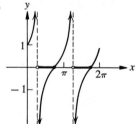

$\{x \mid \pi/4 < x \leq 3\pi/4\} \cup \{x \mid 5\pi/4 < x \leq 7\pi/4\}$.

**27.** $y = -\cot\left(x + (\pi/2)\right)$,

$$y = \frac{-\cos\left(x + (\pi/2)\right)}{\sin\left(x + (\pi/2)\right)}$$

$$= \frac{-\cos x \cos (\pi/2) + \sin x \sin (\pi/2)}{\sin x \cos (\pi/2) + \cos x \sin (\pi/2)}$$

$$= \frac{\sin x}{\cos x} = \tan x.$$

**29.** Greater.                                  **31.** Greater.

**Exercise 3.4** (page 110)

**1.**

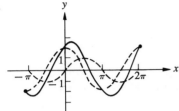

**5.**

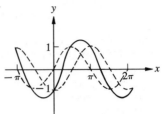

**3**

**9.**

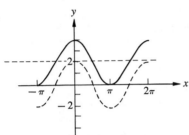

**13.**

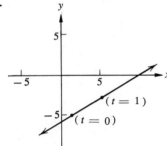

$2x - 3y = 19.$

**15.** $x^2 + y^2 = 4.$

**17.**

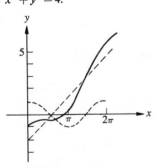

**21.**

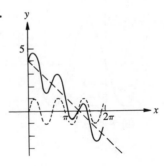

**Exercise 3.5** (page 115)

**1.** $F^{-1} = \{(x, y) \,|\, x = \tan y\}$
$= \{(x, y) \,|\, y = \arctan x\};$
domain of $F$: $\{x \,|\, x \in R, \, x \neq (\pi/2) + k\pi, \, k \in J\}$,
range of $F$: $\{y \,|\, y \in R\}$;
domain of $F^{-1}$: $\{x \,|\, x \in R\}$,
range of $F^{-1}$: $\{y \,|\, y \in R, \, y \neq (\pi/2) + k\pi, \, k \in J\}$.

**3.** $F^{-1} = \{(v, w) \mid v = \csc w\}$
   $= \{(v, w) \mid w = \text{arccsc } v\};$
   domain of $F$: $\{v \mid v \in R, v \neq k\pi, k \in J\},$
   range of $F$: $\{w \mid |w| \geq 1\};$
   domain of $F^{-1}$: $\{v \mid |v| \geq 1\},$
   range of $F^{-1}$: $\{w \mid w \in R, w \neq k\pi, k \in J\}.$

**5.** $F^{-1} = \{(s, t) \mid s = \sin 2t\}$
   $= \{(s, t) \mid t = \frac{1}{2} \arcsin s\}.$

**7.** $F^{-1} = \{(x, y) \mid x = \cot (y/3)\}$
   $= \{(x, y) \mid y = 3 \text{ arccot } x\}.$

**9.** $F^{-1} = \{(k, l) \mid k = \sec l\}$
   $= \{(k, l) \mid l = \text{arcsec } k\}.$

**11.** $F^{-1} = \{(m, n) \mid m = 3 \cos 2n\}$
   $= \{(m, n) \mid n = \frac{1}{2} \arccos (m/3)\}.$

**13.** $F^{-1} = \{(u, v) \mid u = \sec (v + (\pi/2))\}$
   $= \{(u, v) \mid v = \text{arcsec } u - (\pi/2)\}.$

**15.** $\cos y = 1/\sqrt{2}, 0 \leq y \leq \pi; \pi/4.$

**17.** $\cot y = 1, 0 < y < \pi; \pi/4.$

**19.** $\sec y = -\sqrt{2}, 0 \leq y \leq \pi; 3\pi/4.$

**21.** $\sin y = 2, -\pi/2 \leq y \leq \pi/2;$ no value.

**23.** $\tan y = -1, -\pi/2 < y < \pi/2; -\pi/4.$

**25.** $\cos y = 0.8829, 0 \leq y \leq \pi; \approx 0.49.$

**27.** $\sin y = 0.8542, 0 \leq y \leq \pi; \approx 1.02.$

**29.** $\dfrac{1}{\sqrt{2}}.$

**31.** $-\sqrt{3}.$

**33.** $\dfrac{2}{\sqrt{3}}.$

**35.** $\dfrac{\sqrt{3}}{2}.$

**37.** $-2.$

**39.** $-\sqrt{3}.$

**41.** $\approx 0.9249.$

**43.** $\approx 4.193.$

**45.** $\approx -1.557.$

**47.** $\dfrac{\pi}{4}.$

**49.** $\dfrac{\pi}{4}.$

**51.** $\approx 1.16.$

**53.** $\sqrt{1 - x^2}.$

**55.** $\dfrac{x}{\sqrt{1 - x^2}}.$

**57.** $x.$

**59.** $\dfrac{1 - \sqrt{3}}{2\sqrt{2}}.$

**61.** $0.$

*Chapter 3 Review* (page 119)

**1.**

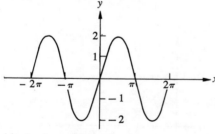

(a) $A = 2.$     (b) $P = 2\pi.$

**3.** (a) $A = 3.$     (b) $P = 4\pi.$

**5.**

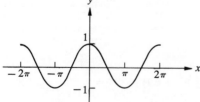

(a) $A = 1.$     (b) $P = 2\pi.$     (c) $|C| = \pi/2$ (left).

**7.** (a) $\{y \mid y \in R\}$. (b) $x = \pi/2, x = 3\pi/2$. (c) $\{0, \pi, 2\pi\}$.

**9.**

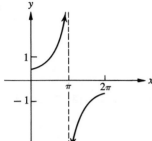

(a) $\{y \mid |y| \geq \frac{1}{2}\}$. (b) $x = \pi$. (c) $\phi$.

**13.**

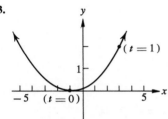

**15.** $F^{-1} = \{(x, y) \mid x = 2 \cos 3y\}$
$= \{(x, y) \mid y = \frac{1}{3} \arccos (x/2)\}$;
domain of $F$: $\{x \mid x \in R\}$,
range of $F$: $\{y \mid |y| \leq 2\}$,
domain of $F^{-1}$: $\{x \mid |x| \leq 2\}$,
range of $F^{-1}$: $\{y \mid y \in R\}$.

**17.** $\cos y = 1/\sqrt{2}, 0 \leq y \leq \pi/2$; $\pi/4$. **19.** $\cot y = -1, \pi/2 \leq y < \pi; 3\pi/4$.

**21.** $1/\sqrt{2}$. **23.** 1. **25.** $\approx 1.018$. **27.** $\pi/4$.

**29.**

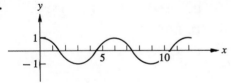

**31.** $\{(\pi/4, 1), (5\pi/4, 1)\}$.

**33.**

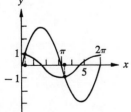

$\{\approx 0.3, \approx 3.4\}$.

**35.** $\sqrt{1-x^2}$.

**37.** $\dfrac{\sqrt{3}-1}{2\sqrt{2}}$.

**39.**

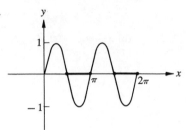

$\{x \mid \pi/2 \leq x \leq \pi\} \cup \{x \mid 3\pi/2 \leq x \leq 2\pi\}.$

**41.** $\{x \mid 0 < x < \pi/2\} \cup \{x \mid \pi < x < 3\pi/2\}.$

**43.**

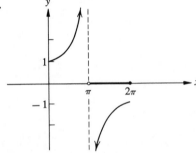

$\{x \mid \pi < x \leq 2\pi\}.$

## Chapter 4

**Exercise 4.1** (page 128)

**1.** $m^\circ(\alpha) = \dfrac{180}{\pi}\, m^R(\alpha); \quad m^R(\alpha) = \dfrac{\pi}{180}\, m^\circ(\alpha).$

**3.** $-60.0°.$            **9.** $-292.5°.$            **15.** $-0.52^R.$            **21.** $-3.52^R.$

**5.** $-240.0°.$           **11.** $10.3°.$            **17.** $1.31^R.$

**7.** $288.0°.$            **13.** $-88.2°.$           **19.** $0.75^R.$

**23.**

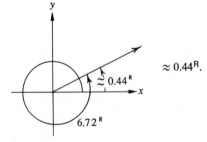

$\approx 0.44^R.$

**25.** $3.41°$

**27.**

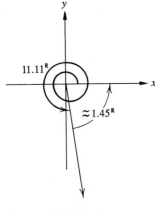

**29.** 72°.

$\approx 1.69^{R}$.

**31.** $(21 + k \cdot 360)°$, $k \in J$.

**33.** $\left(\dfrac{-5\pi}{6} + k \cdot 2\pi\right)^{R}$, $k \in J$.

**35.** $\left(\dfrac{-7\pi}{6} + k \cdot 2\pi\right)^{R}$, $k \in J$.

**37.** (a) $\dfrac{2\pi r}{8}$.    (b) $\dfrac{\pi}{4}^{R}$; 45°.

**Exercise 4.2** (page 131)

**1.** 15.5 feet.

**3.** $\dfrac{-3\pi}{2}$ inches.

**5.** $\dfrac{5}{6}\pi^2$ inches.

**7.** 126.7°.

**9.** 281.0°.

**11.** 36 feet per second.

**13.** 360 radians per minute.

**15.** $-54\pi$ inches per second.

**17.** $\dfrac{-12}{\pi}$ degrees per second.

**19.** $\approx 3.48$ inches.

**21.** $\approx 9.39$ radians per minute.

**23.** $\approx 53,500$ miles per hour.

**27.** $\approx 11.6$ square feet.

**Exercise 4.3** (page 135)

**1.** Quadrant III.    **3.** Quadrant I.    **5.** Quadrant II.    **7.** Quadrant I.

**9.**

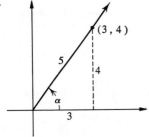

$\cos \alpha = \frac{3}{5}$,    $\sec \alpha = \frac{5}{3}$,

$\sin \alpha = \frac{4}{5}$,    $\csc \alpha = \frac{5}{4}$,

$\tan \alpha = \frac{4}{3}$,    $\cot \alpha = \frac{3}{4}$.

**11.** $\cos \alpha = -\frac{12}{13}$,    $\sec \alpha = -\frac{13}{12}$,

$\sin \alpha = -\frac{5}{13}$,    $\csc \alpha = -\frac{13}{5}$,

$\tan \alpha = \frac{5}{12}$,    $\cot \alpha = \frac{12}{5}$.

**13.**

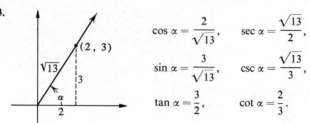

$$\cos \alpha = \frac{2}{\sqrt{13}}, \qquad \sec \alpha = \frac{\sqrt{13}}{2},$$

$$\sin \alpha = \frac{3}{\sqrt{13}}, \qquad \csc \alpha = \frac{\sqrt{13}}{3},$$

$$\tan \alpha = \frac{3}{2}, \qquad \cot \alpha = \frac{2}{3}.$$

**15.** $\cos \alpha = -\dfrac{\sqrt{3}}{2}, \qquad \sec \alpha = -\dfrac{2}{\sqrt{3}},$

$\sin \alpha = -\dfrac{1}{2}, \qquad \csc \alpha = -2,$

$\tan \alpha = \dfrac{1}{\sqrt{3}}, \qquad \cot \alpha = \sqrt{3}.$

**17.**

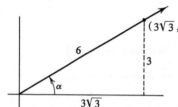

$$\cos \alpha = \frac{\sqrt{3}}{2}, \qquad \sec \alpha = \frac{2}{\sqrt{3}},$$

$$\sin \alpha = \frac{1}{2}, \qquad \csc \alpha = 2,$$

$$\tan \alpha = \frac{1}{\sqrt{3}}, \qquad \cot \alpha = \sqrt{3}.$$

**19.** Undefined for $\alpha = ((\pi/2) + k \cdot \pi))^R$, $k \in J$. Equal to 0 for $\alpha = (0 + k \cdot \pi)^R$, $k \in J$.

**21.** Undefined for $\alpha = ((\pi/2) + k \cdot \pi))^R$, $k \in J$. No secant function values are equal to 0.

**Exercise 4.4** (page 139)

**1.** $\tan \alpha = y/x.$     **5.** $1/2.$     **9.** $2/\sqrt{3}.$     **13.** $-2/\sqrt{3}.$

**3.** $\csc \alpha = 1/y.$     **7.** $\sqrt{3}.$     **11.** $-\sqrt{3}/2.$     **15.** $-\sqrt{3}.$

**17.**

| Function | Circular Functions | | Trigonometric Functions | |
| --- | --- | --- | --- | --- |
| | Domain | Range | Domain | Range |
| Cosine and sine | $x \in R.$ | $y \in R, \lvert y \rvert \leq 1.$ | $\alpha \in A.$ | $\cos \alpha, \sin \alpha \in R.$ |
| Tangent | $x \in R, x \neq \dfrac{\pi}{2}, \dfrac{3\pi}{2}.$ | $y \in R.$ | $\alpha \in A, \alpha \neq \dfrac{\pi^R}{2}, \dfrac{3\pi^R}{2}.$ | $\tan \alpha \in R.$ |
| Secant | $x \in R, x \neq \dfrac{\pi}{2}, \dfrac{3\pi}{2}.$ | $y \in R, \lvert y \rvert \geq 1.$ | $\alpha \in A, \alpha \neq \dfrac{\pi^R}{2}, \dfrac{3\pi^R}{2}.$ | $\lvert \sec \alpha \rvert \geq 1, \sec \alpha \in R.$ |
| Cosecant | $x \in R, x \neq 0, \pi.$ | $y \in R, \lvert y \rvert \geq 1.$ | $\alpha \in A, \alpha \neq 0^R, \pi^R.$ | $\lvert \csc \alpha \rvert \geq 1, \csc \alpha \in R.$ |
| Cotangent | $x \in R, x \neq 0, \pi.$ | $y \in R.$ | $\alpha \in A, \alpha \neq 0^R, \pi^R.$ | $\cot \alpha \in R.$ |

Chapter 4

**Exercise 4.5** (page 143)

**1.**

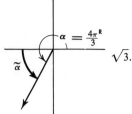

$\sqrt{3}.$

**3.** $-\sqrt{2}.$

**5.**

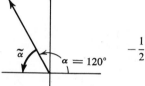

$-\dfrac{1}{2}.$

**7.** $-\dfrac{1}{2}.$  **11.** $\approx 0.9872.$  **15.** $\approx 0.0407.$  **19.** $\approx 1.124.$

**9.** $\approx 0.7379.$  **13.** $\approx 0.7774.$  **17.** $\approx 0.6862.$

**21.**

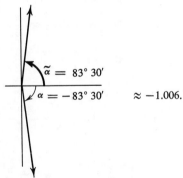

$\approx -1.006.$

**23.** $\approx -2.605.$

**25.**

$\tilde{\alpha} = 86° 30'$    $\approx -16.38.$

**27.** $\approx 0.0634.$

**29.**

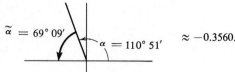

$\tilde{\alpha} = 69° 09'$    $\alpha = 110° 51'$    $\approx -0.3560.$

**31.** $\approx 0.8205.$    **37.** $-30°$ or $330°.$    **43.** $\approx 30° 40'.$    **49.** $\approx -115° 20'.$

**33.** $45°.$    **39.** $135°.$    **45.** $\approx 47° 20'.$

**35.** $30°.$    **41.** $\approx 16° 10'.$    **47.** $\approx 173° 30'.$

### *Chapter 4 Review* (page 146)

**1.** $72.0°.$    **7.** $(131 + k \cdot 360)°, \; k \in J.$    **13.** Quadrant III.

**3.** $1.31^R.$    **9.** $\approx 6.82$ feet.

**5.** $0.95^R.$    **11.** 12 inches per minute.

**15.** $\cos \alpha = -\frac{4}{5},$    $\sec \alpha = -\frac{5}{4},$    **17.** $-1/\sqrt{2}.$
      $\sin \alpha = \frac{3}{5},$    $\csc \alpha = \frac{5}{3},$    **19.** $\approx 0.6095.$
      $\tan \alpha = -\frac{3}{4},$    $\cot \alpha = -\frac{4}{3}.$    **21.** $60°.$

### Chapter 5

### Exercise 5.2 (page 154)

**1.** $\sin 2x.$    **9.** $\frac{1}{2} \sin 2x.$    **27.** $\pm \dfrac{5}{\sqrt{34}}.$

**3.** $\cos 2x.$    **11.** $2 \cos 10x.$    **37.** $\frac{1}{2} \sin 7x + \frac{1}{2} \sin x.$

**5.** $\cos 6x.$    **13.** $\sin x.$    **39.** $\dfrac{3}{2} \cos \dfrac{5\pi}{6} + \dfrac{3}{2} \cos \dfrac{\pi}{6}.$

**7.** $\cos 8x.$    **25.** $\dfrac{7}{25}.$    **41.** $2 \sin \dfrac{\pi}{2} \cos \dfrac{\pi}{3}.$

### Exercise 5.3 (page 158)

**11.** $\sqrt{3}.$    **13.** $0.$    **17.** $2x\sqrt{1 - x^2}.$    **19.** $\pm \sqrt{\dfrac{1 - \sqrt{1 - x^2}}{2}}.$

### Exercise 5.4 (page 161)

**1.** $\left\{ \dfrac{\pi}{6}^R \right\}.$    **3.** $\left\{ \dfrac{\pi}{4}^R \right\}.$    **5.** $\left\{ \dfrac{\pi}{2}^R \right\}.$    **7.** $\left\{ \dfrac{\pi}{4}^R, \dfrac{\pi}{2}^R \right\}.$    **9.** $\left\{ 0^R, \dfrac{\pi}{2}^R \right\}.$

**11.** $\{135°, 315°\}.$    **17.** $\{30°, 210°, 150°, 330°\}.$

**13.** $\{270°\}.$    **19.** $\phi.$

**15.** $\{60°, 300°, 180°\}.$    **21.** $\{90°, 270°, 45°, 315°, 135°, 225°\}.$

**23.** $\{x \,|\, x = \dfrac{2\pi}{3} + k \cdot 2\pi\} \cup \{x \,|\, x = \dfrac{4\pi}{3} + k \cdot 2\pi\} \cup \{x \,|\, x = k\pi\}, \quad k \in J.$

**25.** $\{x \,|\, x \approx 0.41 + k \cdot 2\pi\} \cup \{x \,|\, x \approx 2.73 + k \cdot 2\pi\}, \quad k \in J.$

**27.** $\{x \,|\, x \approx 1.25 + k \cdot \pi\} \cup \{x \,|\, x = (\pi/3) + k\pi\}, \quad k \in J.$

**29.** $\{x \,|\, x \approx 0.93 + k \cdot 2\pi\}, \quad k \in J.$

**31.** $\{x \,|\, x = k \cdot 2\pi\} \cup \{x \,|\, x = (\pi/2) + k \cdot 2\pi\}, \quad k \in J.$

**33.** $\{x \,|\, x = (\pi/2) + k \cdot 2\pi\} \cup \{x \,|\, x = (7\pi/6) + k \cdot 2\pi\}, \quad k \in J.$

**Exercise 5.5** (page 164)

1. $\left\{\dfrac{\pi^R}{6}\right\}$.

9. $\{10°, 50°, 130°, 170°, 250°, 290°\}$.

3. $\{0^R\}$.

11. $\{15°, 75°, 135°, 195°, 255°, 315°\}$.

5. $\left\{\dfrac{\pi^R}{8}\right\}$.

13. $\{180°\}$.

7. $\{0^R\}$.

15. $\{60°, 180°, 300°\}$.

17. $\left\{x \mid x = \dfrac{\pi}{2} + k\pi\right\} \cup \left\{x \mid x = \dfrac{\pi}{6} + k \cdot 2\pi\right\} \cup \left\{x \mid x = \dfrac{5\pi}{6} + k \cdot 2\pi\right\}$,    $k \in J$.

19. $\{x \mid x = k\pi\} \cup \left\{x \mid x = \dfrac{\pi}{6} + k \cdot \dfrac{\pi}{3}\right\}$,    $k \in J$.

21. $\left\{x \mid x \approx 1.28 + k \cdot \dfrac{\pi}{2}\right\} \cup \left\{x \mid x = \dfrac{\pi}{8} + k \cdot \dfrac{\pi}{2}\right\}$,    $k \in J$.

23. $\left\{x \mid x = \dfrac{\pi}{12} + k \cdot \dfrac{2\pi}{3}\right\} \cup \left\{x \mid x = \dfrac{\pi}{4} + k \cdot \dfrac{2\pi}{3}\right\}$,    $k \in J$.

25. $\left\{(0, 90°), (0, 270°), \left(\dfrac{\sqrt{3}}{4}, 30°\right), \left(\dfrac{-\sqrt{3}}{4}, 150°\right)\right\}$.

27.

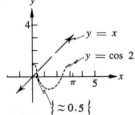

*Chapter 5 Review* (page 166)

9. $\left\{\dfrac{\pi^R}{3}\right\}$.

11. $\{22\frac{1}{2}°, 112\frac{1}{2}°, 202\frac{1}{2}°, 292\frac{1}{2}°\}$.

13. $\dfrac{7}{9}$.

15. $\left\{(0, 0°), (0, 180°), \left(-\dfrac{\sqrt{3}}{8}, 120°\right), \left(\dfrac{\sqrt{3}}{8}, 240°\right)\right\}$.

## Chapter 6

**Exercise 6.1** (page 172)

1. $A \approx 33° 40'$; $B \approx 56° 20'$; $c \approx 3.61$; $\mathscr{A} = 3.00$.
3. $A \approx 62° 00'$; $B \approx 28° 00'$; $b \approx 7.98$; $\mathscr{A} \approx 59.9$.
5. $A = 39°$; $a \approx 14.8$; $b \approx 18.3$; $\mathscr{A} \approx 135$.
7. $A = 67° 50'$; $c \approx 11.9$; $b \approx 4.48$; $\mathscr{A} \approx 24.6$.
9. $\approx 22° 20'$.    11. $\approx 356$ feet.    13. $\approx 53° 10'$.    15. $\approx 133.8$ feet.

**Exercise 6.2** (page 178)

1. $\gamma = 85°$; $b \approx 14.1$; $c \approx 14.6$.
3. $\alpha = 42° 50'$; $b \approx 51.9$; $c \approx 84.6$.
5. $\beta = 52° 50'$; $a \approx 0.307$; $c \approx 0.527$.
7. $\gamma = 77° 30'$; $a \approx 7.03$; $b \approx 13.3$.

**9.**

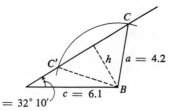

$a > b$; one triangle

**11.**

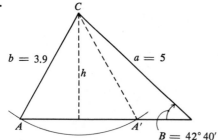

$c \sin \alpha < a < c$; two triangles.

**13.** $c \sin B = b$; one right triangle.

**15.**

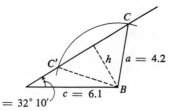

$a \sin B < b < a$; two triangles.

**17.** $B \approx 21° \, 00'$;   $C \approx 129° \, 00'$;   $c \stackrel{.}{\approx} 10.9$.
**19.** $C \approx 50° \, 40'$;   $C' \approx 129° \, 20'$;   $B \approx 97° \, 10'$;   $B' \approx 18° \, 30'$;   $b \approx 7.82$;   $b' \approx 2.50$.
**21.** $A = 60° \, 00'$;   $C = 90° \, 00'$;   $a \approx 27.7$.
**23.** $A \approx 60° \, 20'$;   $A' \approx 119° \, 40'$;   $C \approx 77° \, 00'$;   $C' \approx 17° \, 40'$;   $c \approx 5.61$;   $c' \approx 1.75$.
**25.** $\approx 1.51$.            **27.** $\approx 14.5$.            **29.** $\approx 3.49$.            **31.** $\approx 3,050$ feet.
**39.** $\beta \approx 12° \, 10'$;   $\gamma \approx 122° \, 50'$;   $a \approx 3.35$.
**41.** $\alpha \approx 1° \, 00'$;   $\beta \approx 160° \, 40'$;   $c \approx 2.34$.

**Exercise 6.3** (page 182)

**1.** $a \approx 4.37$;   $\beta \approx 60° \, 00'$;   $\gamma \approx 49° \, 10'$.
**3.** $c \approx 1.90$;   $\alpha \approx 103° \, 10'$;   $\beta \approx 42° \, 00'$.
**5.** $b \approx 1.42$;   $\alpha \approx 17° \, 50'$;   $\gamma \approx 20° \, 30'$.
**7.** $\alpha \approx 95° \, 40'$;   $\beta \approx 54° \, 40'$;   $\gamma \approx 33° \, 30'$.
     Does $\alpha + \beta + \gamma = 180$? Very "close." $\alpha + \beta + \gamma = 179° \, 50'$. Rounding off error $10'$.
**9.** $\alpha \approx 4° \, 00'$;   $\beta \approx 31° \, 40'$;   $\gamma \approx 144° \, 20'$.
**11.** $\approx 42° \, 40'$.                **13.** $\approx 10.9$.                            **21.** $\approx 5.94$.
**23.** $\alpha \approx 37° \, 00'$;   $\beta \approx 53° \, 00'$;   $\gamma \approx 90° \, 00'$.  Does $\alpha + \beta + \gamma = 180°$?  Yes.
**25.** $\alpha \approx 58° \, 40'$;   $\beta \approx 24° \, 20'$;   $\gamma \approx 97° \, 00'$.  Does $\alpha + \beta + \gamma = 180°$?  Yes.

**Exercise 6.4** (page 190)

*Note:* Scale of the geometric vectors used in the following answers has been changed from
     the scale used on page 190.

**1.** $\approx 2.3$;   $\approx 90°$.            **3.** $\approx 3.5$; ` $\approx -63°$.            **5.** $\approx 2.3$;   $\approx -90°$.

**7.**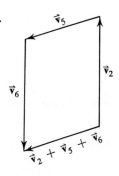

**11.**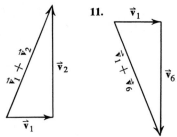

**15.**

**19.**

**27.**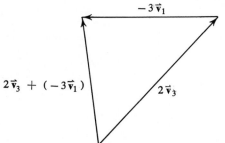

**31.**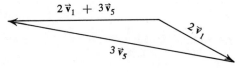

**35.**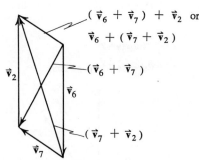

**39.**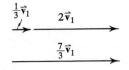

**41.** $c_1 \approx 4$;  $c_2 \approx 2$;  $\|\vec{v}\| \approx 4.5$.
**43.** $c_1 \approx -2$;  $c_2 \approx -2$;  $\|\vec{v}\| \approx 3$.

**Exercise 6.5** (page 197)

**1.** $\|\vec{v}_x\| = 5$;   $\|\vec{v}_y\| = 5$.

**3.** $\|\vec{v}_x\| = 0$;   $\|\vec{v}_y\| = 15$.

**5.**

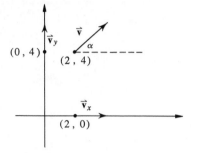

$\|\vec{v}\| = 2\sqrt{2}$;   $\alpha = 45°$.

**7.** $\|\mathbf{v}\| = 17$;   $\alpha \approx 28° \, 00'$.
**9.** $\|\vec{v}_3\| = 5$;   $\alpha \approx 53° \, 10'$;   $\beta \approx 36° \, 50'$.
**11.** $\|\vec{v}_3\| \approx 27.9$;   $\alpha \approx 43° \, 50'$;   $\beta \approx 31° \, 20'$.   *Note:* $\alpha + \beta = 75° \, 10'$.   Rounding off error 10′.
**13.** $\approx 0.967$ miles per hour.
**15.** Drift angle is $\approx 11° \, 20'$ to the right; ground speed is $\approx 612$ miles per hour; and the course is $\approx 146° \, 20'$.
**19.** $\vec{v}_4 = \dfrac{\vec{v}_1 + \vec{v}_3}{2}$.

*Chapter 6 Review* (page 203)

**1.** $A = 39°$;   $a \approx 8.81$;   $b \approx 10.9$;   $\mathscr{A} \approx 48.0$.
**3.** $A \approx 53° \, 40'$;   $B = 36° \, 20'$;   $b \approx 3.97$;   $\mathscr{A} \approx 10.7$.
**5.** $\alpha = 66° \, 10'$;   $b \approx 1.53$;   $c \approx 1.75$;   $\mathscr{A} \approx 1.22$.
**7.** Two triangles are possible.
   $A \approx 50° \, 40'$;   $C \approx 101° \, 50'$;   $c \approx 9.12$;   $\mathscr{A} \approx 15.1$.
   $A' \approx 129° \, 20'$;   $C' \approx 23° \, 10'$;   $c' \approx 3.66$;   $\mathscr{A} \approx 6.08$.
**9.** $c \approx 2.07$;   $\alpha \approx 51° \, 10'$;   $\beta \approx 70° \, 40'$;   $\mathscr{A} \approx 1.85$.
**11.** $\alpha \approx 69° \, 50'$;   $\beta \approx 75° \, 00'$;   $\gamma \approx 35° \, 10'$;   $\mathscr{A} \approx 14.5$.
**13.** $F_x \approx 13.2$ lbs;   $F_y \approx 7.6$ lbs.
**15.** $\|\vec{v}_3\| \approx 8.45$;   $\alpha \approx 35° \, 40'$;   $\beta \approx 44° \, 20'$.

## Chapter 7

**Exercise 7.1** (page 208)

**1.**

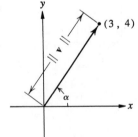

$\|\mathbf{v}\| = 5$;   $\alpha \approx 53° \, 10$.

**3.** $\|\mathbf{v}\| = \sqrt{2}; \quad \alpha = 45°.$

**5.**

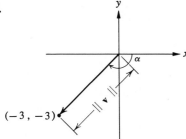

$\|\mathbf{v}\| = 3\sqrt{2}; \quad \alpha = -135°.$

**7.** $\|\mathbf{v}\| = 4; \quad \alpha = 0°.$

**9.**

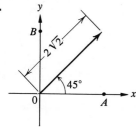

$\mathbf{v} = (2, 2).$

**11.** $\mathbf{v} = (0, 1).$

**13.**

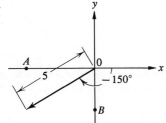

$\mathbf{v} = \left(-\dfrac{5}{2}\sqrt{3}, -\dfrac{5}{2}\right).$

**15.** $\mathbf{v}_1 + \mathbf{v}_2 = (5, 1).$

**17.**

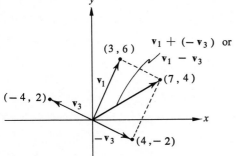

$\mathbf{v}_1 - \mathbf{v}_3 = (7, 4).$

**19.** $c_1(\mathbf{v}_1 + \mathbf{v}_2) = (10, 2).$

**21.**

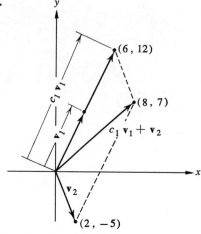

$$c_1 v_1 + v_2 = (8, 7).$$

**23.** $c_1 v_2 + c_2 v_3 = (16, -16).$

**25.**

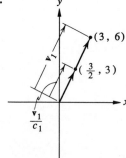

$$\frac{v_1}{c_1} = \left(\frac{3}{2}, 3\right).$$

**33.** (See Problem 26.) The norm of $(2/\sqrt{29}, -5/\sqrt{29})$ equals 1; this vector has a length of 1 unit. This suggests that a vector of unit length, collinear with a given vector, can be found by dividing the given vector by its norm.

**Exercise 7.2** (page 212)

**1.**    y                        $(3, 0).$

**3.** $(1, 1).$

**5.**    y                        $(5, -1).$

**7.** $-4\mathbf{i} + 7\mathbf{j}$.

**9.**

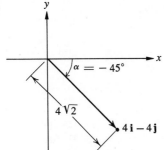

$\sqrt{3}\mathbf{i} + \mathbf{j}$.

**11.** $3\mathbf{i} - 3\sqrt{3}\mathbf{j}$.

**13.**

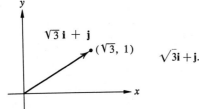

$4\sqrt{2}$;   $\alpha = -45°$.

**15.** $\sqrt{5}$;   $\alpha \approx 26° \, 30'$.

**17.**

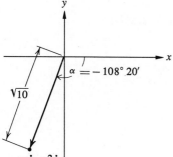

$\sqrt{10}$;   $\alpha \approx -108° \, 20'$.

**19.** $\sqrt{13}$;   $\alpha \approx -33° \, 40'$.

**21.** $\left(\dfrac{4}{5}, \dfrac{3}{5}\right)$.     **23.** $\left(\dfrac{7}{\sqrt{85}}, \dfrac{-6}{\sqrt{85}}\right)$.     **25.** $\dfrac{5}{3\sqrt{3}}\mathbf{i} + \dfrac{\sqrt{2}}{3\sqrt{3}}\mathbf{j}$.

**27.**

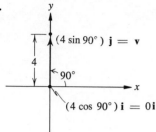

**29.** $v = \dfrac{5}{2}i + \dfrac{5\sqrt{3}}{2}\,j.$

$v = 4j.$

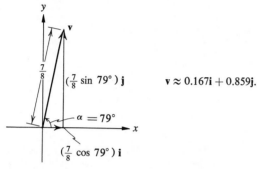

$v \approx 0.167i + 0.859j.$

**Exercise 7.3** (page 218)

| | | | |
|---|---|---|---|
| **1.** 48. | **5.** 45. | **9.** $36\sqrt{2}$. | **13.** 156° 40′. |
| **3.** −2. | **7.** 13. | **11.** $\frac{3}{4}$. | **15.** 33° 50′. |

**17.**

**19.** $\frac{7}{41}\sqrt{41}$;  $\frac{7}{13}\sqrt{13}$.

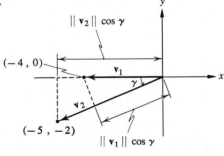

$\frac{20}{29}\sqrt{29}$; 5.

**21.**

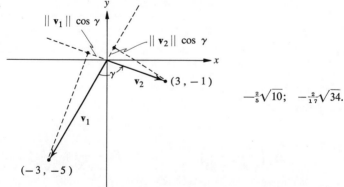

$-\frac{2}{5}\sqrt{10}$;  $-\frac{2}{17}\sqrt{34}$.

**29.** 6.　　　　**31.** ±3.　　　　**33.** ±$\sqrt{30}$.　　　**39.** $-\frac{2}{3}$.　　　**41.** 0.

**47.**

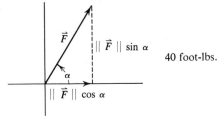

40 foot-lbs.

**Exercise 7.4** (page 225)

**1.** $(3, 20°)$; $(-3, 200°)$; $(-3, -160°)$; $(3, -340°)$.
**3.** $(5, -90°)$; $(-5, -270°)$; $(-5, 90°)$; $(5, 270°)$.
**5.** $(-4, 60°)$; $(4, 240°)$; $(4, -120°)$; $(-4, -300°)$.

**7.** $(2\sqrt{3}, 2)$.　　　　　**15.** $\left(2\sqrt{2}, \frac{\pi}{4}^R\right)$　　　**23.** $\rho = 3 \sin \theta$.

**9.** $\left(-\dfrac{3}{\sqrt{2}}, -\dfrac{3}{\sqrt{2}}\right)$.　　**17.** $\left(1, \dfrac{5\pi}{6}^R\right)$　　**25.** $y = -1$.

**11.** $\left(\dfrac{3\sqrt{3}}{2}, -\dfrac{3}{2}\right)$.　　**19.** $\rho \cos \theta = 4$.　　**27.** $x^2 + y^2 = 4y$.

**13.** $(4, 0^R)$.　　　　　**21.** $\rho = 4$.　　　**29.** $x^2 - 8y - 16 = 0$.

**31.**

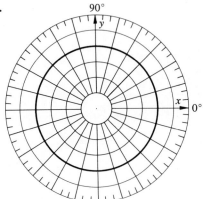

**35.**

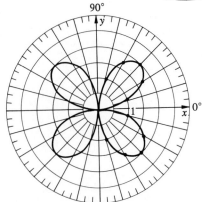

Curve is called a four-leaved rose.

**37.** $(0, 0°)$ and $(\approx 0.7, 45°)$.

**39.**

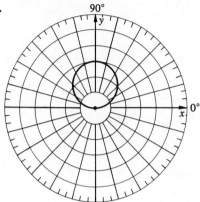

$(0, 0°)$, $(\approx 1.7, 45°)$, and $(\approx 0.3, 225°)$.

**41.** $(0, 0°)$ and $(1/\sqrt{2}, 45°)$; since $\rho$ is a distance measured from the origin, its measure may be zero for more than one value of $\theta$. In this problem, $\rho = 0$ when $\theta = 0°$ in $\rho = \sin \theta$, and $\rho = 0$ when $\theta = 90°$ in $\rho = \cos \theta$.

**43.** $(0, 0°)$, $(\approx 1.71, 45°)$, and $(\approx 0.29, 225°)$.

*Chapter 7 Review* (page 230)

**1.**

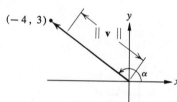

$5$; $143°\ 10'$.

**3.** $\mathbf{v} = (-5\sqrt{2}, -5\sqrt{2})$.

**5.**

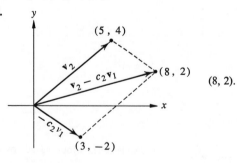

$(8, 2)$.

**7.** $(-1, 3)$.

**9.**

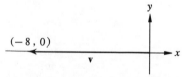

$(-8, 0)$.

**11.** $7\mathbf{i} - 9\mathbf{j}$.

**13.**

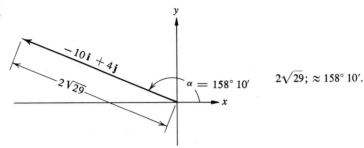

$2\sqrt{29}; \approx 158° \ 10'.$

**15.** $\left(\dfrac{4}{\sqrt{17}}, \dfrac{-1}{\sqrt{17}}\right).$

**17.**

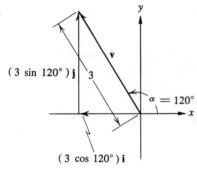

$\mathbf{v} = -\tfrac{3}{2}\mathbf{i} + \tfrac{3}{2}\sqrt{3}\mathbf{j}.$

**19.** $-54.$

**21.** $16\sqrt{2}.$

**23.** $136° \ 50'.$

**25.** $\dfrac{3}{101}\sqrt{101}; \dfrac{3\sqrt{2}}{10}.$

**29.** $1.$

**31.** $(-2, 70°); (-2, -290°); (2, -110°).$

**33.** $(1, \sqrt{3}).$

**35.** $\left(2, \dfrac{7\pi^{\text{R}}}{6}\right).$

**37.** $\rho^2(\cos^2 \theta + 3 \sin^2 \theta) = 5.$

**39.**

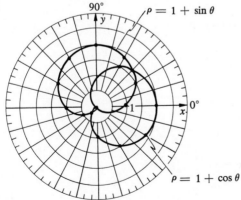

$\rho = 1 + \sin \theta$

$\rho = 1 + \cos \theta$

## Chapter 8

**Exercise 8.1** (page 236)

**1.** $(-2, 8)$.         **9.** $(0, -12)$.         **17.** $(-8, 6)$.

**3.** $(-2, -5)$.       **11.** $(-42, -6)$.      **19.** $(270, 432)$.

**5.** $(-4, 6)$.         **13.** $(-72, 45)$.       **21.** $x = 3, y = 0$.

**7.** $(-17, 21)$.       **15.** $(-4, 16)$.        **23.** $x = 0, y = \frac{3}{2}$.

**25.** $z = (0, 1)$  or  $z = (0, -1)$.          **29.** $x = \dfrac{a}{a^2 + b^2}$ ; $y = \dfrac{-b}{a^2 + b^2}$.

**27.** $z = (1, 0)$  or  $z = (-1, 0)$.

**Exercise 8.2** (page 242)

**9.** $(12, -7)$.     **13.** $(9, -4)$.      **17.** $(\frac{1}{3}, -2)$.      **21.** $(\frac{-42}{53}, \frac{12}{53})$.

**11.** $(5, 1)$.       **15.** $(\frac{21}{13}, -\frac{11}{26})$.    **19.** $(\frac{-3}{41}, \frac{-14}{41})$.    **23.** $(\frac{62}{65}, \frac{-24}{65})$.

**Exercise 8.3** (page 246)

**1.** $-8 + 7i$.      **13.** $-6i$.       **25.** $40 - 78i$.     **37.** $\frac{-19}{89} - \frac{23}{89}i$.

**3.** $-4$.          **15.** $3 + 5i$.     **27.** $5 + (-\sqrt{5} - 1)i$.   **39.** $\frac{3}{5} - \frac{6}{5}i$.

**5.** $7 - \sqrt{2}i$.     **17.** $-3 + 2i$.    **29.** $-64$.         **41.** $-i$.

**7.** $(0, 6)$.        **19.** $x = 1, y = \frac{1}{6}$.   **31.** $-5 + 12i$.    **43.** $-1$.

**9.** $(5, -9)$.     **21.** $x = 0, y = 0$.   **33.** $\frac{3}{2} - \frac{9}{4}i$.

**11.** $(8, 1)$.      **23.** $1 - 4i$.      **35.** $\frac{8}{85} + \frac{36}{85}i$.

**45.** $\{-3 + 2i, -3 - 2i\}$.     **47.** $\{-1 + 2\sqrt{2}i, -1 - 2\sqrt{2}i\}$.     **53.** When $b^2 - 4ac < 0$.

**Exercise 8.4** (page 250)

**1.**

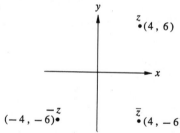

**3.** $z = (-1, 0)$, $-z = (1, 0)$, and $\bar{z} = (-1, 0)$.

**5.**

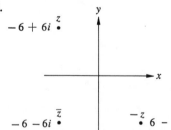

**7.** $z = 5$, $-z = -5$, and $\bar{z} = 5$.

**9.**

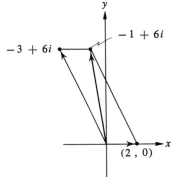

**13.** $3\sqrt{10}$; $\approx 288°\ 30'$.      **17.** $\sqrt{29}$; $\approx 158°\ 10'$.      **21.** 7; 0°.

**15.** 4; 0°.                    **19.** $2\sqrt{2}$; 225°.             **23.** 5; 270°.

**25.** (a) $a \neq 0$, $b = 0$.    (b) $a = 0$, $b \neq 0$.    (c) $b > 0$.    (d) $b < 0$.    (e) $a > 0$.    (f) $a < 0$.

**27.**

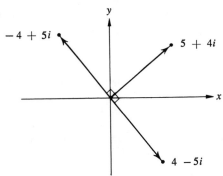

**31.**

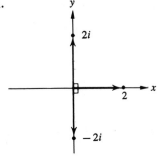

**Exercise 8.5** (page 255)

**1.** $10(\cos 210° + i \sin 210°)$.           **5.** $4\sqrt{2}(\cos 225° + i \sin 225°)$.

**3.** $4(\cos 120° + i \sin 120°)$.            **7.** $8(\cos 180° + i \sin 180°)$.

**9.**

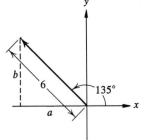

$-3\sqrt{2} + 3\sqrt{2}i$.

**11.** $-\frac{5}{2} - \frac{5}{2}\sqrt{3}i$.

**13.**

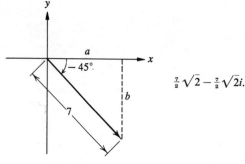

$\frac{7}{2}\sqrt{2} - \frac{7}{2}\sqrt{2}i.$

**15.** $\frac{4}{5}\sqrt{3} - \frac{4}{5}i.$

**21.** $35\sqrt{2} + 35\sqrt{2}i.$

**27.** $\frac{3}{2} - \frac{\sqrt{3}}{2}i.$

**17.** $\frac{3}{2}\sqrt{2} - \frac{1}{2}\sqrt{2}i.$

**23.** $\sqrt{3} - i.$

**29.** $4\sqrt{2} + 4\sqrt{2}i.$

**19.** $-\frac{9}{2}.$

**25.** $-\frac{1}{3} + \frac{\sqrt{3}}{3}i.$

**37.** $-1 + 0i = -1.$

**Exercise 8.6** (page 262)

**1.** $0 - 125i$   or   $-125i.$

**7.** $0 + 32i$   or·   $32i.$

**13.** $\approx 0.000422 - 0.000171i.$

**3.** $-\frac{1}{8} + 0i$   or   $-\frac{1}{8}.$

**9.** $\frac{1}{64}i.$

**15.** $8 + 8i.$

**5.** $-1 + 0i$   or   $-1.$

**11.** $-\frac{1}{64}.$

**17.** $\approx -0.683 + 0.183i.$

**19.**

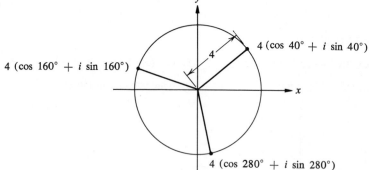

$4(\cos 40° + i \sin 40°), 4(\cos 160° + i \sin 160°), 4(\cos 280° + i \sin 280°).$

**21.** $3^{1/5}(\cos 54° + i \sin 54°), 3^{1/5}(\cos 126° + i \sin 126°),$

$3^{1/5}(\cos 198° + i\sin 198°), 3^{1/5}(\cos 270° + i \sin 270°),$

$3^{1/5}(\cos 342° + i \sin 342°).$

**23.**

$$\frac{1}{(4\sqrt{2})^{1/3}} \; [\cos\,(-225°) + i\,\sin\,(-225°)]$$

$$\frac{1}{(4\sqrt{2})^{1/3}} \; [\cos\,(-345°) + i\,\sin\,(-345°)]$$

$$\frac{1}{(4\sqrt{2})^{1/3}} \; [\cos\,(-105°) + i\,\sin\,(-105°)]$$

$$\frac{1}{(4\sqrt{2})^{1/3}} \; [\cos\,(-105°) + i\,\sin\,(-105°)],$$

$$\frac{1}{(4\sqrt{2})^{1/3}} \; [\cos\,(-225°) + i\,\sin\,(-225°)], \quad \frac{1}{(4\sqrt{2})^{1/3}} \; [\cos\,(-345°) + i\,\sin\,(-345°)].$$

**25.** $\{(\cos 60° + i\,\sin 60°), (\cos 180° + i\,\sin 180°), (\cos 300° + i\,\sin 300°)\}$,

or $\quad \left\{ \dfrac{1}{2} + \dfrac{\sqrt{3}}{2}\,i, \; -1, \dfrac{1}{2} - \dfrac{\sqrt{3}}{2}\,i \right\}.$

**27.** $\{(\cos 0° + i\,\sin 0°), (\cos 120° + i\,\sin 120°), (\cos 240° + i\,\sin 240°)\}$,

or $\quad \left\{ 1, \; -\dfrac{1}{2} + \dfrac{\sqrt{3}}{2}\,i, \; -\dfrac{1}{2} - \dfrac{\sqrt{3}}{2}\,i \right\}.$

**29.** $\left\{ \dfrac{1}{2^{1/3}} \,(\cos 110° + i\,\sin 110°), \dfrac{1}{2^{1/3}} \,(\cos 230° + i\,\sin 230°), \right.$

$$\left. \dfrac{1}{2^{1/3}} \,(\cos 350° + i\,\sin 350°) \right\}.$$

### *Chapter 8 Review* (page 267)

**1.** $(-3, -7)$.      **9.** $-4 + i$.      **17.** $-3 + 3i$.

**3.** $(-9, -123)$.      **11.** $7 + 5i$.      **19.** $\frac{7}{5} - \frac{13}{10}i$.

**5.** $(-15, -1)$.      **13.** $x = -\frac{1}{3}, y = -\frac{2}{3}$.      **21.** $i$.

**7.** $(-\frac{7}{6}, \frac{5}{6})$.      **15.** $-8 + 8i$.      **23.** $\left\{ \dfrac{1 + \sqrt{7}i}{2}, \dfrac{1 - \sqrt{7}i}{2} \right\}.$

**25.**

**27.** $(5, -1)$.      **29.** $\sqrt{29}; \approx 248° \, 10'$.      **31.** $8(\cos 270° + i\,\sin 270°)$.

**33.**

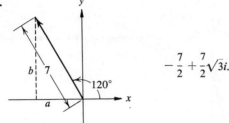

$$-\frac{7}{2}+\frac{7}{2}\sqrt{3}i.$$

**35.** $-\dfrac{\sqrt{2}}{3}-\dfrac{\sqrt{2}}{3}\,i.$    **37.** $-\dfrac{8}{9}.$    **39.** $-64.$

**41.** $5(\cos 110° + i \sin 110°),\ 5(\cos 230° + i \sin 230°),$
$5(\cos 350° + i \sin 350°).$

**43.** $z = (0,\ \sqrt{2})$  or  $z = (0,\ -\sqrt{2}).$

## Appendix

**A.1** (page 271)

**1.** $\log_2 64 = 6.$    **9.** $9^{1/2} = 3.$    **17.** $1.$

**3.** $\log_{32} 2 = \frac{1}{5}.$    **11.** $5^1 = 5.$    **19.** $1.$

**5.** $\log_5 \frac{1}{25} = -2.$    **13.** $2.$    **21.** $0.$

**7.** $2^4 = 16.$    **15.** $\frac{1}{2}.$    **23.** $-2.$

**25.** $\log_{10} x + \log_{10} y + \log_{10} z.$

**27.** $2 \log_{10} x + \log_{10} y.$

**29.** $\log_{10} x - \frac{1}{2} \log_{10} (x - 1).$

**31.** $\log_{10} 2 + \log_{10} \pi + \frac{1}{2} \log_{10} l - \frac{1}{2} \log_{10} g.$

**33.** $\log_{10} \dfrac{xy}{z}.$    **35.** $\log_{10} \dfrac{x^{3/4}}{y^{3/8}}.$    **37.** $\log_{10} \dfrac{\pi r^2}{s^{1/3}}.$

**A.2** (page 275)

**1.** $0.6425.$    **13.** $0.5410.$    **25.** $0.858.$    **37.** $79.3.$

**3.** $1.9196.$    **15.** $1.0306.$    **27.** $0.0247.$    **39.** $62.$

**5.** $2.3856.$    **17.** $9.5731 - 10.$    **29.** $7.525.$    **41.** (a) $6.$  (b) $3.$

**7.** $9.7931 - 10.$    **19.** $6.53.$    **31.** $9,423.$

**9.** $7.7782 - 10.$    **21.** $25.4.$    **33.** $0.01537.$

**11.** $6.6990 - 10.$    **23.** $2,000.$    **35.** $2.52.$

**A.3** (page 279)

**1.** $27.04.$

**3.** $7.595.$

**5.** $2.873.$

**7.** $0.363.$

**9.** $51.23.$

**11.** $2.687.$

**13.** $\left\{\dfrac{\log_{10} 5}{\log_{10} 3}\right\}$  or  $\{\approx 1.46\}.$

**15.** $\left\{\dfrac{\log_{10} 290}{\log_{10} 5} - 1\right\}$  or  $\{\approx 2.52\}.$

**17.** $\left(-\dfrac{\log_{10} 21}{\log_{10} 3}\right)$  or  $\{\approx -2.77\}.$

**19.** $\approx \$14{,}790.$

**21.** $\approx 12.$

**23.** $\approx 8{,}264.$

**25.** $\approx 40.25.$

**27.** $\approx 12.11$.

**29.** $\approx 1.7$.

**31.** 2.81.

**A.4** (page 284)

**1.** $\approx 9.7586 - 10$.

**3.** $\approx 9.8066 - 10$.

**5.** $\approx 9.6028 - 10$.

**7.** $\approx 9.4996 - 10$.

**9.** $\approx 9.9824 - 10$.

**11.** $15° 20'$.

**13.** $73° 00'$.

**15.** $46° 00'$.

**17.** $\approx 1.97$.

**19.** $\approx 1.25$.

**21.** $\approx 0.360$.

**23.** $\approx 30.4$.

**25.** $\alpha \approx 36° 50'$; $\beta \approx 53° 10'$; $c \approx 15.0$.

**27.** $\alpha = 21° 50'$; $c \approx 10.3$; $a \approx 3.85$.

**29.** $\gamma = 99°$; $b \approx 11.4$; $c \approx 14.9$.

**31.** Only one triangle is possible. $\beta \approx 16° 50'$; $\gamma \approx 140° 20'$; $c \approx 20.7$.

**33.** Two triangles are possible. $a \approx 7.75$, $\alpha \approx 132° 40'$, $\gamma \approx 30° 10'$; $a' \approx 2.37$, $\alpha' \approx 13° 00'$, $\gamma' \approx 149° 50'$.

**35.** $\alpha \approx 52° 30'$; $\gamma \approx 97° 30'$; $b \approx 2.52$.

**37.** $\alpha \approx 35° 40'$; $\beta \approx 92° 40'$; $c \approx 7.00$.

**39.** $\alpha \approx 56° 20'$; $\beta \approx 30° 00'$; $\gamma \approx 93° 40'$. Does $\alpha + \beta + \gamma = 180°$? Yes.

**41.** $\alpha \approx 95° 00'$; $\beta \approx 23° 40'$; $\gamma \approx 61° 20'$. Does $\alpha + \beta + \gamma = 180°$? Yes.

*Review* (page 286)

**1.** $\log_3 81 = 4$.

**3.** $4^3 = 64$.

**5.** 3.

**7.** 1.

**9.** $\log_{10} 2 + 3 \log_{10} x + \log_{10} y$.

**11.** $\log_{10} \dfrac{x^2}{\sqrt[3]{y}}$.

**13.** 0.8531.

**15.** 0.0133.

**17.** $\approx 40.2$.

**19.** $\left\{\dfrac{\log_{10} 14}{\log_{10} 2}\right\}$ or $\{\approx 3.81\}$.

**21.** $\approx 9.9774 - 10$.

**23.** $\approx 9.3713 - 10$.

**25.** $66° 20'$.

**27.** $\alpha \approx 39° 40'$; $\beta \approx 50° 20'$; $a \approx 8.30$.

**29.** $\alpha = 32° 20'$; $a \approx 4.23$; $c \approx 3.77$.

**31.** $\approx 40.4$.

**33.** 4.

**35.** When $x < 10^6$.

**App.**

# Index

# Symbols and Notation

The section where the symbol or notation is first used is shown in parentheses.

| | | |
|---|---|---|
| $\{a, b\}$ | the set whose elements (or members) are $a$ and $b$ | (1.1) |
| $=$ | is equal to or equals ($\neq$, is not equal to) | (1.1) |
| $\in$ | is a member of or is an element of ($\notin$, is not a member of or is not an element of) | (1.1) |
| $\{x \mid \ldots\}$ | the set of all $x$ such that ... | (1.1) |
| $\cup$ | the union of | (1.1) |
| $\cap$ | the intersection of | (1.1) |
| $\varnothing$ | the null set or the empty set | (1.1) |
| $R$ | the set of real numbers | (1.2) |
| $N$ | the set of natural numbers | (1.2) |
| $J$ | the set of integers | (1.2) |
| $Q$ | the set of rational numbers | (1.2) |
| $H$ | the set of irrational numbers | (1.2) |
| $0.123\overline{123}$ | repeating decimal | (1.2) |
| $\sqrt{a},\ a \geq 0$ | the positive square root of $a$ for $a > 0$; zero for $a = 0$ | (1.2) |
| $\approx$ | approximately equal to | (1.2) |
| $<$ | is less than | (1.2) |
| $\leq$ | is less than or equal to | (1.2) |
| $>$ | is greater than | (1.2) |
| $\geq$ | is greater than or equal to | (1.2) |
| $\lvert a \rvert$ | the absolute value of $a$ | (1.2) |
| $A \times B$ | the Cartesian product of $A$ and $B$ | (1.3) |
| $(a, b)$ | the ordered pair of numbers whose first component is $a$ and whose second component is $b$ | (1.3) |
| $f, F, \mathscr{F}, \mathscr{R}$, etc. | a function, $f$, $F$, $\mathscr{F}$, $\mathscr{R}$, etc., or a relation, $f$, $F$, etc. | (1.3) |
| $f(x)$ | the value of the function $f$ at $x$ | (1.3) |
| $f(a)$ | the value of the function $f$ when $x$ is replaced by $a$ | (1.3) |
| $P(x, y)$ | the name of a point with coordinates $x$ and $y$ | (1.4) |
| $\overline{P_1 P_2}$ | line segment with end points designated by $P_1$ and $P_2$ | (1.4) |
| $l(\overline{P_1 P_2})$ | the length of line segment $\overline{P_1 P_2}$ | (1.4) |
| $l(\overrightarrow{P_1 P_2})$ | the directed length of a line segment from $P_1$ to $P_2$ | (1.4) |
| $\pm$ | plus or minus | (1.5) |
| $\mathscr{R}^{-1}$ | $\mathscr{R}$ inverse or the inverse of $\mathscr{R}$ | (1.6) |
| $[x]$ | greatest integer, not greater than $x$ | (2.1) |
| $\mathscr{A}$ | area | (2.2) |
| cos | cosine function | (2.2) |
| sin | sine function | (2.2) |
| $\tilde{s}, \tilde{x}$ | the reference arc for arc of length $s$, $x$ | (2.3) |
| tan | tangent function | (2.5) |
| sec | secant function | (2.6) |
| csc | cosecant function | (2.6) |
| cot | cotangent function | (2.6) |
| $2!, 3!$, etc. | $2 \cdot 1,\ 3 \cdot 2 \cdot 1$, etc. | (2.7) |